Hazmatology
The Science of Hazardous Materials

Hazmatology: The Science of Hazardous Materials,
Five-Volume Set
9781138316072

Volume One - Chronicles of Incidents and Response
9781138316096

Volume Two - Standard of Care and Hazmat Planning
9781138316768

Volume Three - Applied Chemistry and Physics
9781138316522

Volume Four - Common Sense Emergency Response
9781138316782

Volume Five - Hazmat Team Spotlight
9781138316812

Applied Chemistry and Physics

Robert A. Burke

CRC Press is an imprint of the
Taylor & Francis Group, an **informa** business

CRC Press
Taylor & Francis Group
6000 Broken Sound Parkway NW, Suite 300
Boca Raton, FL 33487-2742

© 2021 by Taylor & Francis Group, LLC
CRC Press is an imprint of Taylor & Francis Group, an Informa business

No claim to original U.S. Government works

Printed on acid-free paper

International Standard Book Number-13: 978-1-138-31652-2 (Hardback)

This book contains information obtained from authentic and highly regarded sources. Reasonable efforts have been made to publish reliable data and information, but the author and publisher cannot assume responsibility for the validity of all materials or the consequences of their use. The authors and publishers have attempted to trace the copyright holders of all material reproduced in this publication and apologize to copyright holders if permission to publish in this form has not been obtained. If any copyright material has not been acknowledged, please write and let us know so we may rectify in any future reprint.

Except as permitted under U.S. Copyright Law, no part of this book may be reprinted, reproduced, transmitted, or utilized in any form by any electronic, mechanical, or other means, now known or hereafter invented, including photocopying, microfilming, and recording, or in any information storage or retrieval system, without written permission from the publishers.

For permission to photocopy or use material electronically from this work, please access www.copyright.com (http://www.copyright.com/) or contact the Copyright Clearance Center, Inc. (CCC), 222 Rosewood Drive, Danvers, MA 01923, 978-750-8400. CCC is a not-for-profit organization that provides licenses and registration for a variety of users. For organizations that have been granted a photocopy license by the CCC, a separate system of payment has been arranged.

Trademark Notice: Product or corporate names may be trademarks or registered trademarks, and are used only for identification and explanation without intent to infringe.

Library of Congress Cataloging-in-Publication Data
LoC Data here

Visit the Taylor & Francis Web site at
http://www.taylorandfrancis.com

and the CRC Press Web site at
http://www.crcpress.com

Dedication

Volume Three

For Jan David Kuczma, Sr.

Jan Kuczma was my first instructor at the National Fire Academy (NFA) in 1981 for Hazmat 1, which is now called Chemistry for Emergency Response. Jan and I clicked and became friends and, more important, a mentor for me. When I became an instructor for Chemistry for Emergency Response, I saw Jan at the NFA on numerous occasions during the 1990s. Jan was the one who encouraged me to write my first book, Hazmat Chemistry for Emergency Responders. *Later he told me that I should not be a one-book author. So, to date I have authored seven books; later* Hazmatology: The Science of Hazardous Materials *becomes my ninth book project.*

Jan was both a friend and a mentor at NFA; our friendship extended into our personnel lives away from our careers. He helped me many times during my career and was like a father to me. Many people have influenced me over the years, but I would not be where I am today without having known Jan Kuczma. Jan passed away on June 16, 2013. My life has a void that can never be filled. Thank you, Jan!

Contents

Preface xiii
Acknowledgments xv
Author xix

Applied Chemistry and Physics 1
Applied Chemistry 1
Applied Physics 2
Introduction 2
What Constitutes a Hazardous Materials Incident? 6
When Hazardous Materials Are Present 7
Actors in Hazardous Materials Incidents 8
    Chemicals 8
    Incompatible Chemicals 8
    Inert Materials 9
    Container Damage 11
        Incident Caused Container Damage 11
        Post-Incident Container Damage 15
    Boiling Liquid Expanding Vapor Explosion 16
    Weather 18
    Geography 19
Applied Chemistry 21
    States of Matter 21
        Solids 22
        Liquids 23
        Gases 24
        Temperature and Pressure 24
        Physical State and Chemical Hazards 25
    Periodic Table of Elements 25
        Elements 26
    Into the Atom 29
        Atomic Weight 30
        Atomic Number 30
        Metals and Nonmetals 31

　　　　Element Families 31
　　Compounds, Mixtures, Solubility and Miscibility.............................. 34
　　　　Chemical Compounds 34
　　　　Mixtures.. 37
　　　　Solubility.. 38
　　　　Miscibility.. 40
　　Common and Site-Specific Hazardous Materials............................... 40
　　Recognition-Primed Decision Model 41
　　Incident Frequency/Risk Model... 42
　　　　High-Frequency, Low-Risk Incidents 42
　　　　Low-Frequency, Low-Risk Incidents 42
　　　　High-Frequency, High-Risk Incidents.. 42
　　　　Low-Frequency, High-Risk Incidents 43
　　Families of Compounds 46
　　　　Salt Families and Hazards 46
　　　　Salts—LF/LR... 46
　　　　Nonsalt Compounds.. 55
　　　　Hydrocarbons—HF/HR 57
　　　　Hydrocarbon Naming Systems 58
　　　　Hydrocarbon Families 63
　　　　Hydrocarbon Derivatives 72
　　　　Hydrocarbon Derivative Families.. 76
　　　　Polar Compounds... 76
　　　　Complex Hydrocarbon Derivative Compounds.......................... 80
Nine DOT Hazard Classes 84
Class 1 Explosives (LF/HR) 84
　　DOT Definition of Explosive... 84
　　Forbidden Explosives... 86
　　Explosive Families of Compounds (Figure 3.70) 86
　　　　Inorganic Explosives LF/HR.. 86
　　　　Metal Azides LF/HR 88
　　　　Aliphatic Explosive Compounds (Nitro Hydrocarbon
　　　　　Derivatives) LF/HR... 88
　　　　Aromatic Explosive Compounds 92
　　　　Explosive Chemicals Not Considered to Be Explosives............. 94
　　　　Military Explosives 95
　　Historic Incidents Involving Common Explosives............................ 97
　　Homeland Security Monitoring of Ammonium Nitrate................. 108
　　Chemical Notebook 109
　　Categories of Chemical Explosions 109
　　　　Physical Explosion.. 109
　　　　Physical/Chemical Explosion 110
　　　　Chemical Explosion.. 110
　　　　Explosives Tetrahedron 110

- Types of Chemical Explosions 111
  - Detonation 111
  - Deflagration 111
  - Yield vs. Order 112
- Explosive Effects 112
  - Phases of Explosions 112
  - Overpressure 114
  - Dust Explosion 114
- Historic Dust Explosion Incidents 117
  - Chicago, IL August 5, 1897 Grain Elevator Explosion 117
  - Milwaukee, WI April 22, 1926 Sawdust Explosion 117
  - Corpus Christi, TX April 7, 1981 Grain Elevator Explosion 117
  - Bellwood, NE April 7, 1981 Grain Elevator Explosion 118
  - Nuclear Explosions 118
- Hazard Class 2 Compressed Gases (LF/HR) 119
  - Division 2.1 Flammable Gases LF/HR 119
  - Division 2.2 Nonflammable Gases LF/LR 120
  - Division 2.3 Poison Gases LF/HR 121
  - Toxicological Terms 121
  - Common Class 2 Hazardous Materials 124
    - Liquefied Compressed Gases 124
    - Liquefied Petroleum Gases—LF/HR 124
    - Class 2 Multiple Hazard Bad Actors 126
  - Historic Chlorine Incidents 129
    - World War I Usage 129
    - Henderson, NV, 1991 130
    - Sun Bay, FL, 1998 130
    - St. Louis, MO, 2002 130
    - Atlanta, GA 130
    - North Carolina, 2003 131
    - Glendale, AZ, 2003 131
  - Cryogenic Liquids LF/HR 131
    - Chemical Notebook 132
  - Expansion Ratios 134
  - Hazards to Responders 135
    - Chemical Notebook 135
  - Historic Anhydrous Ammonia Incidents 142
    - Crete NE February 19, 1969 Derailment and Anhydrous Ammonia Release 142
    - Tecumseh, NE March 2014 Anhydrous Ammonia Release 143
  - Confined Space Gases 144
    - Chemical Notebook 144
  - Compressed Gas Containers 146
    - Containers 147

- Cryogenic Containers 152
- Tube Banks 154
- Bulk Containers 156

Hazard Class 3 Flammable Liquids (HF/LR) 157
- Chemistry of Fire 158
  - Alkyl Halide (Halogenated Hydrocarbons) 159
  - Ether (Oxide) 160
  - Amine 160
  - Ketone 161
  - Aldehyde 162
  - Alcohol 163
  - Ester 164
- Plastics and Polymerization 164
  - Polymers 165
  - Polymer Family Tree 166
  - Manufacturing Plastics 166
- Combustion Products 168
- Hazards to Responders 169
- Organic Acid 170
- Flammable Liquid Containers 171
- Nonpressure/Low-Pressure Rail Cars 173
- Fixed Facility Flammable Liquid Bulk Storage Tanks 174
  - Cone-Roof Tanks 174
  - Open Floating Roof Tanks 175
  - Internal Floating Roof Tank 176
  - Horizontal Tanks 177
  - Emergency Response to Ethanol Spills and Fires 178
  - Crude Oil and Its Response Challenges 182
  - Propane suspended in oil may have caused railcar explosions 185

Hazard Class 4 Flammable Solids (LF/LR) 193
- Division 4.1 Flammable Solids—LF/LR 193
  - Chemical Notebook 193
- Division 4.2 Spontaneously Combustible—LF/LR 196
  - Liquids 196
  - Solids 196
  - Chemical Notebook 197
- Phenomenon of Spontaneous Combustion 199
  - Chemical Notebook 203
- Division 4.3 Dangerous When Wet—LF/LR 205
  - Chemical Notebook 206

Hazard Class 5 Oxidizers (LF/HR) 206
- Hypergolic Combustion 207
  - Hypergolic Propellant 207
  - Oxidizers 207

- Chemical Notebook.......... 209
- Chemical Notebook.......... 210
- Division 5.2 Organic Peroxide LF/HR.......... 211
  - Chemical Notebook.......... 213
- Containers for Oxidizers.......... 213
- Hazard Class 6 Poisons and Infectious Substances HF/HR.......... 214
  - Division 6.1 Poisons.......... 214
    - Routes of Exposure.......... 216
    - Cyanides and Isocyanates.......... 218
    - Chemical Notebook.......... 218
    - Sulfur Compounds.......... 218
    - Chemical Notebook.......... 219
    - Alkyl Halide.......... 219
    - Aldehyde.......... 220
    - Ester.......... 220
    - Chemical Notebook.......... 221
    - Organic Acid.......... 221
    - Chemical Notebook.......... 221
  - Military Chemical Agents.......... 221
    - Nerve Agents.......... 221
    - Chemical Notebook.......... 222
  - Division 6.2 Infectious Substances LF/LR.......... 222
    - Biological Notebook.......... 224
- (WMD) Unit of the FBI developed guidelines for emergency responders..........233
  - Pesticides.......... 235
    - Signal Word Toxicity Comparison.......... 240
    - Protective Measures.......... 241
    - Synthetic Opioids and the Dangers to Emergency Responders.... 241
    - Illicit Uses.......... 242
    - Clandestine Manufacture.......... 243
  - Naloxone/Narcan.......... 244
  - Weapons Grade Narcotic.......... 244
    - Protecting Personnel.......... 245
  - Chemistry of Clandestine Methamphetamine Drug Labs.......... 245
    - Detection.......... 248
    - Drug Lab Chemicals.......... 248
  - Homemade Ammonia.......... 252
    - Cleanup Concerns.......... 252
- Class 7 Radioactive (LF/LR).......... 253
  - Types of Radiation.......... 254
  - Isotopes.......... 256
  - Radiation Exposure.......... 256
    - Chemical Notebook.......... 259

Uranium Compounds 260
    Chemical Notebook 260
Radium Compounds 261
    Chemical Notebook 261
Containers for Radioactive Materials 263
Hazard Class 8 Corrosive (HF/LR) 264
    Organic Acid 268
        Chemical Notebook 268
    Inorganic Acid 269
        Chemical Notebook 269
    Base 270
    Dilution vs. Neutralization 271
    Corrosive Containers 274
Hazard Class 9 Miscellaneous Hazardous Materials (LF/LR) 276
    Chemical Notebook 277
    Batteries 278
    Solid Materials 278
        Chemical Notebook 279
    Molten Materials 280
Applied Physics (Physical Characteristics) 281
    Combustion Analysis 281
        Hydrocarbons 281
    Boiling Point 282
    Vapor Pressure 283
    Vapor Content 283
    Vapor Density 284
    Polarity 284
    Molecular Weight 285
    Branching 285
    Sublimation 286
    Pyrophoric Materials 286
    Volatility 289
    Flash Point 289
    Flash Point Solids 289
    Fire Point 290
    Ignition Temperature 291
    Relationship of Physical Characteristics 292
    Flammable Range or Explosive Limits 293
    Specific Gravity 296
    Critical Temperature and Pressure 296
    Hazards of the "Invisible Force" 297

Glossary 299
Bibliography 317
Index 319

Preface

Effectively dealing with hazardous materials incidents cannot be accomplished without knowing chemical, physical and family effects of these materials. The key is to know how to apply these effects to a hazardous materials incident keeping responders and the public out of harm's way. Because of the hazardous materials standard of care, it may not always be possible to achieve this goal. Just remember that emergency responders did not cause the incident to occur. Our job is to make the situation better than if we had not responded at all. Sadly, in some cases that is not what happens, because sometimes we do not effectively apply chemistry and physics to a hazardous materials incident.

Focusing on the common hazardous materials in our communities and utilizing the tools from our Hazardous Materials Tool Box as needed, we can effectively and safely deal with hazardous materials. We need to be flexible and use common sense. Looking at historical hazardous materials incidents and understanding how chemistry and physics, along with actors present in the incident, affected the outcome can guide us to deal safely with those common hazardous materials. It is not what I like to call the "exotic" hazardous materials that are faced time and time again, it is the common materials. If we find out the history of all those hazardous materials history has shown we will face on a regular basis, we can avoid pitfalls and everybody goes home.

It is not the focus of this volume to teach chemistry and physics to emergency responders; by reading this volume, you cannot help but pick up on some of the concepts. Reading this volume will not teach you to make chemicals. This volume is intended to build recognition and understanding tools for your Hazmat Tool Box. The purpose of this approach is to bring chemistry and physics concepts down to the level that most emergency responders will understand. Having an understanding will help responders to apply chemistry and physics concepts to hazardous materials incidents in a scientific-based approach, using a risk management system and, most important, common sense.

Acknowledgments

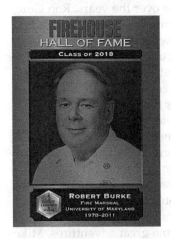

I thank the many fire departments and members across the United States and Canada that I have visited and became friends with during my visits to their departments over the years. I also thank the firefighters from classes I have attended as a student and taught for the National Fire Academy, Maryland Fire and Rescue Institute and Community College of Baltimore County since 1988. Learning is a two-way street, and I have learned much from the students as well. I thank the many friends I have met during the 40 plus years in the fire, EMS, hazardous materials and emergency management fields. There are those who I have not seen for a while; some are no longer with us, but once a friend, always a friend.

I express my thanks to *Firehouse Magazine* for allowing me to write stories about hazardous materials for 33 years and counting. During those years, I have had the pleasure of writing under every editor of the magazine including founder Dennis Smith who gave me the chance to be published for the first time. I also thank Firehouse editors, Janet Kimmerly, Barbara Dunleavy, Jeff Barrington, Harvey Eisner, Tim Sendelbach and Peter Mathews for their support over the years. When I read my first copy of *Firehouse Magazine* in the late 1970s, I was hooked. My dream was to someday go to Baltimore to attend a Firehouse Expo. Never did I dream I would not only attend an expo but teach at numerous expos, write for the magazine and in 2018 be inducted into the Firehouse Hall of Fame. To be placed in a fraternity with sixteen of the people who had an enormous impact on the fire service and who I looked up to my entire career was very humbling.

Several people have been my mentors and have impacted my life and career. When I worked with the State Fire Marshall of Nebraska, Wally Barnett allowed me to accomplish things in the State Fire Marshal's Office

Brent Boydston, Chief Bentonville, AR Fire Department.

that that I otherwise would not have. Because of his ability to let his employees reach their potential, I was able to write for *Firehouse Magazine*, become a contributing editor, teach for the National Fire Academy and other things too numerous to mention. He was proud when I gave him a copy of my first book. I owe much of my success in the fire service to the opportunities Wally gave me. Jan Kuczma and Chris Waters at the National Fire Academy have been mentors to me over the years. Ron Gore, retired Captain from the Jacksonville, FL Fire Department and Owner of Safety Systems, has had a large impact on my life and career. The Jacksonville Hazmat Team was the first emergency services hazmat team in the United States. Ron Gore is the Godfather of Hazmat response in the United States.

Former student of mine and current Chief of the Bentonville, AR Fire Department Brent Boydston has been a great friend to me and my family over the years. Rudy Rinas, Gene Ryan and John Eversole of the Chicago Fire Department have been fellow classmates and students. Mike Roeshman and Bill Doty of the Philadelphia Fire Department both former students and retired as Hazmat Chief Officers have remained friends. I used to ride with Bill and together we had some great adventures. Mike showed me Philadelphia historical areas, like the spot where Ben Franklin flew his kite and his post office, which is so obscure today in downtown Philadelphia. I also stood on the spot where Rocky stood at the top of

Mike Roeshman Retired Hazmat Chief Philadelphia Fire Department.

the steps in the movie. These adventures enjoyed in Philadelphia would not have happened without Bill and Mike.

Just outside of Philadelphia in Delaware County, Tom Micozzie, Hazmat Coordinator for Delaware County, was also a former student and a great friend. We had many adventures together, and I will never forget his introduction to me of the Galati at Rita's Italian Ice! Rita's Italian Ice was started by a retired Philadelphia firefighter and not long ago one opened up in Lincoln, NE.

Thanks to Richmond In Fire Chief Jerry Purcell, who I met during a visit to Richmond to do a Firehouse story on their 1968 explosion in downtown. As a result of

Acknowledgments

William, "Bill" Doty retired Hazmat Chief Philadelphia Fire Department.

the Richmond story being published, I was able to locate and become friends with blast survivor Jack Bales. More recently I visited to do another story on their hazmat team and propane training. Thanks to new friend Ron Huffman who traveled to Richmond to conduct the propane training utilizing water injection to control liquid propane leaks. The article appeared in the September 2019 *Firehouse Magazine*.

Thanks to Tod Allen, Fire Chief in Crete Nebraska who I met when I was researching a train derailment in Crete for another friend Kent Anderson. We have become good friends. Tod is the apparatus operator on Truck one at Station 1 for the Lincoln Nebraska Fire Department. He invited me to come and ride with him, and many adventures later I still go there on a regular basis. I thank all of my friends past and present on "B" Shift at Station 1 for making me feel at home and showing me a good time whenever I am there. Thanks to friend Captain Mark Majors for sharing his experiences with Nebraska Task Force 1 Urban Search and Rescue Team (USAR) and Captain Francisco Martinez Lincoln Hazmat. Finally, I thank Chief Michael Despain and assistant Chief Patrick Borer for their friendship and hospitality while visiting the Lincoln Fire Department on many occasions. This is only the short list—I would have to write a separate book to thank all of you I have met and for the impact you have had on my life over the past 40+ years. You know who you are; I appreciate your friendship and assistance and consider your selves thanked again.

Chief Jerry Purcell Richmond, IN Fire Department.

During my year-long book writing adventure that led to *Hazmatology: The Science of Hazardous Materials*, I met and spoke to many people and made new friends. I thank my cousin Dustin Schroeder, Senior Captain at Houston Station 68, and the firefighters and others I met. I also thank Kevin Okonski, Hazmat at Houston Station 22; Ludwig Benner, former NTSB Investigator and developer of several incident management models; Bill Hand, Houston; Richard Arwood; Charles Smith, Memphis; Kevin Saunders, Motivator; Chief Jeff Miller, Butte, MT; and all of the Nebraska Regional Team leaders and members.

I express my thanks to my cousin Jeanene and her husband Randy for coming all the way from Montana to be with me at the Firehouse Hall of Fame induction. I am also grateful to Brent Boydston, James Rey Milwaukee, Wilbur Hueser and Saskatoon in Canada for the hospitality and tour, and Captain Oscar Robles, Imperial, CA. The list just goes on and on, and there is no room here for everyone, but the rest of you know who you are and I want you to know how much your assistance is appreciated. You are all considered friends, and I hope we will talk and or meet again. Finally, thanks to librarians and historians across the country for your assistance in research, thanks for the memories!

Robert Burke

Author

Robert A. Burke was born in Beatrice and grew up in Lincoln, Nebraska; graduated from high school in Dundee, Illinois; and earned an AA in Fire Protection Technology from Catonsville Community College, Baltimore County, Maryland (now Community College of Baltimore County) and a BS in Fire Administration from the University of Maryland. He has also pursued his graduate work at the University of Baltimore in Public Administration. Mr Burke has attended numerous classes at the National Fire Academy in Emmitsburg, Maryland, and additional classes on firefighting, hazardous materials and Weapons of Mass Destruction at Oklahoma State University; Maryland Fire and Rescue Institute; Texas A & M University, College Station, Texas; the Center for Domestic Preparedness in Anniston, Alabama; and others.

Mr. Burke has over 40 years' experience in the emergency services as a career and volunteer firefighter, and has served as a Lieutenant for the Anne Arundel County, Maryland Fire Department; an assistant fire chief for the Verdigris Fire Protection District in Claremore, Oklahoma; Deputy State Fire Marshal in the State of Nebraska; a private fire protection and hazardous materials consultant; and an exercise and training officer for the Chemical Stockpile Emergency Preparedness Program (CSEPP) for the Maryland Emergency Management Agency; and retired as the Fire Marshal for the University of Maryland. He has served on several volunteer fire companies, including West Dundee, Illinois; Carpentersville, Illinois; Sierra Volunteer Fire Department, Chaves County, New Mexico; Ord, Nebraska; and Earleigh Heights Volunteer Fire Company in Severna Park, Maryland, which is a part of the Anne Arundel County, Fire Department, Maryland.

Mr. Burke has been a Certified Hazardous Materials Specialist (CFPS) by the National Fire Protection Association (NFPA) and certified by the

National Board on Fire Service Professional Qualifications as a Fire Instructor III, Fire Inspector, Hazardous Materials Incident Commander, Fire Inspector III and Plans Examiner II. He served on the NFPA technical committee for NFPA 45 Fire Protection for Laboratories Using Chemicals for 10 years. He has been qualified as an expert witness for arson trials as well.

Mr. Burke retired as an adjunct instructor at the National Fire Academy in Emmitsburg, Maryland in April 2018 after 30 years. He taught Hazardous Materials, Weapons of Mass Destruction and Fire Protection curriculums. He taught at his Alma Mater Community College of Baltimore County, Catonsville Campus and Howard County Community College in Maryland. He has had articles published in various fire service trade magazines for the past 33 years. Mr. Burke is currently a contributing editor for *Firehouse Magazine*, with a bimonthly column titled "Hazmatology," and he has had numerous articles published in *Firehouse, Fire Chief, Fire Engineering* and *Nebraska Smoke Eater* magazines. He was inducted into the Firehouse Hall of Fame in October 2018 in Nashville, TN. Mr Burke has also been recognized as a subject matter specialist for hazardous materials and been interviewed by newspapers, radio and television about incidents that have occurred in local communities including Fox Television in New York City live during a tank farm fire on Staten Island.

Mr. Burke has been a presenter at Firehouse Expo in Baltimore, MD and Nashville, TN numerous times, most recently in 2017. He gave a presentation at the EPA Region III SERC/LEPC Conference in Norfolk, Virginia, in November 1994 and a presentation at the 1996 Environmental and Industrial Fire Safety Seminar, Baltimore, Maryland, on DOT ERG. He was a speaker at the 1996 International Hazardous Materials Spills Conference on June 26, 1996, in New Orleans, Louisiana; a speaker at the Fifth Annual1996 Environmental and Industrial Fire Safety Seminar in Baltimore, Maryland, sponsored by Baltimore City Fire Department; and at LEPC, an instructor for Hazmat Chemistry, August 1999, at Hazmat Expo 2000 in Las Vegas, Nevada. He also delivered a Keynote presentation at the Western Canadian Hazardous Materials Symposium Saskatoon, Saskatchewan, Canada, in 2008.

Mr. Burke has developed several CD-ROM-based training programs, including the Emergency Response Guide Book, Hazardous Materials and Terrorism Awareness for Dispatchers and 911 Operators, Hazardous Materials and Terrorism Awareness for Law Enforcement, Chemistry of Hazardous Materials Course, Chemistry of Hazardous Materials Refresher, Understanding Ethanol, Understanding Liquefied Petroleum Gases, Understanding Cryogenic Liquids, Understanding Chlorine and Understanding Anhydrous Ammonia. He has also developed the "Burke Placard Hazard Chart." He has published seven additional books titled

Hazardous Materials Chemistry for Emergency Responders (1st, 2nd and 3rd Editions, *Counterterrorism for Emergency Responders* 1st, 2nd and 3rd editions, *Fire Protection: Systems and Response* and *Hazmat Teams Across America*.

Currently, Mr. Burke serves on the Homestead LEPC in Southeast Nebraska. He also manages a Hazardous Materials section at the Nebraska Firefighters Museum and periodically rides with friends on "B" shift at Station 1, Lincoln Fire Department. He can be reached via email at robert.burke@windstream.net, on Facebook at https://www.facebook.com/RobertAb8731 and through his website: www.hazardousmaterialspage.com.

Volume Three

Applied Chemistry and Physics

> Science makes us safer. It is essential to start with science.

Applied Chemistry

Applied chemistry is the application of the principles and theories of the science of chemistry and physics to answer specific questions or solve real-world problems during hazardous materials response. We are not using chemistry to become chemists as we are not going to make chemicals. Our real-world problem is that some hazardous material has escaped its container and the hazmat team has been activated to remedy the situation. The question to be answered is, how do we mitigate an escaped hazardous material safely? Knowing the chemistry of the escaped hazardous material will help responders understand the dangers that the material poses to us as emergency responders and to the public that we need to protect (Figure 3.1).

Figure 3.1 Knowing the chemistry and physics (physical characteristics) of the escaped hazardous material will help responders understand the dangers of those materials. (Courtesy Anne Arundel County, MD, Fire Department.)

Applied Physics

Applied physics is the application of the science of physics to help emergency responders tackle their hazardous materials adversaries. What we need to know before we decide to confront an escaping hazardous material is as follows: what are the physical characteristics of the material(s) and how does the hazardous material behave when it is outside the container? Volume 3 provides you with tools for your Hazmat Tool Box to help answer the questions about the escaped hazardous material and how it will behave.

Introduction

Applied chemistry and physics in this volume will focus on materials we are most likely to encounter in the real world (Figure 3.2). We touch on "exotic" chemicals and weapons of mass destruction (WMDs) for informational purposes. There is really nothing exotic about WMD materials. They are generally explosive, toxic, radioactive, and infectious substances. In reality, history shows us that explosives are the only hazardous materials that have been used to any extent in WMD incidents in the United States. The same explosives that were used by terrorists are involved with accidents at agricultural facilities. We have dealt with hazardous materials in these hazard classes before, so WMDs are really nothing new. Even though it was unpopular in the beginning when we started hearing about WMD, it has always been my belief that WMDs are hazardous materials and should be treated as such. Former Deputy Commissioner of

Figure 3.2 Applied chemistry and physics in this volume focus on materials we are most likely to encounter in the real world.

Operations for the Chicago Fire Department Gene Ryan calls WMD materials "hazardous materials with attitude." Therefore, in applied chemistry and physics, WMD materials will be treated as to their chemical and physical makeup rather than the special circumstances in which they are used.

We are more likely to encounter hazardous materials in everyday activities then in a terrorist attack. Looking at incidents that have occurred in the past, those hazardous materials that have killed the most emergency responders in accidental releases are common hazardous materials (Figure 3.3). Specifically most responders have been killed or injured by LPG, propane, butane, anhydrous ammonia, chlorine, gasoline, ammonium nitrate and other explosives in the past 137 years! Those same hazardous materials are also responsible for civilian deaths. Because these materials are so common, responders may not always understand the dangers. Commonplace chemicals do not, however, mean that they are not dangerous. Chlorine is one of the most toxic materials you are likely to encounter. You will notice these materials except for fuels are in the two most dangerous U.S. Department of Transportation (DOT) Hazard Classes: explosives and compressed gases. But if you know they are in your community, you can plan and prepare for them if in the unlikely event they escape their containers. Knowing about the common materials that are killers does not mean that we ignore other common materials, that is, the rest of the DOT Hazard Classes; if your skills are honed on the killers, you can use those same skills on the rest of the DOT Hazard

Figure 3.3 Looking at incidents that have occurred in the past, those hazardous materials that have killed the most emergency responders in accidental releases are common hazardous materials. (Courtesy Kansas City, MO, Fire Department.)

Classes. What I do not advocate is that we waste any time on "exotic" chemicals that we are not likely to ever see. Does that mean we can never encounter them? No, but statistically it is not very likely. What I want you to be able to do is effectively respond to the everyday hazardous materials in your communities and do that response safely. Studying previous incidents, some that have resulted in fatalities among emergency responders and some that went well can help us learn and prevent death and injury in the future. If we do not heed the lessons learned from the past, we are doomed to commit the same errors again in the future. Responders should be aware of the hazardous materials that are used, stored and transported in and through their jurisdictions. Focus should be on the types of materials that are a part of our technologies that bring about the standard of living and our everyday life. Understanding the dangers of these materials, responders can handle them safely and effectively. These are the materials we are most likely to encounter. In fact, 50% of all hazardous materials incidents involve hydrocarbon fuels. Hazardous materials that are also quite common but in smaller quantities in use include the following hazard classes: corrosives, liquid and solid toxic materials, flammable solids and oxidizers. The most likely hazardous materials that will be encountered outside of fixed facilities are in highway transportation. According to Interstate Commerce Commission (ICC) and the National Transportation Board (NTSB), since 1959 there have only been 44 train derailments that involved the release of hazardous materials (NTSB). Most of those derailments have been what I like to call once in a lifetime or career events. Very few responders will ever encounter a derailment of that magnitude.

Just because we have both regulations and standards that make up part of the Hazardous Materials Standard of Care doesn't mean we have to do the same things with all hazardous materials incidents all the time. Ludwig Benner, former NTSB investigator, calls that "a cookbook approach." What he advocates is an adaptive approach using a hazard risk analysis and decision-making models (Ludwig Benner).

Hazmatology Point: *We need to be able to exercise flexibility that is customized to fit the circumstances of any given incident. For example, at a lawn and garden center on the East Coast, a pint of pesticide was dropped by a customer and it broke. The fire department was called as well as the hazmat team. They suited up and made a Level A entry. Look at the time lost from the retailer, the time the fire department and hazmat team were tied up. There was a general lack of the application of common sense. The pesticide is approved for consumer use. Consumers take it home, open the bottle and mix it with water. Common sense should play a major part in deciding what we need to do to mitigate escaped hazardous materials.*

All incidents need to be kept in context; they are not all alike even if the chemicals are the same. Procedures that worked previously do not mean that they will work every time. Quantities and intended uses of hazardous materials as well as hazards will vary widely, and each incident has a different set of circumstances (actors). We can, however, review and study previous incidents and make use of lessons learned that led to unfavorable outcomes. This includes incidents that we have responded to personally or historical incidents that others have experienced.

Volume 3 is intended to provide emergency response personnel with a combination of science, history, adaptability and common sense as it applies to the hazardous materials encountered in daily responses. We cannot train for every specific hazardous materials incident that could happen. In general, we use the same "Hazmat Tool Box" for all hazardous materials responses. Your hazmat tools are both cognitive skills and physical tools. Not every tool will be needed or used in every incident. We need to make sure that we know how to use all of the tools and select the appropriate tools that are needed to safely stabilize an incident. We do not have to use tools that are not needed, just because we have them and we can. With that in mind, we need to make sure that our personnel can implement those skills effectively no matter what hazardous material is involved or how big the incident turns out to be.

Hazardous materials response teams are not unlike sports teams. We all have to do our job well for the outcome of an incident to be successful. Members of the Houston hazmat team told me during a visit that they were in essence all equal team captains. Their senior captain by rank was more like a quarterback than a coach. Everyone knows their strengths and how to do their job. In case of sports, it makes the difference between winning or losing. With hazmat response, it can mean the difference between life and death.

Keep in mind that you are trained to deal with all hazardous materials with the same basic procedures. Common hazardous materials responses can be overwhelming, depending on a responder's level of training and resources to deal with an incident. Do not let that interfere with what needs to be done. Responders at all levels have a job to do, even if it is setting up zones and denying entry. Evacuation or sheltering in place may also be appropriate. If in doubt, protect the public and shelter in place or evacuate as dictated by the circumstances of an incident. Levels of training and protective equipment dictate what you can do safely. Do not exceed those levels.

Planning to deal with hazardous materials based on your ability will identify how to overcome your training and resource limitations. Few jurisdictions will experience "Career Events." No matter what the incident size or hazards of a material, you do the best you can within your limits and know who to call for advanced assistance. Have a general plan for all hazardous materials incidents. You may need a hazmat team or resources from all levels of government. Handling any hazardous materials incident is much like learning to walk; do not rush into an incident, rather take it "one step at a time."

What Constitutes a Hazardous Materials Incident?

Effectively dealing with hazardous materials incidents cannot be accomplished without knowing something about the chemicals involved (Figure 3.4). Knowing the physical, chemical and family effects of hazardous materials along with monitoring and testing of materials can change the potential outcome of incidents. At the very least, responders need to know the hazard class of chemicals they encounter before they can safely begin operational planning. The key to successful handling hazardous materials releases is to know how to apply these characteristics to incidents keeping responders and the public out of harm's way. Remember that emergency responders do not cause incidents to occur. Our job is to make the situation better than it would have been if we had not responded at all. There are many incidents presented in Volume 1, which show that this has not always happened in the past and on a smaller scale continues to this day. Some would call this Monday morning quarterbacking. What is intended is examining lessons learned. Hundreds of brave responders gave the ultimate sacrifice over the years dealing with incidents involving hazardous materials. Learning from these incidents only honors their sacrifice and forever honors them. They are true heroes, and by learning from what happened shows that they did not die in vain and likely saved many lives in future incidents.

Figure 3.4 Effectively dealing with hazardous materials incidents cannot be accomplished without knowing something about the chemicals involved.

First and foremost when responding to any incident, responders must be able to determine if a hazardous materials incident has occurred. You must be able to identify hazardous materials incidents, or all of your training and experience really doesn't really matter. Identification of suspected hazardous materials and the hazard to responders and the public is paramount to a successful response outcome. Conducting a risk analysis is almost impossible if you do not know what the risk(s) are. Understanding the concepts of Applied Chemistry and Physics may help guide the identification and risk analysis process.

Applied Chemistry and physics provides responders with identification and risk management tools. The concepts presented work in the application of chemistry and physics when dealing with hazardous materials. This information may help keep responders from being injured or killed. It is important that responders learn to utilize the science of hazardous materials and common sense to realistically determine the risks involved in dealing with hazardous materials. Intervening in a hazmat incident is not always necessary or advised, just because we can is not a good reason to do so. Crucial to safety and survival, it is necessary to conduct a risk–benefit analysis to determine what actions need to be taken or can be taken safely and to know when no intervention actions are necessary at all.

When Hazardous Materials Are Present

Once determined that a hazardous materials incident does exist, then identification of the hazard is the next most important task following initial protective actions. This can be particularly difficult with no readily available clues. Knowledge of materials identification or, at the very least, knowing hazard class or the family it belongs to will expedite the mitigation of an incident. Identification includes recognition of clues to the presence of hazardous materials and determination of the hazards. Part of your basic hazardous materials training teaches you about clues to the presence of hazardous materials.

Once hazards are known, an effective risk analysis can be initiated. Risk analysis is followed by determination of goals to safely stabilize the incident. Using Applied Chemistry and Physics will help responders to understand the hazardous materials, their hazards and how to safely undertake the stabilization of the incident. Safe stabilization of an incident should be your ultimate goal. Remember that offloading, flaring, overpacking and cleanup are generally not an emergency response function. We never take hazardous materials away from an incident scene. If we do, we become the owners of the materials and are responsible for the proper storage, transportation and disposal including the costs. Private contractors are usually coordinated by state and local departments of environments and the U.S. Department of Environment and many times find funding for these tasks. They are responsible for removing and properly disposing of hazardous materials from an incident scene.

Advances in technology and miniaturization of monitoring instruments have resulted in much easier identification of unknown hazardous materials on the street. Drones and robots have made it much safer to monitor, take samples and conduct on-scene surveillance without placing personnel in harm's way. Life safety is the number one incident priority when dealing with hazardous materials. Responder life safety is the number one life safety priority.

Actors in Hazardous Materials Incidents

When an accident occurs, which results in the escape of hazardous materials from their containers, a number of actors are present in addition to the hazardous materials that can affect a positive or negative outcome. Each actor can have an effect on the hazardous material(s) or containers involved. Actors include the chemical(s) themselves, incompatible chemicals, normally inert materials such as water or nitrogen, containers, damage to containers, weather, geography and others. Responders need to study and understand the impacts of these actors on a hazardous materials incident as well as make them a part of the risk analysis. Do not become so involved in what the hazardous materials are doing that you do not evaluate the presence of all potential actors and what the hazardous materials might do.

Chemicals

Hazardous materials incidents are all about the chemicals involved. Factors that determine how the incident may play out, with or without responder intervention, are physical state, hazard class, physical characteristics, chemical characteristics, compatibilities and incompatibilities (Figure 3.5). These are all actors that can play a part in the successful or unsuccessful outcome of the incidents. Chemical actors will be discussed in detail in this volume in the appropriate section. All of these factors need to be considered and thoroughly researched in order to make a scientific judgment regarding the role the chemical will play in the incident.

Incompatible Chemicals

Let's face it some chemicals just do not like each other. They are incompatible with other individual chemicals or families of chemicals. Storing them together is asking for trouble; that we have control of. When they come together during an incident, it will be beyond our control.

Applied Chemistry and Physics

Figure 3.5 Hazardous materials incidents are all about the chemicals involved.

Using them to neutralize each other in cases of corrosives is within our control and may be an option in small amounts. Reactions that may occur when incompatible chemicals come together may range from mild to explosive. Actors need to be thoroughly researched before we determine operational tasks. Reactions may occur before you arrive. You may encounter a witches' brew of chemicals that no chemist in the world can tell you what the hazards are. Chemicals may mix together that are not normally mixed together and we do not know what they will do or what the hazards are. Nonetheless, these chemicals are actors in incidents and we need to be aware of them.

Inert Materials

Inert materials generally do not have hazards associated with them. They are present in transportation or fixed facilities because they provide safety for the hazardous material so it can be transported and stored without incident. Acetone is used to dissolve acetylene so that it does not detonate in the container. When transporting acetylene, sometimes end users will lay acetylene tanks on their side. When this happens, acetylene can separate from acetone and detonate. Hazardous materials (polymers) subject to polymerization are shipped and stored with materials called inhibitors. Inhibitors prevent the polymerization reaction from occurring under normal circumstances.

However, in an accident, the inhibitor can become separated from the polymer and the polymerization reaction can occur. Once the reaction starts, it cannot be stopped. It will continue until the fuel is consumed. Even if you had a tanker truck full of inhibitor and could apply it to the polymer, it wouldn't stop the reaction at that point.

Kerosene, a flammable liquid, is sometimes used to prevent lithium, potassium and sodium metals from reaching air. They are water-reactive materials and can react with moisture in the air. However, in a fire situation, the kerosene will burn causing additional fuel for a fire.

Phosphorus, on the other hand, is not water reactive, and it is not a metal. It is shipped under water to prevent it from reaching the air. If the water leaks off in an accident or there is a container failure, phosphorus will catch fire when exposed to air.

CASE STUDIES

A train derailed in Brownson, NE, on April 2, 1978. One tank car contained yellow phosphorus (Figure 3.6). Phosphorus is an air-reactive material that spontaneously ignites when exposed to air. Water is used to cover phosphorus in transportation and storage so that it cannot reach air and ignite. During the Brownson incident, the tank car overturned and water spilled from the tank exposing some of the phosphorus to air and the phosphorus ignited immediately.

Figure 3.6 One tank car contained phosphorus. (Courtesy Sidney, NE, Volunteer Fire Department.)

Responders contacted CHEMTREC and were told that the phosphorus would not explode. CHEMTREC was correct. However, the water became a dangerous actor in the incident outcome, and it was not identified as a potential actor by responders or CHEMTREC. As burning phosphorus heated the water inside the liquid tank car, enough steam, a gas, built up the pressure in the tank; when the tank could no longer hold the pressure, it ruptured into several pieces. Phosphorous did not explode; the container exploded from the over pressurization caused by steam. This was basically a boiler-type explosion, and the water turned out to be the bad actor in this incident. Additional incidents have occurred in Gettysburg, PA, involving a truck carrying phosphorous, which was shipped under water in 55-gallon drums. One of the drums leaked, releasing the water covering the phosphorus. When the phosphorus reached air it ignited and burned an causing a chain reaction fire which spread from one drum to another. (*Firehouse Magazine*).

Container Damage

Incident Caused Container Damage

Containers for hazardous materials are generally designed to meet specifications required by local, state or Federal laws and regulations. Design standards provide guidance to make sure that containers will hold the chemicals and pressures so the materials can be transported, stored and used safely. However, when an accident occurs involving these containers, these design features for safety may not work. For example, relief valves are designed to relieve increased pressure caused by an increase in ambient temperature. They are not designed to relieve pressure from flame impingement caused by an accident. Pressure containers may fail under fire conditions or changes in ambient temperature once a tank is damaged in an accident.

When an accident happens involving a pressure container, responders need to assess what I like to call the "mechanism of injury". This term is used in EMS response to assess patient injures, and this same concept can be applied to assess container damage. Damage to containers may not always be visible or accessible. Looking at the mechanism of injury can help estimate the potential damage that has occurred. Because of the potential danger to responders, it is best to view the exposed areas of suspected damaged tanks from a safe distance with binoculars. Drones and robots may also be helpful to safely view the damaged containers from a distance. A well-trained drone operator can get a better view of incident scenes and damaged containers than response personnel making entry or flying over with a conventional aircraft.

Drone use is increasing across the fire service. If you do not use drones, you may want to consider them in the future. If you are unable to approach the accident scenes because of safety issues, you can still make a good estimate of damage using the mechanism of injury and erring on the side of safety.

CASE STUDY

Waverly, TN, experienced a train derailment right in the middle of town on a Wednesday. LPG cars were part of the derailment and were involved with a pileup of other rail cars. It was a winter day with some snow on the ground and temperatures in the 30s. Nothing of an emergency situation developed during the derailment—no explosions, fires or leaks. Emergency responders were called and much of the town was evacuated. After it was determined that no emergency existed, the evacuation was called off. Responders and the public were on or near the incident scene, and it was pretty much business as usual in Waverly. As is the norm for railroads, they wanted to get the track open and commerce moving again as soon as possible.

On Friday, the temperatures rose into the 50s in Waverly. Cranes were on-site along with railroad crews, and they were in the process of clearing the track. Preparations were being made to transfer the LPG from tank car 83013 when propane started leaking followed shortly by a massive explosion (Figure 3.7). Waverly's fire and police chief were killed—the police chief instantly and the fire chief later in the hospital. Sixteen people were killed and 43 injured from the explosion and fires, and $1,800,000 in 1978 dollars in damage. The National Transportation Safety Board (NTSB) investigated and said that the probable cause was the release and ignition of liquefied petroleum gas from a tank car rupture.

The rupture resulted from stress propagation of a crack, which may have developed during movement of the car for transfer of product, or from increased pressure within the tank. The original crack was caused by mechanical damage during the derailment. It is believed that the increase in ambient temperature on Friday caused the buildup of pressure in the tank. The damaged area of the tank

Figure 3.7 Preparations were being made to transfer the LPG from tank car 83013 when LPG started leaking followed shortly by a massive explosion. (Courtesy Waverly, TN, Fire Department.)

could not withstand the increased pressure, and the tank started leaking and quickly came apart explosively. The police chief, fire chief, 2 firemen and civil defense worker were killed along with 11 civilians (*Firehouse Magazine*).

CASE STUDY

On July 2016, 8:15 a.m., an 18-wheeler which included an MC331 propane tanker went around a curve and hit a concrete embankment at 55–60 mph and skidded 200 feet on the concrete roadway. Friction between the steel tank and the concrete left scrape marks on the tanker's side (Figure 3.8). Considering the mechanism of injury (damage) to the tank, hazmat team members on arrival did a visual inspection of the tank and air monitoring to determine if there was a leak. At the time of the incident, the temperature in the container was 85°F, and the tank pressure was 130. Weather reports indicated that summer temperatures in Houston would reach 110°F–115°F during the day. That temperature increase would raise the pressure inside the tank. The decision was made to keep the tank cool with a Ventura using a hose stream, which would also provide a secondary cooling effect.

Figure 3.8 Friction between the steel tank and the concrete left scrape marks on the tanker's side. (Courtesy of Houston Fire Department.)

Figure 3.9 A tent was fashioned from the ever innovative firefighting tool, the salvage cover. The tank was tented and the hose line was placed to begin the cooling process. (Courtesy Houston Fire Department.)

This accident occurred on an elevated section of the expressway, and there was no water supply. 1000′ of 5″ hose was used to supply a ladder pipe that was used to hook the hose to, like an artificial standpipe. A tent was fashioned from the ever innovative firefighting tool, the salvage cover. The tank was tented and the hose line was placed to begin the cooling process (Figure 3.9). The entire well-planned process worked as intended and kept the tank cool until it could be safely offloaded; the accident scene was investigated and cleared of debris. No further damage or injury occurred.

Incidents have also occurred in compressed gas containers that have had improper maintenance, lack of maintenance or improper repairs. These situations will likely go undetected as the accidents frequently occur during filling of the containers. Containers fail during filling and investigation usually identifies improper repair or lack of maintenance (Houston Fire Department).

CASE STUDY

Two plant workers were killed in Verdigris, OK, June 10, 1979 while filling a railcar with anhydrous ammonia. The force of the explosion sent one man through a chain link fence, which was the cause of death. Vapor clouds from the explosion defoliated trees and turned vegetation brown. Investigation found that the tank car was welded in South America, and the improper weld created a weak point in the tank; while the tank was being filled, the tank gave way at the weakened weld. The plant workers were the only people killed and no one else was injured.

Post-Incident Container Damage

Container damage may occur as a result of post-incident issues. Flame impingement on an undamaged pressure or liquid tank can cause that tank to come apart as an incident progresses. Compressed gas containers have relief valves designed to relieve pressure increases caused by temperature changes. Watching relief valve function during flame impingement on an undamaged container can provide information about the pressures inside the tank. Someone should be assigned to watch the relief valve operation throughout the incident. Changes in the height of the flame or sound being emitted from the discharge should be noted. The higher the flame goes and the louder the sound becomes, the closer you are to a boiling liquid expanding vapor explosion (BLEVE).

Location of the flame impingement on the pressure container is critical. Flame impingement on the liquid level of the tank does not cause structural damage to the steel shell below the liquid level.

However, if the flame impingement is above the liquid level, BLEVE is imminent. The National Fire Protection Association (NFPA) says that once fame impingement on the tank is above the liquid level, you have between 8 and 30 minutes, with an average of 15 minutes or less before BLEVE will occur. This is problematic in terms of tactical choices. There are some critical unknowns. When did the flame impingement begin? How long after the flame impingement began was the call made to emergency responders? How long did it take for them to arrive on scene? How much time is left to figure out what to do? This is one of those Low-Frequency, High-Risk Don't Have Time to think types of incidents. These are the ones that killed firefighters in Kingman, AZ; Dumas, TX; Milwaukee, WI and others. You do not take chances with pressure tank fires where flame impingement on the tank is involved. This is a loser and where the binoculars and tennis shoes come into use. Evacuate to the distances listed in the *Emergency Response Guidebook* (ERG) and secure the area. The only safe actions to take would be taking photographs.

Boiling Liquid Expanding Vapor Explosion

BLEVE was mentioned several times in the previous paragraph. Liquefied compressed gas containers carry liquids that are all already above their boiling points. These liquefied gases are kept in liquid form by pressure inside the tank and the confines of the tank itself, which has been engineered to hold the liquid and the pressures involved under normal conditions. Flame impingement quickly weakens the steel to the point it will no longer hold back the pressure building up in the tank, and thus, a BLEVE occurs (Figure 3.10). Because liquefied gas in the tank is kept in liquid form by the pressure, once that pressure escapes the tank, all of the liquefied gas contents immediately return to the gas state tearing the tank apart and forming a gas cloud. If there is an ignition source, a fireball will form burning the gas vapors. All of this occurs explosively within milliseconds of the tank failure. Tank parts can be rocketed thousands of feet from the blast. Once the fireball goes away, there is no more fire unless ordinary combustibles are ignited by the fireball. However, all of the liquefied gas is consumed by the fire, and there is no gas left to burn.

Liquid tanks do not BLEVE, but they rupture. First of all, they are liquids and never have been gases. Tanks that are used to transport, store and use liquids are engineered to hold the liquid contents under limited pressure. When flame impingement occurs, the liquid protects the metal and absorbs the heat. Since the liquids were not boiling previous to flame

Applied Chemistry and Physics

Figure 3.10 Fame impingement quickly weakens the steel to the point it will no longer hold back the pressure building up in the tank, and thus, a BLEVE occurs. (Courtesy Crescent City, IL, Volunteer Fire Department.)

impingement, it takes a period of time to cause the liquid to boil (the temperature of which varies according to altitude). Once the liquid does start to boil, pressure builds up within the tank, which is not designed to hold pressure. Eventually, the tank ruptures at some point weakened by flame impingement, and all of the pressure is released (flammable vapors from the boiling of the liquid).

Flammable vapors will likely ignite from an ignition source and burn as a large fireball above the tank. These fireballs are not nearly as large as those from a liquefied gas. Once the vapors are consumed by the fire, the liquid in the tank continues to burn along with any ordinary combustibles ignited by the fireball. There is no longer pressure buildup in the tank because there is an opening in the tank to release the pressure. The tank did not come apart in the rupture so no tank pieces of any significance were rocked anywhere. Multiple ruptures in multiple tanks can occur in train derailments because of flammable liquid pools burning under multiple tanks. This is a Low Frequency, Low Risk incident where you have enough time to plan operational tasks.

CASE STUDY

About 8:36 p.m., central daylight time, on Friday, June 19, 2009, eastbound Canadian National Railway Company freight train traveling at 36 mph derailed at a highway/rail grade crossing in Cherry Valley, IL. Derailed tank cars were breached or lost product and caught fire. At the time of the derailment, several motor vehicles were stopped on either side of the grade crossing waiting for the train to pass. As a result of the fire that erupted after the derailment, a passenger in one of the stopped cars was fatally injured (Figure 3.11), two passengers in the same car received serious injuries and five occupants of other cars waiting at the highway/rail crossing were injured. Two responding firefighters also sustained minor injuries. The release of ethanol and the resulting fire prompted a mandatory evacuation of about 600 residences within a 1/2-mile radius of the accident site. Monetary damages were estimated to total $7.9 million (NTSB).

Figure 3.11 As a result of the fire that erupted after the derailment, a passenger in one of the stopped cars was fatally injured.

Weather

Weather conditions including wind directions and speed can play a major role on the scene of a hazardous materials incident. This role may be helpful or may cause negative issues when trying to mitigate an incident. High winds during a gas release may help disperse and dilute the gas to the point it is no longer harmful. On the other hand, lower winds may move a toxic or flammable cloud beyond the incident scene and place other areas in harm's way. Sunny days can cause an increase in pressure in uninsulated containers and containers with dark colors, which absorb heat. An increase in pressure can cause a damaged portion of a tank to fail. Lighter colored tanks can reflect the heat of the sun. Leaks from tanks when it is raining can spread flammable liquids, which float on top of run-off water increasing the size of the incident scene. Water can also dissolve water-soluble gases and create another class of hazardous material, such as ammonium hydroxide from an ammonia plume. Ammonium hydroxide while not as hazardous as anhydrous ammonia; it is corrosive and can cause injury in contact with skin and corrode metal.

Solid materials are potentially one of easiest hazardous materials to deal with on a sunny, windless day. However, if the solid is water reactive, you do not want it to be raining or snowing. Moisture contacting a water-reactive solid material can release a gas with hazards far worse than the solid itself. Some water-reactive solid materials can react to high humidity. For example, a spill of the solid calcium carbide exposed to moisture in the form of rain or snow releases acetylene gas. Acetylene gas is highly flammable with a very wide flammable range. In the green section of the DOT Emergency Response Guide, there is a list of water-reactive solid materials and the toxic gases released when they become wet (Figure 3.12). Responders need to be aware of the atmospheric conditions when arriving on the scene of a hazardous materials incident. Watching for changing weather conditions is also necessary as in some parts of the country, weather can change quickly. Someone should be assigned to monitor on-scene conditions and keep up with forecast changes. These weather conditions can be a good or bad actor when it comes to successful mitigation of hazardous materials release.

Geography

Geography changes from one part of the country to the other. It changes from one part of a state to another. It changes from one part of a local jurisdiction to another. That is a fact we can do little if anything about when a hazardous materials incident occurs. However, we need to recognize how geography will affect our hazardous materials incident. Hills or mountains will determine from which direction you respond to an incident. They may also cause you problems or help you when it comes to vapor cloud travel. A hill or mountain, even a big ditch, would help contain a vapor cloud of a heavier than air vapor. Depending on the elevation of the obstruction, it could keep a toxic vapor cloud from reaching a population on the other side. On the downside, inclines can carry vapors and liquids down to locations you may not want them to go.

Rivers or creeks would spread the spill of a flammable hydrocarbon fuel beyond the source. Hydrocarbon fuels in general float on the surface

TABLE OF WATER-REACTIVE MATERIALS WHICH PRODUCE TOXIC GASES

Materials Which Produce Large Amounts of Toxic-by-Inhalation (TIH) Gas(es) When Spilled in Water

ID No.	Guide No.	Name of Material	TIH Gas(es) Produced	
1162	151	Dimethyldichlorosilane	HCl	
1242	139	Methyldichlorosilane	HCl	
1250	155	Methyltrichlorosilane	HCl	
1295	139	Trichlorosilane	HCl	
1298	155	Trimethylchlorosilane	HCl	
1340	139	Phosphorus pentasulfide, free from yellow and white Phosphorus	H_2S	
1340	139	Phosphorus pentasulphide, free from yellow and white Phosphorus	H_2S	
1360	139	Calcium phosphide	PH_3	
1384	135	Sodium dithionite	H_2S	SO_2
1384	135	Sodium hydrosulfite	H_2S	SO_2
1384	135	Sodium hydrosulphite	H_2S	SO_2
1397	139	Aluminum phosphide	PH_3	
1412	139	Lithium amide	NH_3	
1419	139	Magnesium aluminum phosphide	PH_3	
1432	139	Sodium phosphide	PH_3	
1433	139	Stannic phosphides	PH_3	
1541	155	Acetone cyanohydrin, stabilized	HCN	
1680	157	Potassium cyanide	HCN	
1689	157	Sodium cyanide	HCN	
1714	139	Zinc phosphide	PH_3	
1716	156	Acetyl bromide	HBr	
1717	132	Acetyl chloride	HCl	
1724	155	Allyl trichlorosilane, stabilized	HCl	
1725	137	Aluminum bromide, anhydrous	HBr	

Chemical Symbols for TIH Gases:

Br_2	Bromine	HF	Hydrogen fluoride	PH_3	Phosphine
Cl_2	Chlorine	HI	Hydrogen iodide	SO_2	Sulfur dioxide
HBr	Hydrogen bromide	H_2S	Hydrogen sulfide	SO_2	Sulphur dioxide
HCl	Hydrogen chloride	H_2S	Hydrogen sulphide	SO_3	Sulfur trioxide
HCN	Hydrogen cyanide	NH_3	Ammonia	SO_3	Sulphur trioxide

Use this list only when material is spilled in water.

Figure 3.12 In the green section of the DOT Emergency Response Guide, there is a list of water-reactive solids and the toxic gases released when they become wet.

of the water and carry it down stream, in some cases even if it is on fire. Water can facilitate the spread of fire or carry a liquid into an ignition source and cause a fire that could flash back to the source of the spill. Soil composition may contain spills or allow the spilled material to leach through the soil contaminating the soil below and potentially subsurface water supplies. Geographical components can be a good or bad actor, but you need to be aware of the geographical surroundings as you approach a hazardous materials incident scene.

Applied Chemistry

Science of hazardous materials involves the study of chemistry. Chemistry is the study of matter. Most chemicals can be divided into two groups: inorganic and organic. Acids, bases, salts and certain elements are some of the materials included with inorganic chemistry. Organic chemistry, on the other hand, is the study of compounds that all contain carbon and a few other elements such as oxygen, nitrogen, fluorine, chlorine and bromine. Subfamilies of chemicals with similar hazards exist within each general group. For example, ether compounds as a family are generally flammable and anesthetic. Oxysalts are strong oxidizers. Hydrocarbons are flammable to varying degrees. Elements in the alkali metal family on the periodic table are water reactive. These concepts will be covered in more detail in other parts of this volume.

States of Matter

One of the first things you want to determine about a hazardous material is the state of matter you are dealing with. Containers and hazardous classes can be a hint as to what the state of matter may be, or to determine what it is not by process of elimination. Determining the state of matter of a hazardous material helps in determining hazard and risk analysis and operational goals. Matter is defined as "anything that occupies space and has mass." Studying matter provides response personnel with some simple facts about materials and containers based on the state of matter in which a hazardous material is encountered. Matter exists in three basic forms: solid, liquid and gas.

Knowing the physical state of a hazardous material indicates where the material is likely to be and where it is likely to go (Figure 3.13). It is

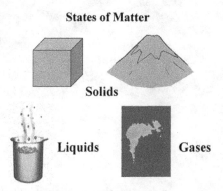

Figure 3.13 The physical state of a hazardous material indicates where the material is likely to be and where it is likely to go.

important that responders know these states of matter well and how to determine them and the advantages and disadvantages of each state. Different states of matter will require different tactics. Also understand that interactions with other actors can cause a state of matter to change from one state to another. Temperature is an issue with gases and liquids. Weather and moisture can be a concern with solid materials.

Solids

Definition of solid. Matter in the solid state maintains a fixed volume and shape. Solids can be transformed into liquids by melting, and liquids can be transformed into solids by freezing. Solids can also change directly into gases through the process of sublimation. Solids are the easiest state of matter to deal with. Unless there is wind or rain, the material will not likely go anywhere beyond the location where it was spilled (Figure 3.14). Tarps can be used to cover the solid to protect it from the wind. Sublimation is a process whereby a solid turns into a gas without entering the liquid phase. One example of sublimation is dry ice. Dry ice turns directly into a gas without becoming a liquid. Snow and ice under certain conditions can turn directly into water vapor (gas) without first becoming a liquid. Camphor and moth balls are also examples of materials that sublime.

Solid materials may be encountered in various forms from dust to solid blocks. Forms can make a difference between little hazard to flammable and explosive hazards. Intermediate steps exist in the process of classifying solids, liquids and gases. Some solids may have varying

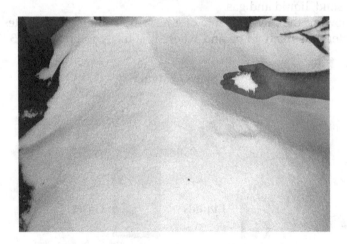

Figure 3.14 Solids are the easiest state of matter to deal with. Unless there is wind or rain, the material will not likely go anywhere beyond the location where it was spilled.

particle sizes from large blocks to filings, chips and dusts. This can affect the flammability of the particle size. Though not considered a hazardous material by the Department of Transportation (DOT), dusts can be explosive if suspended in the air in the presence of an ignition source. Dust explosions have been occurring for over a hundred years and show no sign of slowing down. The Chemical Safety Board (CSB) has undertaken an initiative to reduce dust explosions. Dusts can include grain, flour, starch, saw dust and many others. Particle sizes of vapors may vary from invisible vapors that are very small to mists and droplets that are readily visible. Vapors require detection equipment to determine their presence or some physical characteristics. If the temperature of a liquid is above its boiling point, vapors can be detected in the same manner as the gases.

Liquids

Definition of liquid. It is a state of matter that is not solid. Liquid commonly refers to substances, such as water, oil and alcohol, and the like, that are neither solids nor gases. Water ceases to be a liquid when it is frozen or turned to steam. Liquids are somewhat easier than gases to deal with because you can at least see where the liquid is located in a spill, and based upon terrain you will know where it will likely go (Figure 3.15). Procedures can be deployed for stopping the flow of a liquid keeping it from places you don't want it to go, such as sewers and water supplies. Damming, diking and collection are examples of actions used to control the movement of liquids.

Figure 3.15 Liquids are somewhat easier than gases to deal with because you can at least see where the liquid is located in a spill and based upon terrain you will know where it will likely go. (Courtesy Gwinnett County, GA Fire Department).

Gases

Definition of gas. It is a state of matter that has no definite shape or volume. It flows freely when released into the environment depending on its physical characteristics such as vapor density, vapor pressure and temperature (Figure 3.16). Gas will conform to a container and remain in a properly designed container, unless the container is damaged or breached in an accident. In spite of its lack of organization, a released gas may be controlled or contained if you know the physical characteristics of the gas. For responders, gases are the most difficult to deal with because you don't always know where the gas is or where it is going, and may not be able to stop it. Gases and vapors may be invisible or colorless. Locating them will likely require monitoring instruments.

Temperature and Pressure

Temperature and pressure have a great deal to do with the physical state of a chemical; however, they do not change the chemical properties. Elevated temperatures can be your worst enemy on the scene of a chemical release. Ambient temperature and radiated heat from a fire can change the physical state of a liquid chemical depending on the boiling point of the liquid. Water is a liquid at normal temperatures and pressures. If water exposed to temperatures below 32°F (0°C), it becomes a solid. There are some exceptions: water with varying levels of salt may take a lower temperature to freeze. The solid form of water still maintains all of the chemical properties of liquid water.

If water is heated above 212°F (100°C), it becomes a gas. However, the gas still maintains all of the chemical properties of liquid water.

Figure 3.16 Gas is the most difficult physical state for responders to deal with.

Boiling Point of Water and Atmospheric Pressure (psi) Vs. Altitude

Altitude	Atmospheric Pressure	Boiling Point
Sea Level	14.7 psi	212° F
3300 Feet	13.03 psi	203° F
5280 (1 mile)	12.26 psi	201° F
10,000 Feet	10.17 psi	194° F
13,000 Feet	9.00 psi	188° F

Figure 3.17 Comparison of water boiling points at various elevations.

As temperature increases, more particles have enough energy to escape to the gas phase. This increases the vapor pressure. When vapor pressure equals atmospheric pressure, the liquid boils. Altitude also affects the boiling point of water. At the lowest point in Death Valley, the boiling point would be about 212.0°F/100.3°C, and in deep mines, the boiling point could be even higher. As elevation increases, atmospheric pressure decreases because air is less dense at higher altitudes. Therefore, less heat is required to make the vapor pressure equal to the atmospheric pressure. The boiling point is lower at higher altitudes (Figure 3.17).

Physical State and Chemical Hazards

The hazard presented by a chemical may be affected by the physical state of the material when it is encountered. For example, only gases burn, solids and liquids do not burn, even though they may be listed as flammable. A solid or liquid must be heated until it produces enough vapor to burn. Water has a cooling effect on the skin as a liquid. When water is turned into a gas (steam), it causes burns to the skin. When water is in the solid state, it can freeze the skin and cause thermal burns. It is important to determine the physical state of a hazardous material early in the incident in order to formulate mitigation goals.

Periodic Table of Elements

Chemistry cannot be discussed without examining the Periodic Table of Elements (Figure 3.18). The periodic table is chemistry's method of organizing everything that is known about the chemical universe on one piece of paper. The table reveals the relationship between elements by showing

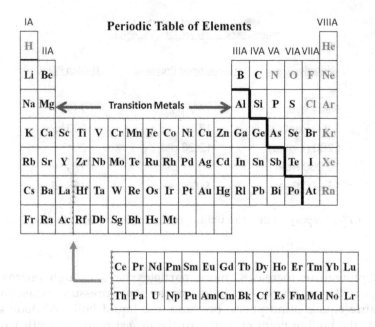

Figure 3.18 Periodic table of elements.

the tendency of their properties to repeat at regular intervals. All chemical compounds are derived from two or more elements or combinations of elements from the periodic table.

Elements

Elements are the building blocks of chemical compounds. Symbols are used to represent each of the elements on the table. The periodic table is composed of a series of blocks representing each element. Within each block is a symbol that represents the name of that element. Also within the block is the atomic number (Figure 3.19) and atomic weight (Figure 3.20). The symbol is a type of shorthand for the element's name. For example, carbon has a symbol of C, the element gold is represented by the symbol Au, chlorine's symbol is Cl and potassium's symbol is K. Each symbol represents one atom of that element. Symbols may be a single letter or two letters together. Single letters are always capitalized. When there are two letters, the first is capitalized and the second is always lowercase. This is important to understand when trying to identify elements and compounds. For example, CO is the molecular formula for the compound carbon monoxide; Co, on the other hand, is the symbol for the element cobalt. There are two totally different materials with quite different hazards.

Symbols and names of elements are derived from a number of sources. They may have been named after the person who discovered the element.

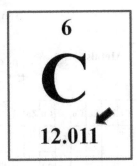

Figure 3.19 C indicates the symbol for the element carbon. Numbers with decimal point indicate the atomic weight of one atom of that element.

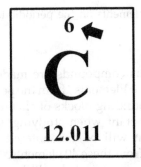

Figure 3.20 The whole number indicates the atomic number of that element.

For example, W which is the symbol for tungsten is named after Wolfram, the discoverer. Other elements are named after famous scientists, universities, cities and states. Es is the symbol for einsteinium, named after Albert Einstein. Cm is the symbol for curium, named after Madam Curie. Bk is the symbol for berkelium, named after the city of Berkeley, California. Cf is the symbol for the element californium, named after the state of California. Other element names come from Latin, German, Greek and English languages. In the case of sodium, Na comes from the Latin word for *natrium*. Au, the symbol for gold, comes from *aurum*, meaning "shining down" in Latin. Cu (copper) comes from the Latin *cuprum* or *cyprium* because the Roman source for copper was the island of Cyprus. Fe (iron) comes from the Latin *ferrum*. Bromine means "stinch" in Greek. Rubidium means red in color. Mercury is sometimes referred to as quick silver. Sulfur is referred to as brimstone in the Bible.

Studying the makeup of chemical compounds can help responders to understand the hazards and identify the materials in the field. Not all of the elements on the periodic table are common or particularly hazardous

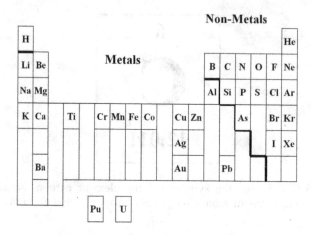

Figure 3.21 There are 38 elements on the periodic table, that we will call the "Hazmat Elements."

to responders. Hazardous compounds are made up of one or more of approximately 40 "Hazmat Elements" from the periodic table (Figure 3.21). These elements are the building blocks of chemical compounds that are hazardous and are important when studying applied chemistry. Most hazardous materials that will be encountered by response personnel include or are produced from these 40 elements.

Hazardous materials personnel should become familiar with these elements and be able to recognize them by symbol and name. Formulas may appear on container labels, MSDS Sheets, literature, reference books and computer data bases. The ability to recognize symbols in a formula could help responders identify chemical families and potential hazards. Elements with atomic number 83 or above are radioactive; many are rare and probably will not be encountered by emergency responders. Manmade elements are the result of nuclear reactions and research. These elements may have existed on earth at one time, but because they are radioactive and many half-lives have passed, they no longer exist naturally.

Elements can be hazardous materials by themselves such as hydrogen, chlorine, potassium, uranium and others. These are considered pure chemicals rather than compounds. Some elements do not exist naturally as single atoms. They chemically bond with another atom of that same element to form "diatomic" molecules. The term "di" simply means two and "atomic" refers to the atom. Therefore, diatomic means two atoms. The diatomic elements are hydrogen, oxygen, nitrogen, chlorine, bromine, iodine and fluorine.

One way of remembering the diatomic elements is by using the acronym HONClBrIF, pronounced honk-le-brif, which includes the elemental symbol for each of the diatomic elements. Oxygen is commonly referred to as O_2 in emergency response. This reference to O_2 is primarily because oxygen is a diatomic element. Two oxygen atoms have covalently bonded together and act as one unit.

Chemical compounds are further organized into chemical families based upon the types of building blocks used to make them up. This chemical family relationship can determine the hazards of a chemical family, which can assist in the identification process. Family effect occurs based upon the types and numbers of elements that make up the family members. That is why it is important to be able to recognize the elements on the periodic table and in a compound, which may be hazardous based upon the type of elements present.

Hazmatology Point: *A periodic table should be a resource available to hazardous materials responders as a reference tool. Periodic tables can be useful on computers and in hard copy to write notes on during an identification process.*

Into the Atom

An atom is the smallest particle of an element that can be found that retains all of its elemental characteristics. The word atom comes from the Greek word meaning "not cut." For example, take a sheet of paper and tear the paper in half. Keep tearing the paper in half until it becomes so small that it cannot be torn any smaller by hand. Then take a knife and cut the paper into smaller pieces. Eventually, it will not be able to be cut any smaller. The atom is like that last piece of paper. You cannot have a smaller piece of an element than an atom. A single atom cannot be altered chemically. To create a smaller part of an element would require that the atom be split in a nuclear reaction. Therefore, a single atom is the smallest particle of an element that would normally be encountered.

An atom is comprised of three major parts: electrons having a (−) negative charge, protons having a (+) positive charge and neutrons having no charge, which are neutral. The atom is like a miniature solar system with the nucleus in the center and the electrons orbiting around the outside (Figure 3.22). These parts are referred to as subatomic particles because they are smaller than the atom itself. Electrons are most important to chemistry, and the nucleus is most important to radioactivity. Orbiting in shells or energy levels around the nucleus are varying numbers of electrons. Electrons are very important in discussing chemistry for hazardous materials responders. Electrons in the outer most shell are the only ones involved in chemical bonding. Elements bond together and form compounds to become stable. Bonding occurs so that each atom of each element will become complete and stable like the noble gases in

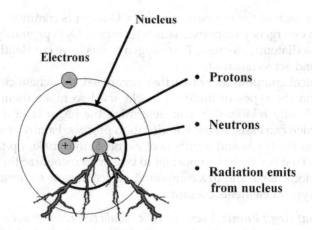

Figure 3.22 The atom is like a miniature solar system with the nucleus in the center and the electrons orbiting around the outside.

family eight of the periodic table. Because the noble gases have a full outer shell of electrons, they do readily bond with other elements to form compounds. They are electrically stable and usually do not, but still have hazards associated with them that can harm response personnel and the public.

> **Hazmatology Point:** *In a confined area, noble gases can displace oxygen in the air and cause asphyxiation even though they do not react chemically.*

Atoms are electrically neutral because they have an equal number of protons and electrons so that the positive and negative charges balance. An atom is held together by the strong attraction between the positive protons and the negative electrons. This is an example of the fact that opposite charges attract.

Atomic Weight

Each element's atomic weight is listed on the periodic table. Atomic weight is the sum of the weights of the protons and neutrons located in the nucleus of an atom. All of the weight of the element occurs in the nucleus. For the purposes of "applied chemistry," electrons have no weight. Atomic weight is located on the periodic table above or below the symbol of an element. It is the number that is not a whole number. Location of the atomic number varies among different versions of periodic tables, so be sure to look for the number with the decimal point.

Atomic Number

The other number on the Periodic Table located above or below the symbol is a whole number and is known as the atomic number. Atomic number is equal to the number of protons in the nucleus. It also equals the total number of electrons in the orbits outside the nucleus of an atom. Protons have a positive charge (+) and electrons have a negative charge (–). There must be an equal number of protons and electrons in an atom of a particular element to maintain an electrical balance or electrical stability. An element is identified by the number of protons, which is the atomic number. The number of protons in an element does not change. If the number of protons is changed, the result is a different element. Protons act as kind of "social security number" to identify a specific element.

Metals and Nonmetals

A stair-step line divides the periodic table into two sections. This line starts under hydrogen and goes over to boron and then stair steps down one element at a time to astatine or radon depending on which periodic table is being used. Elements to the left and below the stair-step line are metals. Metals make up about 75% of all elements. Metals lose their outer shell electrons easily to the nonmetals when forming compounds. Metals conduct heat and electricity very well. Metals are malleable or can be flattened and ductile, which means they can be drawn into a wire. Metals are all solids except for gallium, mercury, francium and cesium which are liquids under "normal" conditions. The 17 elements to the right and above the line are nonmetals. Nonmetals have a strong tendency to gain electrons when forming a chemical bond. Nonmetals may be solids, liquids or gases. They are poor conductors of heat and electricity. Solid nonmetals are either hard and brittle or soft and crumbly.

Element Families

Certain vertical columns on the periodic table contain elements that have similar chemical characteristics in their pure elemental form. These elements have the same number of electrons in the outer shell. Thus, they have similar chemical behaviors. Similar elements are sometimes referred to as families. Families have elements that are hazardous in their elemental form and may or may not be hazardous when forming compounds. For example, sodium is a metal that is very water reactive. Chlorine is a toxic and strong oxidizer. When combined together, they form sodium chloride; table salt is neither water reactive nor exceptionally toxic and isn't an oxidizer.

Families that are important to building compounds include the alkali metals in column 1, the alkaline earth metals in column 2, the halogens in column 7 and transitional metals between the "towers" and the noble or inert gases in column 8. These elements are examples of the family effects of these elements. Alkali metals in column 1 begin with lithium and continue through sodium, potassium, rubidium, cesium and francium. These elements are all solids except for cesium and francium, which are liquids at normal temperatures. Alkali metals are water reactive. The reaction with water is violent producing flammable hydrogen gas and enough heat to ignite the gas. These elements are so reactive that they do not exist in nature as the pure element. Instead, they are found as compounds of the metal such as potassium oxide and sodium chloride. Isotopes of Cesium can be and all isotopes of francium are radioactive. Since these elements are somewhat rare, you are not likely to see them on the street.

Alkaline earth metals in column 2 are less reactive than the alkali metals in column 1. Beryllium does not react with water at all. Others have varying levels of reaction with water. Alkaline earth metals are all solids. In order for them to react, they have to be burning or in a smaller physical form, such as shavings or dust. Magnesium, for example, is extremely water reactive when on fire. Application of water on a magnesium fire will cause violent explosions that can endanger responders. To fight fires involving magnesium, water should be applied from a safe distance using unmanned appliances. Other elements in column 2 are also water reactive to varying degrees, including calcium, strontium, barium and radium, which is radioactive.

Transition metals located in the center of the periodic table are metals with economic value as pure elements, like gold, silver, which when combined with nonmetals form some types of salts. These metals are considered a family because they all have differing numbers of electrons in their outer shell. This gives them differing characteristics when combined with nonmetals. Just an FYI, since we are not creating salts, transition metals are no important for applied chemistry, other their elemental hazards.

Halogens in column 7 are nonmetals; they may be solids, liquids or gases. Fluorine and chlorine are gases at normal temperatures and pressures. Bromine is a solid up to 19.7°F and a liquid up to 136°F and produces vapor readily when above that temperature. Iodine is a solid, and astatine is a radioactive solid; however, such a small amount has ever been found that you are not likely to encounter it. Halogens are all toxic and strong oxidizers; fluorine is a much stronger oxidizer than oxygen. In fact, fluorine is the strongest oxidizer known. In the pure elemental form, halogens do not burn; however, they will accelerate combustion much like oxygen because they are oxidizers. Some halogen compounds were components of former fire extinguishing agents called halon. Halon was phased out in the United States because of the damage caused to the ozone layer.

Elements in column 8 are all gases. For the purposes of applied chemistry, we will consider that they are nonflammable, nontoxic and nonreactive. They do not form compounds easily or of any importance; so for our purposes we will say that they do not form compounds. Family 8 elements are sometimes referred to as the inert or noble gases. Normally they do not react chemically with themselves or any other chemicals.

Helium is used in weather balloons and airships. Neon is used in lighting and beacons. Argon is used in electric light bulbs. Krypton and xenon are used in special light bulbs for miners and in light houses (and it doesn't hurt Superman). Radon is radioactive and is used in tracing gas leaks and treating some forms of cancer. It is also a gas found in homes that can cause harm to the occupants at certain concentrations. Noble gases have a complete outer shell of electrons, two in helium and eight in the rest. It is because of the complete outer shell of electrons that the noble gases are not reactive and do not bond chemically. Elementally, noble gases do have physical hazards. All of them can cause asphyxiation and are usually stored and shipped as cryogenic liquids (Figure 3.23). All of the other elements on the periodic table try to reach the same stable electron configuration.

The reason why chemical reactions occur is that the elements are trying to reach stability. Column 8 elements are all nonflammable and nontoxic gases. However, they can displace the oxygen in the air and cause asphyxiation inside of buildings or in confined spaces. Most of the noble gases are present in the air like nitrogen and oxygen. Inert gases are commonly shipped and stored as compressed gases or cryogenic liquids. If a container of these materials were to leak outside, letting them leak into the air and return to where they came from would be an option if it could be done safely. The periodic table can provide valuable assistance as another tool when trying to identify the physical state as well as hazards of elements.

Figure 3.23 Nobel gases can cause asphyxiation and are usually stored and shipped as cryogenic liquids.

Compounds, Mixtures, Solubility and Miscibility

Chemical Compounds

Just as the atom is the smallest part of an element, a single element is the smallest portion of a chemical compound. Chemical compounds are made up of two or more elements covalently or ironically bonded together. Ionic bonding occurs between metals and nonmetals, and involves the exchange of electrons. Covalent bonding involves the sharing of electrons between nonmetals. Chemical compounds are represented by formulas much like elements are represented by symbols. Two basic groups of chemical compounds are formed from elements: salts and non-salts.

Salts Salt compounds are made up of a metal and a nonmetal (Figure 3.24). They are all solids and generally do not burn. Metals do not bond together. Metals that are combined are melted and mixed together to form an alloy, in which the metals do not react chemically. For example, copper and zinc are melted and mixed together to make brass. Brass is not an element. There is no chemical bond involved, but rather it is just a mixture of zinc and copper.

When outer shell electrons of a metal are given up to a nonmetal element, a salt compound is formed through a chemical bond (Figure 3.25). The outer shell of the metal is now empty so the next shell becomes the outer shell. This shell will have two or eight electrons which are considered a stable configuration just like the noble gases of family 8 on the periodic table. The metal is then stable and satisfied. The nonmetal receives the electrons from the metal and now has eight electrons in its outer shell. The nonmetal is now stable and satisfied. The result is that a compound is formed, a salt, which usually has different characteristics and hazards than the individual elements used to make it up.

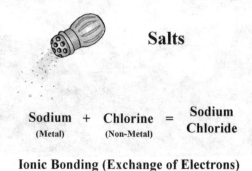

Salts

Sodium + Chlorine = Sodium Chloride
(Metal)   (Non-Metal)

Ionic Bonding (Exchange of Electrons)

Figure 3.24 Salt compounds are made up of a metal and a nonmetal.

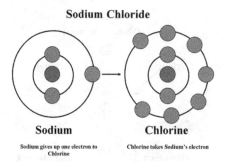

Figure 3.25 When outer shell electrons of a metal are given up to a nonmetal element, a salt compound is formed through a chemical bond.

Author's Note: *This is an oversimplification of the process, but remember that in applied chemistry we are not trying to create compounds. We are trying to understand them so we can recognize salts on the incident scene and know their family hazards.*

For example, magnesium combined with bromine results in a compound called magnesium bromide. Magnesium is water reactive when burning, and bromine is a toxic material and an oxidizer. When combined into magnesium bromide, a medication is formed that is used to treat nervous system disorders.

Nonmetals The second group of compounds is made up totally of nonmetal elements. Nonmetals are comprised of two or more nonmetal elements combining to form a compound. For example, the nonmetal carbon combines with the nonmetal hydrogen to form a hydrocarbon. A typical hydrocarbon might be methane with the molecular formula CH_4. Nonmetals may be solids, liquids or gases. They may burn as well as being toxic, explosive, corrosive and oxidizers. Hazardous materials most frequently encountered by emergency responders are compounds made up of just a few nonmetal elements, namely, carbon, hydrogen, oxygen, sulfur, nitrogen, phosphorus, fluorine, chlorine, bromine and iodine. In elemental form and in compounds, these elements make up the bulk of hazardous materials found by emergency responders in most incidents.

Chemical bonding, in the case of nonmetals, involves electrons that are shared between the nonmetal elements (Figure 3.26). This process of sharing electrons is called covalent bonding. Covalent bonds are pairs of shared electrons which can be represented in "shorthand" by two dots, each representing an electron that is being shared (Figure 3.27). Even more simply, the pair of shared electrons can be represented by a dash. Dashes are the way the bonding process among chemicals in a compound that are

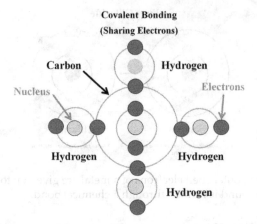

Figure 3.26 Chemical bonding, in the case of nonmetals, involves electrons that are shared between the nonmetal elements.

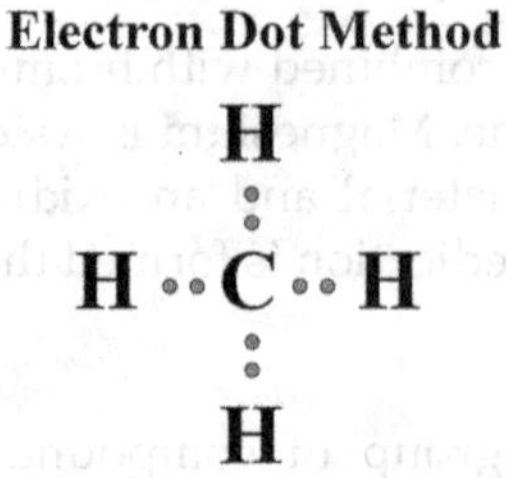

Electron Dot Method

$$\begin{array}{c} H \\ \vdots \\ H \cdot\cdot C \cdot\cdot H \\ \vdots \\ H \end{array}$$

Figure 3.27 Covalent bonds are pairs of shared electrons which can be represented in "shorthand" by two dots, each representing an electron that is being shared.

sharing electrons will be represented in real-life situations (Figure 3.28). Approximately 90% of covalently bonded hazardous materials are made up of carbon, hydrogen and oxygen. The remaining 10% are composed of chlorine, nitrogen, fluorine, bromine, iodine, sulfur and phosphorus. It is still necessary that each atom of each element have two or eight electrons in the outer shell. However, there is no exchange of electrons. When the bonding takes place, each atom of each element brings along its electrons and shares them with the other elements. Chemical compounds become electrically stable through the process of sharing and exchanging electrons.

Compounds that have become electrically stable do not mean they are no longer hazardous. Quite simply, elements combine and chemical reactions occur so that compounds can become stable. Combinations of elements that form compounds create many new hazardous and

Dash Method

Each dash represents a pair of shared Electrons.

Figure 3.28 Even more simply, the pair of shared electrons can be represented by a dash. Dashes are the way the bonding process among chemicals in a compound that are sharing electrons will be represented in real-life situations.

nonhazardous chemicals. In terms of emergency response, the really important thing we get out of covalent bonding is the type of bond. Type of bond determines reactivity. Reactivity concerns us when the hazardous material escapes its container.

> **Hazmatology Point:** *Emergency responders may encounter elemental chemicals that are hazardous when released in an accident. However, most of the hazardous chemicals encountered by emergency responders will be in the form of compounds or mixtures. Compounds and mixtures have a broad range of hazards including explosive, corrosive, flammable, toxic and oxidizers. Covalently bonded compounds also have specific families based upon the types of elements present. These families have particular hazards associated with them. Each member of the family will have similar hazards.*

Mixtures

Chemical compounds may also exist in the form of mixtures. A mixture is two or more compounds combined together without any chemical bonding taking place. Each of the compounds retains its own characteristic properties. For example, gasoline is not a pure compound. You cannot write a formula for gasoline because it is a mixture of compounds. Two types of mixtures exist: homogeneous and heterogeneous. Homogeneous means "the same kind" in Latin. In a homogeneous mixture, every part is exactly like every other part. For example, water has a molecular formula of H_2O. Pure water is homogeneous; it contains no substances other than hydrogen and oxygen. Loosely translated to include mixtures, homogeneous refers to two or more compounds or elements that are uniformly dispersed in each other. A solution is another example of a homogeneous mixture. Heterogeneous means "different kinds" in Latin. In a heterogeneous mixture, different parts of the mixture have different properties. The components in a heterogeneous mixture can be separated mechanically into their component parts. Some examples of heterogeneous mixtures are gasoline, the air we breathe, blood and mayonnaise.

Solubility

Solubility is another term associated with mixing two or more compounds together. The definition of solubility, from the *Condensed Chemical Dictionary*, is "the ability or tendency of one substance to blend uniformly with another, e.g., solid in liquid, liquid in liquid, gas in liquid, and gas in gas." Salts placed into water are totally soluble yielding salt water. Alcohol placed in water is totally soluble. Gases such as hydrogen chloride gas placed in water are soluble and form hydrochloric acid. Solubility may have varying degrees from one substance to another. Temperature can also affect solubility. When researching chemicals in reference sources, relative solubility terms may include very soluble, slightly soluble, moderately soluble and insoluble. Generally speaking, nothing is absolutely insoluble. Insoluble actually means "very sparingly soluble," that is, only trace amounts dissolve.

Compounds of the alkali metals of family 1 on the periodic table are all soluble in water. While they are soluble in water, the reaction with water can be violent before the materials are dissolved in the water. Salts containing the ammonium cation (NH_4), nitrate (NO_3), perchlorate (ClO_4) and chlorate (ClO_3), and organic peroxides containing carbon, hydrogen and two oxygens bonded together are also soluble in water. All binary salt chlorides ($Cl-$), bromide ($Br-$) and iodide ($I-$) are soluble, except those containing the metals silver (Ag), lead (Pb_2) and mercury (Hg_2). All sulfates are soluble, except those with the metals lead (Pb_2), calcium (Ca_2) strontium (Sr_2), mercury (Hg_2) and barium (Ba_2). All hydroxides (OH) and metal oxides (containing O) are insoluble, except those of family 1 on the periodic table and Ca_2, Sr_2 and Ba_2. All compounds that contain phosphate (PO_4), carbonate (CO_3), sulfate (SO_3) and sulfur (S) are insoluble, except those containing metals in family 1 on the periodic table and the ammonium ion (NH_4).

Most hydrocarbon compounds are mixtures and are not soluble, such as gasoline, diesel fuel, fuel oil and so on. When placed together in a container or when spilled in water in the environment, hydrocarbons will form layers on top of the water because they are lighter than water (Figure 3.29). Compounds that are polar such as the alcohol, ketone, aldehyde, ester and organic acid are soluble in water, because water is also a polar compound. Compounds may be soluble in other compounds or chemicals as well. Polar compounds are generally soluble in other polar compounds. There is a rule in chemistry that says "like materials dissolve like materials."

Several factors affect the solubility of a material. One is particle size: the smaller the particle, the more the surface area that is exposed to the

Applied Chemistry and Physics

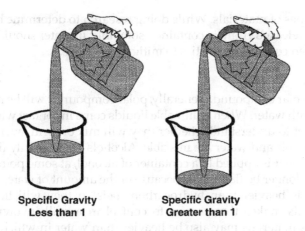

Specific Gravity Less than 1   **Specific Gravity Greater than 1**

Figure 3.29 When placed together in a container or when spilled in water in the environment, hydrocarbons will form layers on top of the water because they are lighter than water.

solvent; therefore, more dissolving takes place over a shorter period of time. For example, placing table salt into water will dissolve fairly quickly. If you place a block of salt into water, like those used to supply salt to animals in a pasture, it would take a long time to dissolve. Salt is soluble in water, but the physical form dictates how long it will take. Higher temperatures usually increase the rate of dissolving. The term "miscibility" is often used synonymously with the term "solubility." If a compound, for example, is miscible in water, it is water soluble.

If it is immiscible, then it is insoluble in water. Acetylene gas is produced when calcium carbide, a salt, is mixed with water. Acetylene gas is very flammable and unstable. When placed in a container, acetylene gas is dissolved in acetone to stabilize it. Characteristics of the acetylene and acetone do not change. They do not bond together or their chemical characteristics do not change, and they are merely occupying the same space without interacting chemically. Many inorganic acids are made when a gas is dissolved in water to form the liquid acid. Anhydrous ammonia is very soluble in water. The term "anhydrous" means "without water."

Solubility of a material in water can be important to hazmat responders as water can be used to control the spread during a release. For example, ammonia or other anhydrous materials can be removed from the air by applying water by hose lines to control the spread of a vapor cloud which may be toxic. Remember, however, that ammonia and water form ammonium hydroxide on the ground, which is a strong base that can also be water reactive. So, only apply water to the gas and not the liquid formed on the ground from the gas. Water may also dilute a soluble material to the point that it is no longer flammable or reduces the hazardous effects of other types of materials. While doing research to determine hazards of a material released from its container, solubility in water should be determined when considering tactics to mitigate a release.

Miscibility

Water is a polar compound. Generally, polar compounds will be miscible or will mix with water. When flammable liquids come in contact with water, it is important to understand whether they will mix or form layers with the water. Alcohols and water are miscible. Alcohols are generally flammable. If enough water is applied to a container of alcohol, at some point the alcohol will no longer be flammable because of the amount of water present.

Water is heavier than hydrocarbon fuels so they will float on the surface. This makes them easier to control in a spill involving water. Hazardous materials may also be heavier than water in which case they will fall to the bottom and the water will float on the top. In this situation, the hazardous materials will be difficult to control or clean up if covered with water, particularly in natural waterways such as streams, rivers and lakes or the oceans and related bodies of water.

Damming a hazardous material when water is present is based on the weight of the hazardous material compared to water. Hazardous materials lighter than water will float on the top and would be contained with an underflow dam. Heavier than water materials would be contained with an overflow dam. Utilizing underflow and overflow damming, the hazardous materials can be diked to prevent further spread of the product while letting the water flow freely requiring less room in the diked area.

Common and Site-Specific Hazardous Materials

Time will be spent in this here pointing out details of the common rather than the "exotic" hazardous materials. Common materials are the ones that have killed and injured responders in the past. In reality, there have been only a few common materials involved in most of these deaths and injuries:

- Ammonium nitrate
- Anhydrous ammonia
- Chlorine
- Combustible dusts
- Liquefied petroleum gases
- Petroleum products such as gasoline and benzene

Common hazardous materials, not the "exotic", are the ones responders are most likely to encounter in transportation and fixed-facility incidents today. Although those materials listed above are the killers in the past, there are many other common hazardous materials we may also have to deal with. Frequent encounters occur between emergency responders and acids, bases, oxidizers, flammable solids, poisonous liquids and solids and water-reactive materials. Even though they may be common materials, it doesn't mean we have extensive experience in handling them. Our skills in recognition and identification along with experience will help us determine what the dangers are.

> **Hazmatology Point:** *Once the hazard is determined, a thorough hazard/risk analysis needs to be conducted before operational tasks are identified. Gordon Graham's video about Risk/Frequency Models is part of the Hazardous Materials Incident Management Course I taught at the National Fire Academy before I retired in April 2018. These models apply to all kinds of risk/benefit analysis situations, not just the fire service and not specifically for hazmat. However, I have modified the concept so that it can be useful in determining risk/benefit in hazardous materials incidents. My opinion is that this really is a good way to look at hazard/risk analysis as it applies to hazardous materials incidents.*

Recognition-Primed Decision Model

Your brain is much like a computer hard drive, which has the capacity of holding maybe a gazillion bytes of information. When you get involved in any type of incident, your brain scans it and stores the information in your on-board computer. It looks for a match of an incident you have been involved with previously. If the hard drive finds a match, it automatically directs the behavior based on the past behavior that ended with a satisfactory result. Simply, the decision maker has an idea of how things work based on the knowledge that has been gained from experience. The options are compared against what is known to work.

Historical incidents, like those in Volume One can be substituted for incidents you actually experience yourself. Reading and learning about them will put you there, and you can experience what went wrong, if anything; then by looking at lessons learned, you can also learn what is known to produce a satisfactory result.

> **Hazmatology Point:** *On March 4, 1996, eighteen years after the Waverly, TN, incident, a similar derailment occurred in Weyuwega, WI, and Assistant Chief Jim Baehnman told me when I visited there that he used his knowledge of the Waverly incident to formulate tactics in Weyuwega. This may have had a direct impact on the Weyuwega incident in terms of safety to emergency personnel and residents. There was not a single serious injury or death as a direct result of the derailment in Weyuwega. Baehnman said, "from the start of the incident, the tone of the incident would be driven by safety and not time."*

Incident Frequency/Risk Model

HR LF	HR HF
LR LF	LR HF

High-Frequency, Low-Risk Incidents

Throughout the fire service, including hazmat response, we strive to do things right. In spite of the complexity and variety of the tasks we are called upon, we usually do them right. This increases personnel safety, limits liability, increases respect and improves customer service. Most of the tasks we perform in the fire service on a daily basis we are really good at doing. We would call those High Frequency (HF) Low Risk (LR) tasks. For example, flammable liquid and fuel incidents make up 50% of all hazardous materials incidents nationally. These are HF/LR incidents because we do them on a regular basis. Hazards and quantities of these types of incidents are limited, so the risk in dealing with them is low. Even if mistakes occur with HF/LR tasks, the consequences are usually not significant. Things you do all the time generally do not cause problems.

Low-Frequency, Low-Risk Incidents

Low Frequency (LF), Low Risk (LR) incidents should not be of concern because the consequences and likelihood are relatively low. Incidents involving radioactive materials would be an example. Design of transportation containers and regulations affecting storage of radioactive materials make the chance of encounters with them very remote.

High-Frequency, High-Risk Incidents

Houston's hazmat team is located in the petrochemical capital of the world. They are a dedicated team; hazmat response and preparation is all they do. "Exotic" chemicals are considered high-frequency incidents in Houston compared to other jurisdictions. Common chemicals that are the killers are considered high-frequency incidents in Houston. Hazmat responders in Houston are used to dealing with these types of events and are able to handle them with a minimum of risk because they do it all the time. As I mentioned earlier, things we do all the time, we usually do pretty well. If, however, errors are made, the consequence is great.

Applied Chemistry and Physics

> ***Hazmatology Point:*** *North Platte Nebraska population about 25,000 does not have a dedicated hazmat team. Team members respond to fires, EMS and other types of calls. On an annual basis, they respond to an average of 52 hazmat calls that include fuel spills, odor complaints, gas leaks and regional response. They do not have high-frequency, high-risk incidents. High frequency high risk incidents only occur in areas where there are high volumes of high-risk hazardous materials. Few departments across the country have to deal with HR/HR incidents on a regular basis, or most cases will never have to deal with them.*

Low-Frequency, High-Risk Incidents

Low-frequency, high-risk incidents are once in a lifetime incidents that I like to refer to as "career" events. Incidents like this are usually well beyond the ability of a single jurisdiction to handle. People that have experienced a career incident have all pretty much conveyed the same synopsis to me. It would have been difficult to plan or prepare for the incident that they experienced. Career incidents are so far from what we normally deal with in hazmat response that it is an overwhelming situation. Now having said that, it doesn't mean we are destitute. We know the drill about dealing with hazardous materials incidents. All of the tools are in your mental tool box. What is missing from the mental tool box is the type of career event that you might be facing. You have no information in your data base from your past experiences.

However, if you had studied previous incidents that others have been involved in and looked at their lessons learned, that would be in your mental database and you can draw from that incident. It is important that we do the things that need to be done, just like learning to walk, "one step at a time (FEMA, NFA)."

> ***Hazmatology Point:*** *HR/LF incidents have the highest probability of going bad with nasty consequences. They have a high probability of causing you grief. Why? Your hard drive is empty. You have not likely experienced an incident like this before. You most likely do not know what to do. These are the incidents you should worry about. To quote Gordon Graham, "The bells of St. Mary's should be going off in your head." There are two types of HR/LF incidents, Discretionary Time (DT)-Time to think it through and No Discretionary Time (NDT)-No time to think it through. Luckily, most, but not all, hazardous materials incidents have time to think them through. If you do have time, then SLOW DOWN! SLOW DOWN, think things through, use common sense there is no need rush.*
>
> *Having spent over 40 years involved in emergency response at various levels, I have found that there is something different about emergency responders. Responders are always rushing in when others are fleeing an emergency. Emergency responders take risks that other people*

would not. Everything has to be done right now. Hazardous materials by nature are dangerous to both responders and the public. When it is possible, and that is most of the time, we as hazmat responders need to slow down and think things through. Do not give in to the temptation, fostered by our culture and traditions, to rush in. This reminds me of a song lyric "Fools rush in where Angels fear to tread." SLOW DOWN.

CASE STUDY

MIAMISBURG, OH, JULY 8, 1986, DERAILMENT PHOSPHORUS FIRE

On July 8, 1986, at 4:25 p.m., a southbound Baltimore and Ohio Railroad Company (B&O) freight train Southland Flyer (FLFR) derailed near Miamisburg, Ohio. Only one car on the train was carrying a considered hazardous material white phosphorus. It derailed, and over turned. Near the phosphorus car were a tank car of animal fat, which was leaking into the creek, and a box car of rolled paper, neither of which was involved in the fire. Phosphorus is air reactive and shipped under water to keep it from spontaneously combusting. Several holes developed in the tank during the impact of the derailment, and the water started leaking off causing the phosphorus to be exposed to air and ignited (Figure 3.30). Minor explosions occurred, which may have been the gas tanks of automobiles on the two car carriers that derailed. Mutual aid was requested from 30 area fire

Figure 3.30 Several holes developed in the tank during the impact of the derailment and the water started leaking off causing the phosphorus to be exposed to air and it ignited. Special Collections & Archives, Wright State University.

departments along with 107 medic units. Police agencies sent personnel from Cincinnati and Dayton along with other communities that sent help as well.

During the first and second days of the emergency, numerous proposals for handling the phosphorous tank were suggested and evaluated. Some of them included direct hose stream attack, plugging, water flooding of the interior of the tank car, foam application, burial, opening the manhole to allow air injection to accelerate the burn rate and use of explosive demolition (it was suggested that the Air Force could use a Fighter from nearby Wright-Patterson Air Force Base to shoot the tank car with a missile). Chief Menker in consultation with the city manager decided to proceed with the suggestion of opening the manhole to accelerate the burn rate.

Initially 10,000 people were evacuated. Wind direction changed in the evening of July 9 and coupled with a wind inversion that kept the phosphorus cloud close to the ground, which prompted a second evacuation affecting approximately 30,000 people. Initially the command post was established on a bridge downwind, south of the derailment. The bridge provided an excellent vantage point for the incident because the tracks ran right under the bridge with a view north of the tracks and of the derailment site. However, the wind changed to the north and the command post had to be moved. Radio communications was the biggest problem as many agencies were on separate frequencies; local mutual aid channels were overloaded with communications. So much so, that remote activation of evacuation sirens could not be accomplished when needed.

Over the 5-day duration of the incident, crews were rotated in and out, and millions of gallons of water were flowed onto the fire. By the time the emergency was declared under control, local hospitals received 569 persons with nonfatal injuries, including 13 emergency response personnel. Of these, 27 were hospitalized due to their injuries. The cost, excluding costs to evacuees, of community disruption or business interruptions was estimated to be over 3.5 million dollars (*Firehouse Magazine*).

> **Hazmatology Point:** *The scope of this incident was well beyond the resources of tiny Miamisburg, yet the responders requested additional resources and managed the incident well with minimal injuries. They used incident command and did a risk analysis to determine the best tactics to use, and the outcome of the response was good. They handled a low-frequency, high-risk incident very effectively, utilizing all of the tools available them.*

By focusing on the likely rather than the unlikely, you will be able to better handle the incidents you may be faced with, and those skills will also prepare you for the unlikely events. The exception I mentioned above involves locations where you have "exotic chemicals" in a fixed facility or transported through your jurisdiction (like Houston). Even though you may have exposure to the exotic chemicals, that does not mean you will have an incident involving them. By knowing they are in your community, you can learn about them and prepare for the unlikely which will make your response more effective if it does occur. Generally, "exotic" materials are few and far between and incidents are rare.

Common hazardous materials and families can be placed into the DOT Hazard Classes, which are generally organized by the hazard and physical state of the materials. Hazard classes are organized from the most dangerous to the least dangerous from top to bottom. Common hazardous materials are shown in the hazard class to which they belong. Each hazard class has been identified as to the hazardous materials included in that class and general frequency in terms of incidents using the Risk Frequency Model (FEMA/NFA).

Families of Compounds

Salt Families and Hazards

There are two basic groups of chemical compounds that are formed from elements: salts and nonsalts. Within each group are families of compounds that have particular hazards associated with each. By understanding these family relationships, responders can determine the general hazards of materials by identifying the family they belong to. This "rule-of-thumb" information can assist responders in identifying hazards of particular hazardous materials upon initial response to an incident. This type of recognition does not, however, eliminate the need to research the chemicals further before mitigation actions are taken.

> **Hazmatology Point:** *Just as the Periodic Table of the Elements is organized into families of elements, compounds can also be placed into families with similar characteristics. One of the basic reasons for looking at a family is that each family has particular hazards associated with it. If you are able to recognize which family a material belongs to by the name or the formula, you should be able to determine the hazard even if you don't know anything else about the specific chemical.*

Salts—LF/LR

Salts are made up with a metal and a non-metal to form a salt compound. There is one exception and that is the ammonium ion (NH_4). Ammonium compounds are High Risk (HR). Why and how the ammonium ion works

Applied Chemistry and Physics

is beyond the scope of Applied Chemistry. Just know that the ammonium ion acts like a metal in a salt compound and does not change the salt family characteristics, but it does add some hazards of its own. Ammonium hydroxide and ammonium nitrate are some examples of the use of the ammonium ion. Salts in general are solids, water soluble, and most do not burn. While salts generally do not burn, they can be oxidizers and support or accelerate combustion. Some salts are toxic and some may be water reactive. A good general rule of thumb when dealing with salts is "Don't touch, and keep them dry." The hazards present in a given salt family, other than the binaries, are a result of reactivity with water.

Binary Salt

Salts have particular hazards depending upon which salt family they belong to. The first family of salt compounds that will be discussed are the binary salts. Binary (meaning two) salts are made up of two elements: a metal and any nonmetal except oxygen. They end in "ide," such as potassium chlor**ide**. So, when a compound is encountered with a metal that ends in "ide," responders would know this is a binary salt. They would realize that it needs to be looked up to determine actual hazards (Figure 3.31). Binary salts are the one type of family in which the family effect of common hazards does not apply except for a few compounds (Figure 3.32). Binary salts as a group have varying hazards. However, knowing that the "ide" identifies the family, you should know that you will have to research the salt to find out the hazards. They may be water reactive, toxic, and in contact with water may form a corrosive liquid and

Binary Salts

Metal + Non-Metal
Not Oxygen
End in "-ide"
Variable Hazard
Except NCHP

Figure 3.31 Binary (meaning two) salts are made up of two elements: a metal and any nonmetal except oxygen.

Salt Hazard Key

WR = Water Reactive
RH = Release Heat
RCL = Corrosive Liquid
RO = Release Oxygen

Figure 3.32 Abbreviations for salt hazards.

release heat. Chemical reactions often release heat. This characteristic is referred to as an exothermic reaction. The hazard of an individual binary salt cannot be determined by the family, other than four types of binary salts.

Varying hazard applies to all binary salts except for nitride, carbide, hydride and phosphide. One way that may be helpful in remembering these four binary salts is by using the first letters of the element to form NCHP, which can be represented by "North Carolina Highway Patrol." These are compounds in which the metal has been bonded with one of the nonmetals nitrogen, carbon, phosphorus or hydrogen. These compounds do have particular hazards associated with them when they are in contact with water. They are all water reactive. When in contact with water, nitrides give off ammonia gas, carbides produce acetylene gas, phosphides give off phosphine gas and hydrides form hydrogen gas. Gases produced in the reaction with water each have their own specific hazards:

- **Nitride:** Ammonia, NH_3, is a colorless gas with an intensely sharp irritating odor. While not considered a flammable gas by the DOT, it is flammable under certain conditions, particularly inside of buildings and in confined spaces. Furthermore, ammonia is toxic by inhalation with a TLV of 25 ppm.
- **Carbide:** Acetylene, H_2N_2, is a colorless, odorless gas with a wide flammable range almost to 100%. It is highly flammable and will burn inside containers.
- **Hydride:** Hydrogen, H_2, is a highly flammable gas that burns with an almost invisible flame and produces little if any smoke. Hydrogen has a wide flammable range from 4%–75%.
- **Phosphide:** Phosphine, PH_3, is a colorless gas with a disagreeable, garlic-like odor. It is spontaneously flammable, toxic by inhalation and a strong irritant. Phosphine has a TLV of 0.3 ppm in air.

In addition to the gases released when these salts contact water, a corrosive base is formed. The corrosive base will be the hydroxide of the metal that is attached to the nonmetal. For example, calcium carbide in contact with water will produce acetylene gas and the corrosive liquid calcium hydroxide. Acetylene gas is a common welding and cutting gas. It is so unstable that it is not shipped in tank car or truck quantity. Instead calcium carbide is shipped to an acetylene-generating facility where acetylene is generated and placed in the familiar small welding tanks. Acetylene is so unstable that it cannot be pressurized so it is dissolved into acetone in the tank to stabilize it.

With remaining binary salts, you have to look them up to determine the hazard of a particular salt. For example, when lithium metal is combined with chlorine, the resulting compound has a metal and a nonmetal other than oxygen, and the name ends in "ide." Therefore, it fits

Applied Chemistry and Physics 49

the definition of a binary salt. If lithium chloride is researched in reference sources, it is found to be soluble in water. It is not water reactive; in fact, lithium chloride doesn't present any significant hazard in a spill. The DOT does not list lithium chloride on its hazardous materials tables.

The next example combines the metal calcium and the nonmetal phosphorous resulting in the compound calcium phosphide. It is a dangerous fire risk. The compound name ends in "ide." Therefore, this is also a binary salt. Lithium fluoride is a strong irritant to the eyes and skin. Potassium bromide is toxic by ingestion and inhalation. Sodium chloride is table salt, a medical concern when ingested in excess, but certainly of no significant hazard to emergency responders. However, if sodium chloride is washed into a farmer's field from an incident, the farmer may not be able to grow crops in that field for many years!

Binary or Metal Oxide The next family of salt compounds is known as the binary or metal oxides. They are also made up of two elements: a metal and a nonmetal; but in this case, the nonmetal can only be oxygen (Figure 3.33). They end in "oxide," such as aluminum oxide. As a group, they are water reactive and, when in contact with water almost, always produce heat and form a corrosive liquid. However, they do not give off oxygen because there is not an excess of oxygen.

Peroxide Peroxides are the next group of salt compounds. They are composed of metal and nonmetal peroxide radical "O_2^{-2}" (Figure 3.34). The prefix "per" in front of a compound or elemental name means that the material is "loaded" with atoms of a particular element. In the case

Binary (Metal) Oxides

Metal + Oxygen
End in –"oxide"
WR = RH, CL

Figure 3.33 Characteristics of binary oxide salts.

Peroxide Salts

- Metal + $(O_2)^{-2}$ Radical
- Ends in "peroxide"
- WR = RH, RO_2, RCL

Figure 3.34 Peroxides are composed of metal and nonmetal peroxide radical "O_2^{-2}."

of the peroxide salts, they are loaded with oxygen. When peroxide salts come in contact with water, heat is produced, a corrosive liquid is formed and oxygen is released. This makes them particularly dangerous in the presence of fire. As was previously mentioned, salts are soluble in water.

When a porous material comes in contact with water that is mixed with an oxidizer salt, it will get wet. However, once it dries, the oxidizer is still in the material. If another fire occurs, peroxide will add oxygen making the fire burn faster and hotter than would be expected. The same is true for bunker gear. If you get the salt solution on your bunker gear, it will soak in, and the next time you are exposed to fire, your bunker gear could erupt in flames. Peroxide salts release oxygen because unlike the oxide salts, there is an excess of oxygen present. The heat produced in the reaction with water can be enough to ignite nearby combustibles. The excess oxygen present can then accelerate the combustion. When sodium metal combines with the peroxide radical to form the compound sodium peroxide, the name ends in "peroxide" so this is a peroxide salt. When in contact with water, sodium peroxide is a dangerous fire and explosion risk, and it is also a strong oxidizing agent.

Hydroxide The next group is known as the hydroxide salts. They are made up of a metal and a nonmetal hydroxide radical $-OH^{-1}$ (Figure 3.35). The name ends in the word "hydroxide." They are water reactive and, when in contact with water, release heat and form a corrosive liquid. For example, if calcium metal is combined with the hydroxide radical, the resulting compound is calcium hydroxide, a hydroxide salt. If the ammonium ion were added to a hydroxide radical, the resulting compound would be ammonium hydroxide.

Calcium hydroxide is a strong corrosive base with a pH of 12.4. It is a skin irritant and inhalation hazard. The TLV is 5 mg/m³ of air. When responders encounter carbides and other water-reactive salts in a spill, they should be careful not to apply water unless absolutely necessary. If rain is forecast, acetylene gas would be generated when the rain contacted the calcium carbide in a spill. Care would have to be taken to cover the material to keep it dry.

Hydroxide Salts

Metal + (OH) $^{-1}$ Radical
Ends in "hydroxide"
WR = RH, RCL

Figure 3.35 Hydroxide Salts are made up of a metal and a nonmetal hydroxide radical $-OH^{-1}$.

Cyanide Salts (Metal Cyanide) Cyanide salts are formed when a metal is attached to the cyanide radical (CN) (Figure 3.36). They are named by using the metal name first and then ending with the word "cyanide." For example, when potassium metal reacts with the cyanide radical, the resulting compound is potassium cyanide. Similarly, when the sodium metal is added to the cyanide radical, the resulting compound is sodium cyanide. These compounds are highly toxic and are absorbed by all routes of entry. Hydrogen cyanide is formed when the cyanide salt comes in contact with any acid.

Oxysalts Oxysalts are another family of salt compounds (Figure 3.37). A radical is an "incomplete" group of nonmetal elements acting together as a unit, but lacking a metal element to create a complete compound. Radicals will not exist by themselves; they will bond quickly with other elements. They are made up of a metal and an "oxy" radical. They end in "ate" or "ite" and may have the prefix "per" or "hypo" depending upon the number of oxygen atoms in the oxysalt. Generally, as a group they do not react with water; they dissolve in water. Some of the "hypo-ites"

Cyanide Salts

M + CN
Ends in Cyanide
Poison and WR=HCN
And CL

Figure 3.36 Cyanide salts are formed when a metal is attached to the cyanide radical (CN).

Oxy Salts

Any Metal + Oxy Radical

-1	-2	-3
FO_3	CO_3	PO_4
ClO_3	SO_4	
BrO_3		
IO_3		
NO_3		
MnO_3		

Figure 3.37 A radical is an "incomplete" group of nonmetal elements acting together as a unit, but lacking a metal element to create a complete compound.

technically do react with water to release chlorine, but the reaction is a mild one. Oxysalts are oxidizers as a family, which means that they release oxygen that will accelerate combustion if fire is present. Just like the peroxide salts, hazard occurs when oxysalts dissolve in water, and the water is soaked into another flammable material such as cardboard packaging or firefighter turnouts. The water will evaporate and the oxysalts will be left in the material. If the material is then exposed to heat or fire, the material will burn very rapidly because of the oxysalts in the material accelerating the combustion.

Nine oxysalt radicals are presented in this group. These are not the only "oxy" radicals, but the ones chosen are considered most important to emergency response personnel. The "oxy" radicals are FO_3 fluorate, ClO_3 chlorate, BrO_3 bromate, IO_3 iodate, NO_3 nitrate, MnO_3 manganate, CO_3 carbonate, SO_4 sulfate and PO_4 phosphate. Now, the fact that all oxysalt radicals end in "ate" or "ite" is a hint when you are dealing with an oxysalt. Some of the hydrocarbon derivatives end in "ate," so you have to look at what is attached. If it is a metal, then it is an oxysalt. If not, then it is a hydrocarbon derivative. This is the type of information you need to collect from the oxysalts. They start with a metal and end in "ate" or "ite." It really doesn't matter how much oxygen is in a particular oxysalt; all you need is to recognize that it is one.

All of the radicals listed above are considered to be in their base state. The base state has to do with the number of oxygen atoms that are present in the compound. The base state is the "normal" number of oxygen atoms present in that oxy radical. When a metal is added to any oxy radical in the base state, the compound ends in "ate" such as sodium phosph**ate**. The compound does not have a prefix on the oxy radical and it ends in "ate"; therefore, it is the base state of the compound. Oxy radicals may be found with varying numbers of oxygen atoms.

There may be more oxygen atoms than the base state or less atoms than the base state. When naming the compounds with an additional oxygen atom, the prefix "per" is used to indicate excess oxygen over the base state, but the ending is still "ate." An example would be sodium **per**sulf**ate**. When potassium metal is combined with the oxy radical perchlorate, the resulting compound is potassium perchlorate. The level of oxygen is one above the base state. Potassium perchlorate is a fire risk in contact with organic materials, a strong oxidizer and a strong irritant. The more oxygen present, the greater the hazard for acceleration of combustion if a fire is present.

When the number of oxygen atoms is one less than the base state of an oxy radical, the ending of the oxy radical name is "ite." An example would be magnesium sulf**ite**. If the metal sodium is combined with the oxy radical phosphate, there is now one less oxygen than the base state. In addition to being an oxidizer, sodium phosphite is also an antidote in mercuric chloride poisoning.

Finally, an oxy radical can have two less oxygen atoms then the base state. The oxy radical name will now have a prefix "hypo" and will end in "ite." An example would be aluminum **hypo**phosph**ite.** In the following example, calcium metal is combined with the oxy radical hypochlorite. The resulting compound is calcium hypochlorite, a common swimming pool chlorinator. Calcium hypochlorite is an oxidizer and a fire risk when in contact with organic materials.

Ammonia Salts Some ammonium compounds are oxysalts. Although ammonia is not a metal, in the case of the ammonium ion, it acts like a metal when attached to the oxy radicals previously mentioned. When ammonia gas is added to water, it readily dissolves and remains as NH_3. One of the hydrogen atoms leaves water, but leaves its electrons behind. The electrons of the hydrogen then attach to the unbonded electrons on nitrogen to complete its duet. This hydrogen is loosely held to the nitrogen and comes off easily. The hydrogen ions in the water are attracted to the negative side of the ammonia molecules where that unbonded pair of electrons is located (you can think of the ammonia as a slightly polar molecule). The ammonium ion is positive because the hydrogen ion contributes no electrons. It is not important to understand why this happens, but rather that the hazards of the compounds are similar to the rest of the oxysalts, that is, they are oxidizers. This is a complex covalent-sharing arrangement and is one of those chemistry concepts that should be accepted rather than explained for the purposes of emergency response.

Ammonium nitrate, NH_4NO_3, is an oxidizer, which may explode under confinement and high temperatures. When mixed with fuel oil, a deflagrating explosive material is created. This mixture was utilized in the bombings at the Murrah Federal Building in Oklahoma City and World Trade Center bombings in the mid-1990s.

Ammonium perchlorate, NH_4ClO_4, is an unstable explosive. It is used in rocket fuel and was the force behind the massive explosion at the PEPCON plant in Henderson, NV, in the 1980s.

As we have found with the salt compounds, most other hazardous materials will have multiple hazards. When these materials are found in transportation or at fixed facilities, they may be placarded or labeled. The DOT only placards the most severe hazard of a hazardous material as determined by DOT rules. That does not mean that there is only one hazard that material may present. The "Burke Placard Hazard Chart" (Figure 3.38) has been developed to assist emergency responders in identifying other potential hazards of materials that may not be indicated by the placards. Across the top of the chart is a listing of potential hazards.

Down the left side of the chart are colors of potential placards and labels that may be found on hazardous materials in transportation or in fixed storage and use. An "X" in one of the columns marks the primary

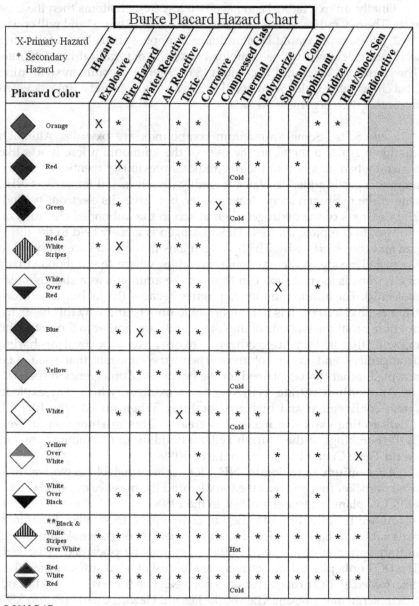

Figure 3.38 DOT only placards the most severe hazard of a hazardous material as determined by their rules. That does not mean that there is only one hazard that material may present.

hazard of a material as is determined by the DOT with a particular placard or label. The "*" in other columns indicates other potential hazards that may also be present. This chart gives first responders a heads-up on "hidden hazards" because they generally do not have any reference books to identify other potential hazards. When used in conjunction with the Emergency *Response Guidebook*, responders may be able to get a better idea of the hazards they may be facing in the initial phases of a hazardous materials incident.

Nonsalt Compounds

The second basic group of chemical compounds are referred to here as non-salt or nonmetal compounds (Figure 3.39). Nonmetal compounds are combinations of nonmetallic elements. These are the elements to the right and above the dividing line between metals and nonmetals on the periodic table. Nonmetals and their compounds may be solids, liquids and gases. Some may burn, some may be toxic and some may also be reactive, corrosive oxidizers, and the noble gases are inert. but can cause asphyxiation. The largest quantities of hazardous materials encountered are made up of nonmetal (non-salt) materials. Nonmetals combine in a covalent bonding process. When elements bond covalently, the bonding electrons are shared between the elements.

Nonmetal compounds are formed from nonmetal elements. For example, when the nonmetal carbon is combined with the nonmetal sulfur, the compound formed is carbon disulfide, which is a poison by absorption and is a highly flammable, dangerous fire and explosion risk; has a wide flammable range from 1%–50% and can be ignited by friction. Carbon disulfide also has a low ignition temperature and can be ignited by a steam pipe or light bulb. Nonmetal compounds can be represented in three ways: chemical name, molecular formula and structural formula. In the above example, carbon disulfide is the chemical name, the molecular formula is **CS_2** and its structural formula is shown in Figure 3.40.

Non-Salts

- **Inorganic Acids**

- **Hydrocarbons**

- **Hydrocarbon Derivatives**

- **Miscellaneous Compounds**

Figure 3.39 The second basic group of chemical compounds are referred to here as non-salt or nonmetal compounds.

Carbon Di-Sulfide

$$S=C=S$$

Figure 3.40 Structural formula of carbon disulfide.

The structural formula illustrates the way the bonding takes place between the atoms of elements in a compound. The molecular formula shows the numbers of each atom in the compound. Responders may encounter the molecular formula along with the name; however, it is unlikely that they will see a structural formula in the field. As can be seen with the structural formula of carbon disulfide, there are two bonds between the sulfur and the carbon. This configuration of bonding is referred to as double bonding. Double bonds are very unstable; which is why carbon disulfide is so very flammable. Most compounds that have this carbon to carbon double bond are unstable. Triple bonds are even more unstable so these compounds are not usually shipped in transportation except for welding bottles of acetylene. Acetylene is the only commercially valuable triple-bonded material you are likely to encounter.

Carbon dioxide CO_2 is another common nonmetal compound. It is a colorless, odorless gas with a vapor density of 1.53, which is heavier than air. It is used as a fire extinguishing agent, in carbonated beverages, aerosol propellant and many other uses. Carbon dioxide can also be found as a liquid cryogenic material with a boiling point of –130°F below zero. It is nonflammable and relatively nontoxic; however, it is a simple asphyxiation hazard. The structural formula of carbon dioxide is shown in Figure 3.41.

Some nonmetal compounds can be separated into families which will have similar hazards among individual members of the family. These families of compounds are called hydrocarbons and hydrocarbon derivatives. Hydrocarbons are primarily flammable liquids and make up the fuels for motor vehicles and heating structures. They are the most frequently released hazardous material making up over 50% of all hazardous materials incidents.

Carbon Di Oxide

$$O=C=O$$

Figure 3.41 The structural formulas of carbon dioxide.

Applied Chemistry and Physics

Hydrocarbons—HF/HR

Author's Note: Within the hydrocarbon family, there are compounds that are not overly hazardous even though they are frequently involved during incidents. However, with the liquefied petroleum gases in this family, their danger outweighs the low risk of the fuels.

Hydrocarbon compounds contain only carbon and hydrogen in their formulas. Carbon has the ability to bond with itself almost indefinitely to satisfy its need for electrons; this happens frequently in the hydrocarbon families. Carbons form straight chained structures with particular physical and chemical characteristics. An example of a straight chained hydrocarbon compound is found in Figure 3.42.

Carbons in compounds can be arranged in configurations other than straight chains. Changing the configuration also changes some of the physical characteristics of the materials. These other configurations will be discussed later. Hydrogens in the compounds are not represented in the naming process. Hydrogens fill out the remaining bonds after the carbons connect to each other. Carbon is in family 4 on the periodic table and has four electrons in its outer shell. These four electrons are available to be shared with electrons from other elements to form an electrically stable compound. Carbon needs four more electrons to be complete. In the nonmetal compounds, the electrons are shared. Hydrogen is in family 1 on the periodic table and has one electron in its outer shell to share. This makes hydrogen a good filler to satisfy the bonding needs of carbon.

When hydrogen shares a bond with one of the electrons of carbon, hydrogen is complete with two electrons. Carbon needs three more, so

$$
\begin{array}{cccc}
H & H & H & H \\
| & | & | & | \\
H-C-C-C-C-H \\
| & | & | & | \\
H & H & H & H \\
\end{array}
$$

Figure 3.42 An example of a straight chained hydrocarbon compound.

$$
\begin{array}{c}
H \\
| \\
H-C-H \quad\quad CH_4 \\
| \\
H \\
\end{array}
$$

Figure 3.43 When hydrogen shares a bond with one of the electrons of carbon, hydrogen is complete with two electrons. Carbon needs three more, so three additional hydrogen atoms are attached to the carbon.

three additional hydrogen atoms are attached to the carbon (Figure 3.43). Once this occurs, both the hydrogen and the carbon are electrically satisfied. This process occurs with other elements also when they attach to carbon. This will be discussed further in hydrocarbon derivative families.

Hydrocarbon Naming Systems

Trivial Understanding the naming of hydrocarbon compounds is a tool that can help responders in the identification of materials at an emergency scene. Naming of the hydrocarbon and hydrocarbon derivative families uses both the trivial and International Union of Pure and Applied Chemistry (IUPAC) naming systems. The trivial naming system uses prefixes indicating the number of carbons in a compound. For example, meth = 1 carbon, eth = 2 carbons, prop = 3 carbons, but = 4 carbons, pent = 5 carbons, hex = 6 carbons, hept = 7 carbons, oct = 8 carbons, non = 9 carbons, and dec/dek = 10 carbons.

Theoretically, carbon chains can go on indefinitely. There are prefixes for as many carbons as you might want to connect together. However, the vast majority of important hazardous materials are formed with prefixes of ten or less carbons. Hazmatology focuses on those with ten carbons or less. The IUPAC naming system is utilized where it fits into the hydrocarbon families. Most of the time the trivial naming system will be used for hydrocarbon compounds with a few exceptions. IUPAC is more predominant with the hydrocarbon derivatives. Endings of the hydrocarbon names reflect the family and type of bond between the carbons in the compound.

IUPAC rules of nomenclature	
Hydrocarbons with more than ten carbons	
$C_{11}H_{24}$	Undecane
$C_{20}H_{42}$	Eicosane
$C_{12}H_{26}$	Dodecane
$C_{21}H_{42}$	Heneicosane
$C_{13}H_{28}$	Tridecane
$C_{22}H_{46}$	Docosane
$C_{14}H_{30}$	Tetradecane
$C_{23}H_{48}$	Tricosane
$C_{15}H_{32}$	Pentadecane
$C_{26}H_{54}$	Hexacosane
$C_{16}H_{34}$	Hexadecane

(Continued)

IUPAC rules of nomenclature	
Hydrocarbons with more than ten carbons	
$C_{30}H_{62}$	Triacontane
$C_{17}H_{36}$	Heptadecane
$C_{31}H_{64}$	Hentriacontane
$C_{18}H_{38}$	Octadecane
$C_{32}H_{66}$	Dotriacontane
$C_{19}H_{40}$	Nonadecane
$C_{33}H_{68}$	Tritriacontane
$C_{40}H_{82}$	Tetracontane
$C_{49}H_{100}$	Nonatetracontane
$C_{50}H_{102}$	Pentacontane
$C_{60}H_{122}$	Hexacontane
$C_{70}H_{142}$	Heptacontane
$C_{80}H_{162}$	Octacontane
$C_{90}H_{182}$	Nonacontane
$C_{100}H_{202}$	Hectane
$C_{132}H_{266}$	Dotriacontahectane

There are four types of formula:

Molecular: C_2H_{10}

Structural:
$$\begin{array}{c} \text{H H H H H} \\ | \; | \; | \; | \; | \\ \text{H}-\text{C}-\text{C}-\text{C}-\text{C}-\text{C}-\text{H} \\ | \; | \; | \; | \; | \\ \text{H H H H H} \end{array}$$

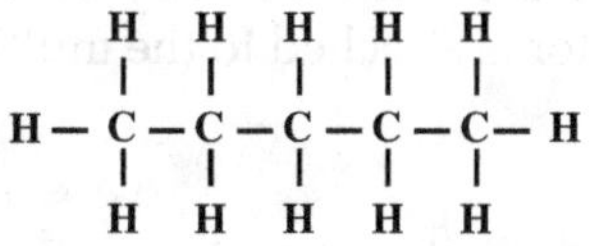

Condensed structural: $CH_3CH_2CH_2CH_3$

Skeleton: C–C–C–C

The naming of all the alkanes is based upon the number of carbon atoms in the longest continuous chain of carbon atoms. If, for example, the longest chain contains four carbon atoms, the compound is called butane. If it has five carbon atoms, it is pentane, and so on.

Names of branched-chain hydrocarbons and hydrocarbon derivatives using the IUPAC system are based on the name of the longest continuous carbon chain in the molecule, with a number indicating the location of a branch or substituent. In order to locate the position of a branch or substituent, the carbon chain is numbered consecutively from one end to the other, starting at the end that gives the lowest numbers to the substituent. For example,

```
      1    2    3    4    5
      H    H    H    H    H
      |    |    |    |    |
  H—  C —  C —  C —  C —  C  — H     Is 2-Methylpentane
      |    |    |    |    |
      H    |    H    H    H
         H—C — H
           |
           O
           |
           H


      1    2    3    4
      H    Cl   H    H
      |    |    |    |
  H—  C —  C —  C —  C  — H     Is 2,2-Dichlorobutane
      |    |    |    |
      H    Cl   H    H
```

The prefixes "di," "tri," "tetra," "penta," "hexa," etc. indicate how many of each substituent is in the molecule. A cyclic (ring) hydrocarbon is designated by the prefix "cyclo."

Double bonds in hydrocarbons are indicated by changing the suffix "ane" to "ene," and triple bonds by changing to "yne." The position of the multiple bond within the structure is indicated by the number of the first or lowest-numbered carbon atoms attached to the multiple bond. For example,

```
      1    2    3    4    5
      H    H    H    H    H
      |    |    |    |    |
  H—  C —  C — C = C —  C  — H     Is 2-Pentene
      |    |         |
      H    H         H


      5    4    3    2    1
      H    H    H
      |    |    |
  H—  C —  C —  C — C ≡ C — H     is 1-Pentyne
      |    |    |
      H    H    H
```

Applied Chemistry and Physics

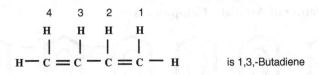

is 1,3,-Butadiene

Most of the hydrocarbon derivative functional groups in organic compounds are designated by either a suffix or a prefix. Rules regarding whether a prefix or a suffix designation is used are as follows:

1. When one such group is present, the suffix will be used.
2. When more than one such group is present, only one will be designated by a suffix, and the others will be designated by prefixes.
3. The order of precedence for deciding which group takes the suffix designation.

Examples are as follows:

```
    H   H
    |   |
H — C — C — O — H        is ethanol, not Hydroxyethane
    |   |
    H   H
```

```
        3   2   1
        H   H   O
        |   |   ||
H — O — C — C — C — O — H    is 3-Hydroxypropanoic acid
        |   |
        H   H
```

In numbering the carbon chain, the lowest numbers will be given preference to

1. Groups named by suffixes
2. Double bonds
3. Triple bonds
4. Groups named by prefixes (groups named by prefixes are listed in alphabetical order)

Nomenclature of Aromatic Compounds

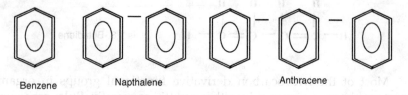

Benzene   Napthalene   Anthracene

Monosubstituted Compounds
Names are derived using prefixes followed by the name "benzene."

Chloro Benzene   Ethyl Benzene   Methyloxy Benzene

Names can be indicated by commonly accepted names.

Disubstituted compounds
Names are derived using prefixes (including commonly accepted names) and numbers or words.

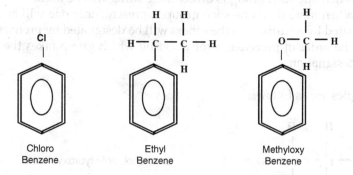

Toluene

Benzoic acid   Benzaldehyde

Phenol   Analine   Benzene Sulfonic acid

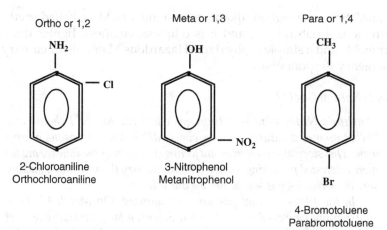

Hydrocarbon Families

Hydrocarbon compounds can be divided into four families: alkanes, alkenes, alkynes and aromatics. With alkane, alkene and alkyne families, the primary hazard is flammability. Vapors of these compounds may be lighter or heavier than air among the gases and heavier than air with the liquids. Most of the flammable liquids have a specific gravity less than 1 and will float on water. As flammable liquids, they may be incompatible with ammonium nitrate, chromic acid, hydrogen peroxide, sodium peroxide, nitric acid and the halogens (F, Cl, Br and I).

Alkane Hydrocarbons Alkanes are all naturally occurring compounds (Figure 3.44). They have single bonds, are saturated with hydrogen, and are considered to be stable even though many of them are flammable and found primarily as an ingredient in mixtures. Individual family members have names beginning with one of the prefixes that indicate the number of carbons in the compound. For example, a one carbon compound begins with the prefix "meth." The prefix "meth" indicates one carbon. Alkanes are all single bonded, so the ending for an alkane

ALKANE HYDROCARBONS (PARAFFINS)

Bond: Single covalent
GF: C_nH_{2n+2}
Name: Prefix + "ane"
Use: Fuels
Hazard: Combustibility; lighter compounds displace O_2, heavier may boil over

Figure 3.44 Alkanes are all naturally occurring compounds.

is "ane." So a one-carbon alkane compound would be called methane. Methane is a natural gas, and it is odorless, colorless, lighter than air, flammable and a simple asphyxiant (Hazardous Materials Chemistry for Emergency Responders).

Why Is Natural Gas Odorized?

Author's Note: *Why is natural gas odorized? Maybe I should ask, "What prompted natural gas to be odorized?" because it was not always done. This story also reinforces my feeling that most of the time we are far more interested in making money with technology than we are in making sure the technology is safe before we put it in use.*

In the 1930s, natural gas was not odorized. On March 18, 1937, an explosion happened at the London School, a large structure of steel and concrete in New London, TX. Originally the architect's plan for the building was to use a boiler and steam distribution system to heat the building. That plan was overruled by the school board, and they opted for 72 gas heaters throughout the building. The school building served all grades from elementary to high school.

Early in 1937 the school canceled their natural gas contract and had plumbers install a tap into a residue gas line in order to save money. Residue gas is the natural gas that is left after thorough and complete natural gas processing. It is free of impurities, moisture and natural gas condensates and is ready to be transported to the end-user market through gas pipelines. This practice, while not explicitly authorized by local oil companies, was widespread in the area. The odorless and colorless natural gas extracted with the oil was seen as a waste product and was flared off. Since there was no value to the natural gas, the oil companies turned a blind eye.

On March 18, 1937, at 3:05 p.m. a shop teacher unplugged an electric sanding machine. Unknown to him, the area was filled with natural gas. Activating the switch ignited the gas/air mixture. The resulting explosion lifted the building from its foundation causing walls to collapse. Victims were buried in a mass of brick, steel and concrete debris. Sound waves from the explosion reached 4 miles away, and a 2 ton concrete slab was hurled 200 ft where it crushed a car. Approximately 600 students and 40 teachers were in the building at the time of the explosion. Only 130 escaped without serious injury. Most of the bodies were either burned beyond recognition or blown to pieces so it was difficult to estimate the number of people killed. Estimates ranged from 296 to 319. As a result of the London School Disaster, the state passed an odorization law, which required that distinctive malodorants be mixed with natural gas for commercial and industrial use so the smell would warn people of any leaks.

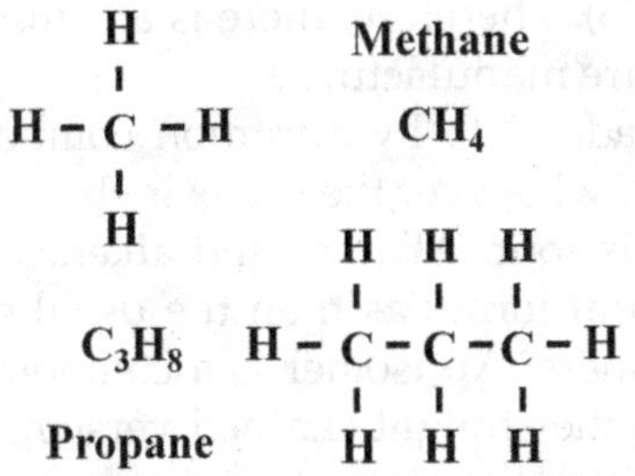

Applied Chemistry and Physics

```
           H          Methane
           |
       H - C - H       CH₄
           |
           H         H  H  H
                     |  |  |
         C₃H₈    H - C - C - C - H
                     |  |  |
        Propane      H  H  H
```

Figure 3.45 Molecular and structural formulas of methane and propane.

A two-carbon alkane would begin with "eth" and end in "ane," forming the compound ethane; a three-carbon alkane would begin with "prop" and end in "ane", forming the compound propane, which is also a gas. Propane is very flammable, odorless, colorless, heavier than air, and a simple asphyxiant. If an explosion occurs in a structure and the damage is at the foundation level, it may be a heavier-than-air gas such as propane. Explosion damage up high such as in an attic may involve a lighter-than-air gas like methane. Molecular and structural formulas of methane and propane are shown in Figure 3.45.

The same procedures apply to other alkane compounds that have four or more carbons hooked together in a chain. They are also named by first using the prefix for the number of carbons and ending in "ane."

- Four-carbon prefix is "but" with the ending "ane," and the compound is called butane.
- Five-carbon prefix is "pent" with the ending "ane," and the compound is called pentane.
- Six-carbon prefix is "hex" with the ending "ane," and the compound is called hexane.
- Seven-carbon prefix is "hept" with the ending "ane," and the compound is named heptane.
- Eight carbon prefix is "oct" with the ending "ane" and the compound is named octane.
- Nine carbon prefix is "non" with the ending "ane," and the compound is named nonane.
- Ten carbon prefix is "dec" with the ending of "ane," and the compound is named decane.

Many of the alkanes are pure compounds and can be found in transportation and fixed fatalities, whereas others are found primarily in mixtures of compounds such as gasoline and diesel fuel. Alkene hydrocarbons are man-made (Figure 3.46). Therefore, there is a commercial value to them, and that is why they are manufactured.

Prefixes may be added to hydrocarbon compounds. These prefixes indicate the number of what ever the prefix is shown in front of. Di is two, tri is three and tetra is four. Alkanes and alkenes are sometimes found with different structural formulas than the usual straight chains. These are referred to as isomers. An isomer is a compound that has the same molecular formula as the straight chained version, but a different structural formula. This concept is also referred to as branching (Figure 3.47). This has the effect of lowering the boiling point of the material. For example, butane has a boiling point of around 31°F. It can be used as a heating fuel like propane. Butane and propane are stored and used at ambient temperatures. However, with the boiling point of 31°F, the areas in the country where butane could be used in the winter would be limited by the winter time temperatures. If, however, butane is branched to form isobutane, this

Figure 3.46 Alkene hydrocarbons are man-made and refined from crude oil.

Isomers or Branched Compounds

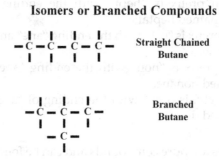

Figure 3.47 An isomer is a compound that has the same molecular formula as the straight chained version, but a different structural formula.

Applied Chemistry and Physics

process lowers the boiling point of butane to around 10°F, making butane more useful as a heating fuel in many colder parts of the country.

The most commonly encountered hazardous material is the hydrocarbon fuel, which makes up over 50% of all hazardous materials releases that occur. Fuels made up of various pure hydrocarbon compounds in a mixture that has different characteristics than the individual compounds are common. Physical characteristics such as boiling points, flash points, ignition temperatures and flammable ranges may be different because of the mixing of the products together. Products such as gasoline, diesel fuel and fuel oil are all used to power our vehicles or heat our homes and businesses. Propane, butane and natural gas are also fuels but are not mixtures; they are pure compounds. Propane and butane may be mixed together to form liquefied petroleum gases. Fuels are transported frequently and stored in large quantities, which cause frequent problems for emergency responders. They are all flammable and some may be toxic or cause asphyxiation by displacing the oxygen in the air.

Another configuration or variation in structure of a straight chained hydrocarbon is the "ringed" compound. All carbons in the ringed compound are hooked together end to end. Ringed compounds are referred to as cyclic, and when a ringed compound is named, the prefix "cyclo" is used in front of the hydrocarbon name. This configuration can be very unstable in the case of three or four carbons, or seven or eight carbons. However, five or six carbons in a ring form a stable compound. Arranging carbons in a circle tends to raise the boiling point of the material (Figure 3.48). For example, hexane is a six-carbon straight chained compound with a boiling point of around 155°F. Cyclohexane, on the other hand, has a boiling point of around 178°F.

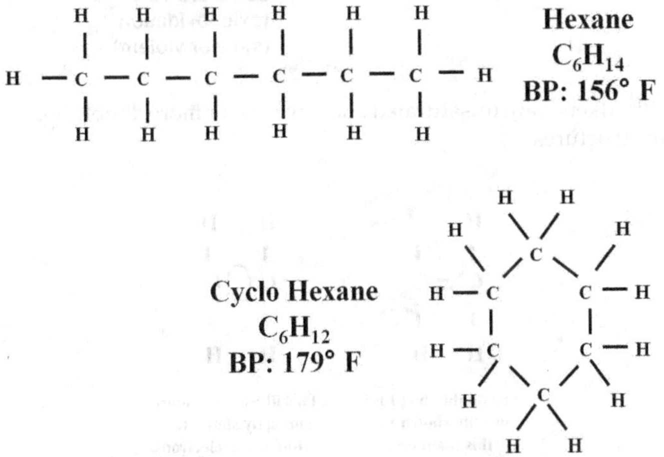

Hexane
C_6H_{14}
BP: 156° F

Cyclo Hexane
C_6H_{12}
BP: 179° F

Figure 3.48 Arranging carbons in a circle tends to raise the boiling point of the material.

Alkene Hydrocarbons   Alkenes are unsaturated and have one or more double bonds between carbons in structures (Figure 3.49). The double bond(s) make the compound unstable, and the oxygen in the air can react with the bond to break it. When a bond breaks, heat is generated. If the heat is confined within the material or a combustible carrier, spontaneous combustion can occur (Figure 3.50).

Another hazard of double bonded materials is polymerization. This is a spontaneous expansion type of chemical reaction. If it occurs within a container, the container may come apart explosively. Once the polymerization reaction starts, it will continue until it is finished regardless of what responders do. Materials that are capable of undergoing polymerization are referred to as monomers. They must have a chemical added to them in transportation and storage to prevent the polymerization from occurring until they are ready for it in a chemical process. This chemical is called an inhibitor. In a chemical process vessel, the inhibitor is removed and a catalyst is added which controls the rate of the chemical reaction.

Alkene (Olefin) Hydrocarbons

Bond: Double covalent.
GF: C_nH_{2n}
Name: Prefix + "ene"
Use: Plastics manufacture
Hazards: Polymerization, some are very toxic, oxidation (slow or violent)

Figure 3.49 Alkenes are unsaturated and have one or more double bonds between carbons in structures.

```
H   H              H   H
|   |              |   |
C = C              C O C
|   |              |   |
H   H              H   H
```

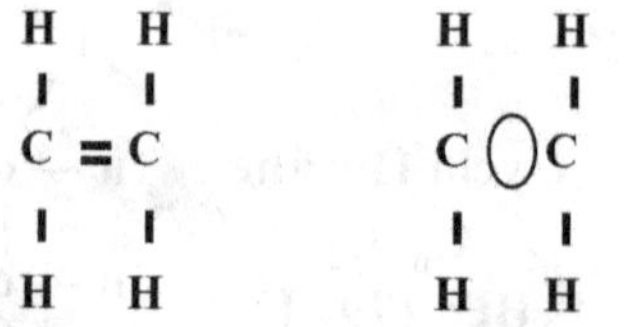

The double bond is usually shown in this manner.

This illustration more correctly shows the out-of-plane electrons.

Figure 3.50 Out of plane electrons allow for oxygen in the air to come in and break the double bond easier.

Because the primary characteristic of an alkene is the double bond that occurs between two carbons, there are no single-carbon alkenes. The first possible alkene has two carbons with the prefix "eth" and the ending "ene," and the compound is called ethene. Alkene compounds may also have a "yl" after the prefix. So, ethene may also be called ethylene however, the ene at the end still identifies the compound as a double bonded alkene. Ethylene is a colorless gas with a sweet odor and taste. It is highly flammable with explosive limits of 3%–36% in air. In addition to being flammable, it is also a simple asphyxiant gas. Ethylene is used in the production of other chemicals such as polyethylene, polypropylene, ethylene oxide, ethylene glycols ethyl alcohol and others.

Alkenes with four carbons begin with the prefix "but." It is possible for alkene compounds to have more than one double bond. A four-carbon compound with two double bonds has the prefix "but" plus an "a" to make the name flow and the ending "ene." However, there must be something in the name to indicate that there are two double bonds. The prefix "di", meaning "two," is inserted in front of "ene" indicating two double bonds. The name of the compound is butadiene. It is a colorless gas with a mild aromatic odor. It is highly flammable with explosive limits of 2%–11% in air. It is a polymer that must be inhibited during transportation and storage. Butadiene is used in the production of plastics. Molecular and structural formulas of ethylene and butadiene are shown in Figure 3.51.

There are other compounds with one, two or even three double bonds. The rules for naming remain the same. Use a prefix for the number of carbons, end in "ene" and use a prefix to indicate the number of double bonds.

Alkyne Hydrocarbons Alkynes are hydrocarbon compounds that have one or more triple bonds in the structure (Figure 3.52). They are also unsaturated and very reactive. Although it is possible to create a different alkyne compound on paper using the rules for the family, there is really only one commercially valuable compound that responders are likely to encounter. Just as with the alkenes, there are no one-carbon alkynes. The first compound has two carbons with the prefix "eth" and the ending "yne",

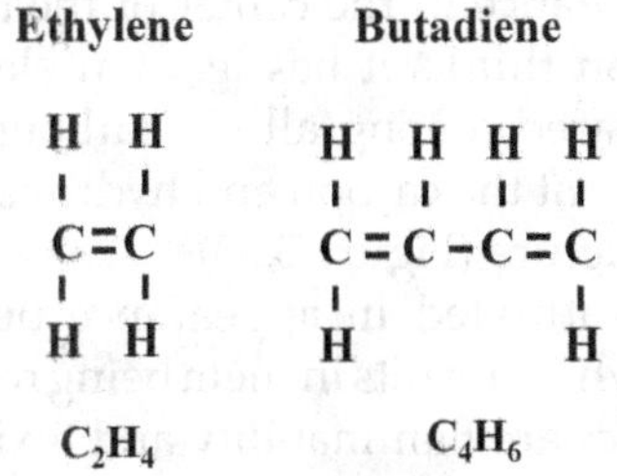

Ethylene   Butadiene

H H   H H H H
| | | | | |
C=C   C=C-C=C
| | | |
H H   H   H

C_2H_4 C_4H_6

Figure 3.51 Molecular and structural formulas of ethylene and butadiene.

Ethyne (Acetylene) -

$$H-C\equiv C-H \qquad C_2H_2$$

Figure 3.52 Alkynes are hydrocarbon compounds that have a triple bonds in the structure.

and the name of the compound would be ethyne. Ethyne, however, is the chemical name of the compound; it is commonly known by its trade name which is acetylene.

Ethyne has two carbons and two hydrogens with a single triple bond between the carbons. Ethyne or acetylene is a highly flammable, colorless gas with an ethereal odor. It has wide explosive limits from 2.5% to 80% in air. It can form explosive compounds with silver, mercury and copper. Acetylene is so unstable that it is dissolved in acetone inside of its container. Acetylene is not shipped in tank car or tank truck quantities. Calcium carbide, a salt, is used to generate acetylene by reacting with water. When the water contacts calcium carbide, acetylene gas is generated. Calcium carbide is shipped in specially designed closed containers that are placed on flatbed rail and highway vehicles. Molecular and structural formulas are shown for ethyne (acetylene) are shown in Figure 3.52.

Aromatic Hydrocarbons The last of the hydrocarbon compounds are the aromatics based on benzene, sometimes referred to as the "BTX" fraction. Unlike the alkanes, alkenes, and alkynes, the naming conventions and endings do not follow any specific rules. It is important that the responder just become aware of the names of these compounds and their characteristics. Aromatics are known by their characteristic ring formation, sometimes called the benzene ring. They are ringed compounds with some variances. Bonding associated with aromatic compounds is called resonant. The concept of resonance is too complicated to discuss here, so just understand that the benzene ring has 6 carbon and 6 hydrogen elements in the compound. There are no double bonds in the benzene ring. Carbons are connected to the additional electrons that each element needs, which rotate in a state of resonance in the center of the compound. They move so fast that each carbon thinks it has its own electron, when in reality electrons are being shared among all six carbons. The benzene ring is sometimes shown without the carbon and hydrogen. It is understood that they are still in the structure (Figure 3.53).

Aromatics are unsaturated in appearance but in reality act like a saturated compound, which results in them being relatively stable. Primary hazards of the aromatics are flammability and toxicity. They are also carcinogenic compounds. While flammable, aromatics burn with incomplete combustion and are very smoky when they burn. The parent member of

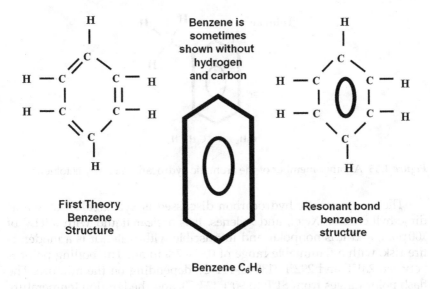

Figure 3.53 The benzene ring is sometimes shown without the carbon and hydrogen. It is understood that they are still in the structure.

Figure 3.54 Benzene is a colorless to light yellow liquid with an aromatic odor.

the aromatic family is benzene as all other aromatic compounds are made from benzene. Benzene is a common industrial chemical. It is a colorless to light yellow liquid with an aromatic odor (Figure 3.54). Its boiling point is around 175°F, its explosive limits are 1.5%–8% in air and it has a TLV of 10 ppm. Not long ago benzene's TLV was thought to be 1,000 ppm until workers who were exposed to benzene daily developed illnesses associated with the exposures.

Another member of the aromatic hydrocarbon family is toluene (Figure 3.55). It also has a benzene ring, but a "methyl radical" is added to distinguish between benzene and toluene. Hazards are similar to benzene which are flammable and toxic with explosive limits of 1.2%–7% in air and a TLV of 100 ppm. Toluene's boiling point is 230°F, and it is a colorless liquid with a benzene-like odor.

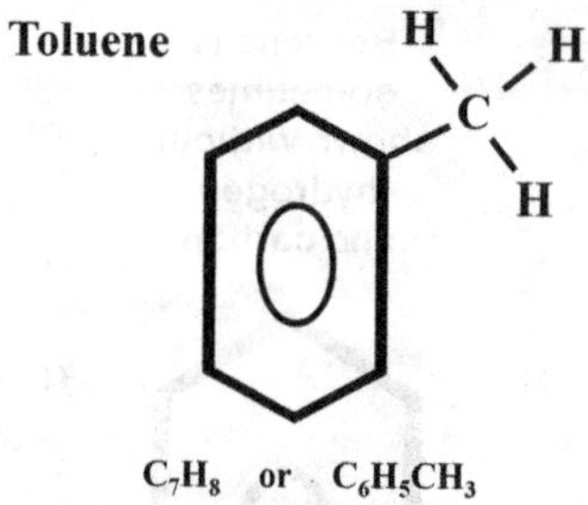

Toluene

C_7H_8 or $C_6H_5CH_3$

Figure 3.55 Another member of the aromatic hydrocarbon family is toluene.

The next aromatic hydrocarbon discussed is xylene, also known as dimethylbenzene. Xylol, and Xylenes. It is a clear liquid with a TLV of 100 ppm, and it is nonpolar and immiscible with water. It is a moderate fire risk, with a flammable range of 0.9%–7% in air. The boiling point is between 281°F and 292°F (138°C–144°C), depending on the mixture. The flash point ranges from 81°F to 90°F (27°C), and the ignition temperature ranges from 867°F to 984°F (463°C–528°C). Xylene is a commercial mixture of the three isomers: *meta*-xylene, *para*-xylene and *ortho*-xylene (Figure 3.56), (Figure 3.57), (Figure 3.58). The last member of the benzene family is styrene (Figure 3.59). It is, however, quite different from the other members of the aromatic family. Styrene is a monomer used in the manufacture of polystyrene. It has a vinyl radical attached to the benzene ring. The double bond in the vinyl radical is reactive. Reaction can occur with the oxygen in the air, with an oxidizer, or it can self-react in storage.

Hydrocarbon Derivatives

Hydrocarbon derivatives are chemical compounds that do not occur naturally. The alkane hydrocarbons occur naturally, whereas other hydrocarbons are man-made. It is important to understand that in the making of hydrocarbon derivatives, the process starts with a hydrocarbon with some other nonmetal element(s) added to create a new compound. In some cases,

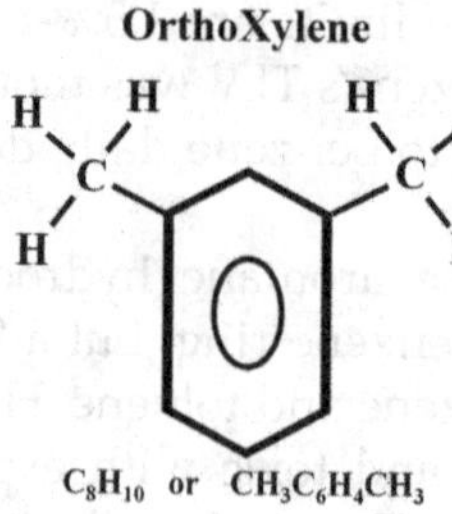

OrthoXylene

C_8H_{10} or $CH_3C_6H_4CH_3$

Figure 3.56 *ortho*-Xylene.

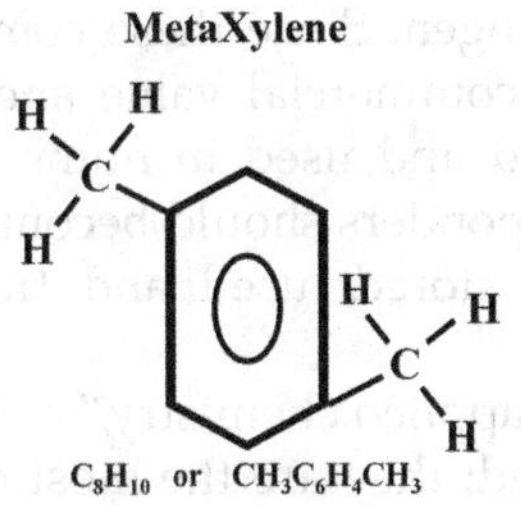

Figure 3.57 meta-Xylene.

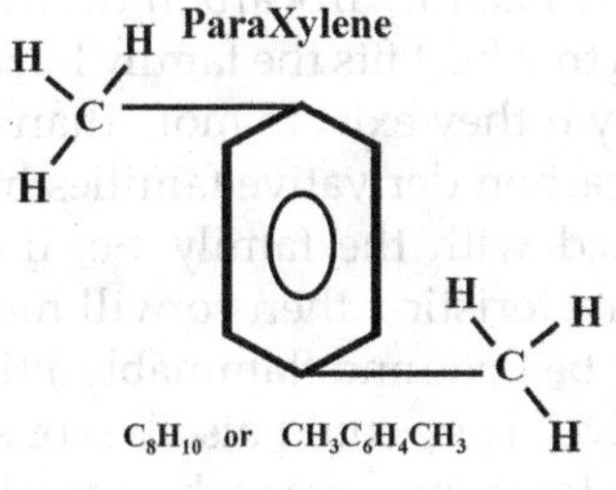

Figure 3.58 para-Xylene.

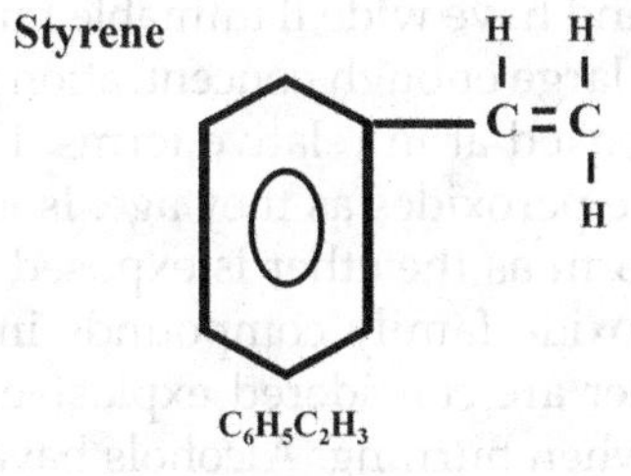

Figure 3.59 Styrene is the last member of the benzene family presented here.

parts of the name of the hydrocarbon, with which the process started, are found in the name of the derivative compound. Some of the characteristics of the hydrocarbon may remain a part of the new compound. In other cases, a completely different compound may be formed with totally different characteristics. For example, in the compound trifluorobromomethane, the compound started with the flammable gas methane. The toxic elements fluorine and bromine are added, and the resulting compound is not flammable or very toxic. In fact the compound is more commonly known as Halon 1301 once used as a common fire extinguishing agent.

Only a few nonmetal elements are added to hydrocarbons to form hydrocarbon derivatives; they are commonly oxygen, fluorine, chlorine, bromine, iodine and nitrogen. Since these compounds do not occur in nature, they have some commercial value and therefore are manufactured, stored, transported and used in numerous industrial and commercial occupancies. Responders should become familiar with the types of hazardous chemicals stored, used and transported through their communities.

For the purposes of "applied chemistry," only 13 hydrocarbon derivative families are presented; they are the most common and most likely compounds to be encountered by emergency responders. Hydrocarbon derivative families include: alkyl halide; nitro; amine; cyanide, ether; isocyanate; organic peroxide; sulfur compounds; alcohol; ketone' ester; aldehyde; and organic acids. Each hydrocarbon derivative family is covered within the hazard class that best fits the family hazards. Some are covered in more than one family if they exist in more than one physical state.

Each of the hydrocarbon derivative families has certain hazards and characteristics associated with the family. So, if one compound in the family has certain characteristics, then so will most of the others. Alkyl halides as a family may be toxic and flammable although all are not. Nitro compounds are explosive! They could also be considered flammable and perhaps in some cases toxic, however who cares! They are going to go BOOM, and that is what will more likely injure or kill response personnel. Amine compounds may be toxic and flammable. Cyanides are toxic. Ethers are flammable and have wide flammable ranges; additionally, they are anesthetics and, in large enough concentrations, toxic, but so is water! So toxicity has to be looked at in relative terms. Ethers are a compound prone to form explosive peroxides as they age. Isocyanates are extremely toxic. Peroxides may form as the ether is exposed to air, particularly the oxygen in the air. Peroxide family compounds including the peroxides that are formed in ether are considered explosive. Sulfur oxides release hydrogen sulfide gas when burning. Alcohols have varying levels of toxicity and are flammable. Ketones are narcotic and flammable. Esters are flammable an will polymerize. Aldehydes are toxic by inhalation, irritants, flammable and carcinogens. Organic acids are corrosive, some are flammable and oxidizers.

Radicals Hydrocarbon derivatives are formed when one or more hydrogen atoms are removed from a hydrocarbon compound (Figure 3.60). When hydrogen is removed, the compound is no longer complete or the same compound. The compound without the hydrogen is considered a "radical." Radicals, as they are not complete compounds, do not exist for very long because of the need to bond with some other elements. Therefore, responders will not encounter tanker trucks, or other containers of methyl or other radicals going down the highway. Because the radicals are not the same compounds as they were, the compound name is changed to

Derivative Prefixes (Radicals)

Single Bonds	Double Bonds
Meth = Methyl/Form	Ethylene = Vinyl
Eth = Ethyl/Acet	Propylene = Acryl
Prop = Propyl	
But = Butyl	
Pent = Pentyl	

Figure 3.60 Hydrocarbon derivatives are formed when one or more hydrogen atoms are removed from a hydrocarbon compound.

indicate that the hydrocarbon is a radical of the original compound. The same compound names are used to identify the number of carbons present in a radical as in the hydrocarbon compounds they came from. For example, if methane, an alkane hydrocarbon with all single bonds, has one hydrogen removed, it is no longer methane. There is still one carbon, so "meth" is used, and a "yl" is added to the "meth" to form the name "methyl." Therefore, a one-carbon radical of methane is the methyl radical.

Alkene Radicals There are really only two alkene radicals that are important to emergency responders: the two- and three-carbon alkenes. When one hydrogen is removed from ethene, the radical formed is called vinyl. A hydrocarbon radical with two carbons and a double bond with the formula C_2H_3 is the "vinyl" radical. There is no rule for knowing that it is just a matter of memorizing the vinyl radical. Another important alkene used for making hydrocarbon derivatives is pentene, with one carbon removed, which becomes the "acryl" radical. So, there are three carbons in the radical with one double bond.

Aromatic Radicals There are two important radicals formed from the aromatic hydrocarbons (Figure 3.61). The first aromatic radical is benzene with one hydrogen removed from the benzene ring which becomes the "phenyl" radical. Phenol is a very common and deadly chemical formed with the "phenyl" radical and an alcohol. The second aromatic radical comes from toluene with one hydrogen removed from the methyl radical on the benzene ring. This radical of toluene is known as the "benzyl" radical. This may be confusing because one would think that the "benzyl" radical would be the radical of benzene. Benzoyl peroxide is a chemical that is an organic peroxide made with the "benzyl" radical. All of the previously mentioned radicals from the alkane, alkene and aromatic hydrocarbons may be used with any of the hydrocarbon derivative families to form compounds.

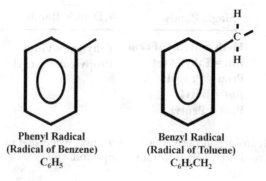

Figure 3.61 Two important radicals formed from the aromatic hydrocarbons.

Hydrocarbon Derivative Families

Each of the hydrocarbon derivative families has certain general hazards associated with the family. The first eight hydrocarbon derivative families—alkyl halide, nitro, ether, amine, cyanide, isocyanate, organic peroxide and sulfur compounds—are considered nonpolar families for the purposes of "applied chemistry." Their hazards include flammability, toxicity, oxidizer and explosive. The last five derivative families—alcohol, ketone, ester, aldehyde and organic acid—are all considered polar compounds. Hazards of the last five hydrocarbon derivatives include flammable, corrosive, narcotic, polymerization and varying degrees of toxicity. With all of the families, there may be different degrees of hazard associated with compounds within a family. Until responders have researched a particular compound, they should use the overall family hazard(s) as a "rule of thumb' until they find out the specific hazards.

Polar Compounds

What Is Polarity? For applied chemistry we will say that A molecule is basically said to be either polar molecule or nonpolar molecule. That translates into either polar compounds or nonpolar compounds as found in hydrocarbon derivative families.

Polar molecules. A polar molecule is usually formed when the one end of molecule is said to possess more number of positive charges and whereas the opposite end of the molecule has more number of negative charges, creating an electrical pole. When a molecule is said to have a polar bond, then the center of the negative charge will be on one side, whereas the center of the positive charge will be on the otherside. The entire molecule will be a polar molecule. Water is a polar material (Figure 3.62). A molecule of water has a slightly negative field on one side and a slightly positive field on the other side. There is a rule in physics

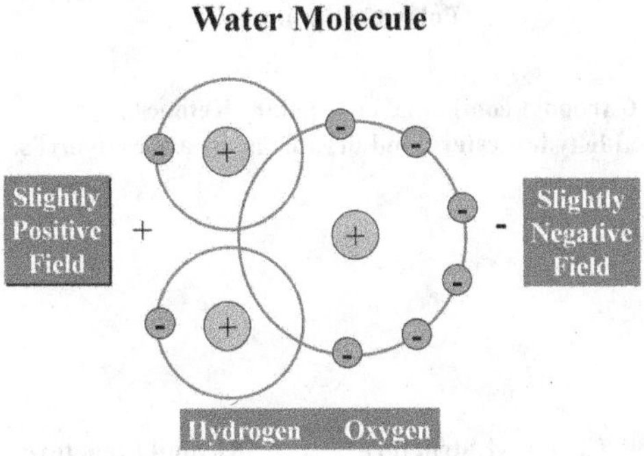

Applied Chemistry and Physics

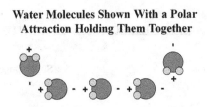

Figure 3.62 Water is a polar compound.

Water Molecules Shown With a Polar Attraction Holding Them Together

Figure 3.63 The negative side of one water molecule is attracted to the positive side of another water molecule.

that says opposite charges attract each other. The negative side of one water molecule is attracted to the positive side of another water molecule, and this continues until all water molecules are attracted to each other (Figure 3.63). That is why water with a molecular formula of a gas can exist as a liquid! A hydrocarbon derivative with a carbonyl or an –O–H in the formula determines a polar compound (Figure 3.64). Polarity raises the boiling point of a liquid. It also determines the type of fire extinguishing agent needed for extinguishment.

Nonpolar Molecules. A molecule that does not have the charges present at the end because electrons are finely distributed and those that symmetrically cancel out each other are the nonpolar molecules. In a solution, a polar molecule cannot be mixed with a nonpolar molecule. For example, consider water and oil. In this solution, water is a polar molecule, whereas oil behaves as a nonpolar molecule. These two molecules do not form a solution as they cannot be mixed up.

Hydrocarbon derivatives are compounds with other elements in addition to hydrogen and carbon. Weight determines the boiling points when comparing hydrocarbon derivatives within the same family. However, when comparing different families, the concept of polarity has to be considered with some of the compounds.

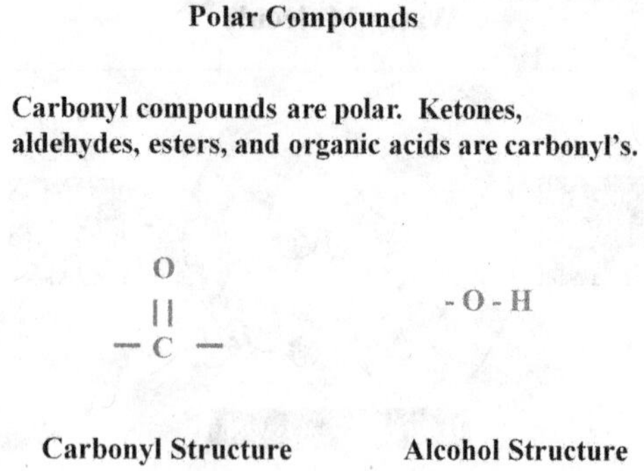

Polar Compounds

Carbonyl compounds are polar. Ketones, aldehydes, esters, and organic acids are carbonyl's.

$$\begin{array}{c} O \\ \| \\ -C- \end{array} \qquad -O-H$$

Carbonyl Structure   Alcohol Structure

Figure 3.64 A hydrocarbon derivative with a carbonyl or an –O–H in the formula determines a polar compound.

POLARITY OF HYDROCARBON DERIVATIVES

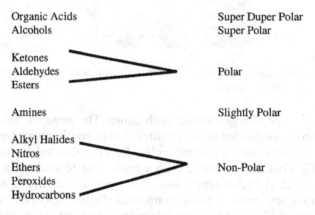

Figure 3.65 There are degrees of polarity among the polar families.

There are degrees of polarity among the polar families (Figure 3.65). For the purposes of "applied chemistry," we use the "scientific terms" such as polar, super polar and super duper polar. Members of the family can be subdivided into three groups: carbonyl, alcohol and organic acid. The carbonyl family is represented in the formula of a compound by the letters "CO." They include ketone, aldehyde and ester. These families are considered polar compounds. Alcohols do not have the carbonyl CO in their formula, so they are not carbonyls. They are, however, polar, because of the "OH" or hydroxyl formulation in the compound. The OH in a hydrocarbon derivative family indicates polarity just as the CO in the

Figure 3.66 Polarity of compounds is important when it comes to firefighting and choosing the right foam.

carbonyl compounds. Alcohol family compounds are considered super polar. Organic acid family members contain both the CO for carbonyl and the OH like the alcohol family. With two forms of polarity, they are considered "super duper polar." In the structural formula of a compound, the carbonyl is represented by the structure in which a single carbon atom is bonded to a single oxygen atom by a double bond. Polarity of compounds is important when it comes to firefighting and choosing the right foam (Figure 3.66).

Normally, a double bond is considered to be reactive or unstable as it is in the alkene and alkyne hydrocarbons. Double bonds are stable within the carbonyl families, except for ester and aldehyde. Ester compounds have the double bond and their hazard is polymerization. Within aldehyde compounds, the double bond has a hydrocarbon radical on each side, protecting the double bond, and the compound is stable (Figure 3.67). Oxygen in the air has difficulty getting to the double bond because radicals are in the way; therefore, the bond is relatively stable. For example, a CO is placed between two ethyl radicals. The radicals protect the double bond from the oxygen in the air making this a stable compound. With aldehyde, there is only one radical, leaving the double bond exposed to oxygen from the air that can break the bond and form peroxides (Figure 3.68).

Hazmatology Point: *Aldehyde has a hidden hazard. It can form explosive peroxides as it ages. It is also a known carcinogen.*

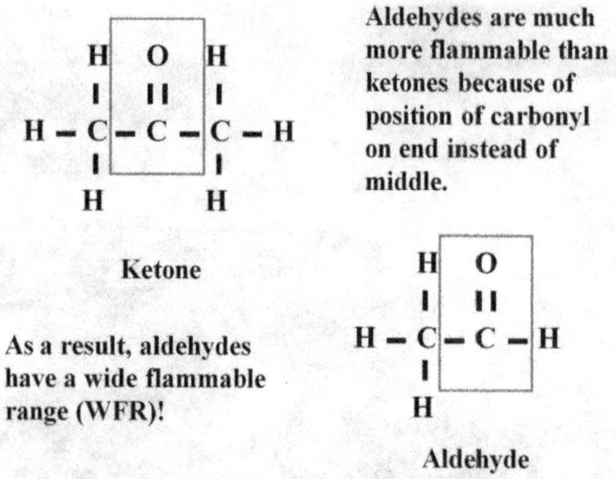

```
        H   O   H
        |   ||  |
    H — C — C — C — H
        |       |
        H       H
```

Aldehydes are much more flammable than ketones because of position of carbonyl on end instead of middle.

Ketone

As a result, aldehydes have a wide flammable range (WFR)!

```
        H   O
        |   ||
    H — C — C — H
        |
        H
```

Aldehyde

Figure 3.67 Within aldehyde compounds, the double bond has a hydrocarbon radical on each side, protecting the double bond, and the compound is stable.

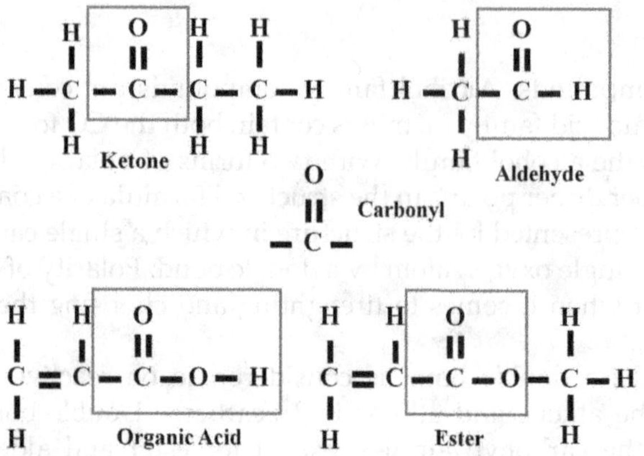

```
    H   O   H   H                H   O
    |   ||  |   |                |   ||
H — C — C — C — C — H        H — C — C — H
    |       |   |                |
    H       H   H                H
      Ketone                      Aldehyde
```

```
         O
         ||
    — C —
```
Carbonyl

```
    H   H   O                     H   H   O       H
    |   |   ||                    |   |   ||      |
    C = C — C — O — H             C = C — C — O — C — H
    |                             |                   |
    H   Organic Acid              H     Ester         H
```

Figure 3.68 Radicals protect the double bond from the oxygen in the air making this a stable compound.

Hydrocarbon and hydrocarbon derivative families all have particular hazards associated with them. If responders become familiar with the hazards of the family, they will have an idea of the hazards facing them during an incident, because all compounds in a particular family will have similar hazards. Each of the 13 hydrocarbon derivative families is placed in the appropriate DOT hazard class and discussed in detail.

Complex Hydrocarbon Derivative Compounds

Rules for naming and identifying hydrocarbon derivative families and determining their hazards are pretty straightforward. There are, however, other compounds that contain two or more hydrocarbon derivative families or several different elements attached to the hydrocarbons. For the purpose of "applied chemistry," let's refer to them as "exotic hydrocarbon derivatives." It is very unlikely that these materials will be found in many communities. Some are "forbidden" in transportation. Few if any incidents of any significance have occurred with these materials. They are presented here as an informational overview. These compounds are not as clear at first glance as to the family they belong to or the hazards that may be associated with them. However, if you look closely at the name or formula if available, there are hints to the component parts and potential hazards.

Ketone and Peroxide Compounds—LF/HR Ketones may have different names that disguise the hydrocarbon derivative family they belong to. Mesityl oxide might be taken as an ether ending in "oxide." It is, however, a ketone with a double bond in the structure. Mesityl oxide is an oily, colorless liquid, with a honey-like odor. The flash point is 90°F with an ignition temperature of 652°F. It is a moderate fire risk but is toxic by inhalation, ingestion and skin absorption. The TLV is 15 ppm in air. Mesityl oxide is used as a solvent, a paint and varnish remover, and an insect repellent.

If there is some information available, such as name or formula, you may be able to notice the components of the material you recognize. For example, in the compound methyl ethyl ketone peroxide, there are two hydrocarbon derivative families, ketone and peroxide (organic). Ketones are flammable and narcotic, whereas peroxides are potentially explosive and oxidizers. By taking precautions for all of the families identified, responders would be looking at the worst-case scenario. As it turns out methyl ethyl ketone peroxide is listed in the *Condensed Chemical Dictionary* as being a "fire risk in contact with organic materials and a strong irritant to skin and tissue." CFR 49 Hazardous Material Table lists it as a "Forbidden" commodity in transportation when the active oxygen in the compound is greater than 9%.

Methyl ethyl ketone peroxide is listed in the *Emergency Response Guidebook* as an organic peroxide/heat and contamination sensitive and a severe irritant. MSDS sheets indicate that the material is stable, does not undergo polymerization, and is a combustible liquid. The point I am trying to make here is if you had taken the worst-case scenario for methyl ethyl ketone peroxide, you would have been in error. However, it is always better to err on the side of safety. Taking precautions for an organic peroxide protects you against this compound. Overprotection, yes, but, overprotection on the side of safety is better when you don't know the exact hazards of a compound. You may get teased, have "egg on your face" and be embarrassed; however, no one has ever died from any of these things. On the other hand, if you don't err on the side of safety, when you haven't verified the hazards, you could be seriously injured or die!

Benzoyl peroxide is another compound that has the peroxide and ketone derivative families in the structure and formula. It is also listed as a UN/DOT Class 5.2 Organic Peroxide. The UN 4-digit identification number is 2085, and Orange Guide 146 is used from the *Emergency Response Guidebook*. When wet, it is stable; when dry with less than 1% water, it may explode spontaneously. It is also highly toxic by inhalation with a TLV of 5 mg/m^3 of air. Benzoyl peroxide is a white granular, crystalline solid that is tasteless and has an odor of benzaldehyde. There is approximately 6.5% active oxygen in the compound. It decomposes explosively above 220°F and has an autoignition temperature of 176°F. Benzoyl peroxide is used as a bleaching agent for flour, fats, oils and waxes and in pharmaceutical and cosmetic compounds in lower concentrations. Initially, treating benzoyl peroxide as a worst-case scenario, organic peroxide would be the appropriate action. Anything less could spell disaster to first responders.

Alcohol Compounds—LF/HR   Some alcohol compounds contain more than one alcohol derivative group. For example, ethylene glycol is a common coolant and antifreeze. It has two alcohol groups attached to the hydrocarbon ethene or ethylene. The compound is a clear, colorless, syrupy liquid that has a sweet taste and is odorless. Its boiling point is approximately 387°F, and the flash point is about 7°F. Ethylene glycol is toxic by ingestion and inhalation with a TLV (vapor) ceiling of 50 ppm. The reported lethal dose by ingestion is 100 cc.

Propylene glycol also has two alcohol groups, but they are attached to propene, also known as proplyene. Its physical characteristics are similar to those of ethylene glycol, except that the flash point is much lower at below –40°F. It is also much less toxic than ethylene glycol. Some of its uses include an antifreeze, a coolant in refrigeration systems, a solvent for food additives, colorings and flavorings, and a deicing fluid for airport runways.

Glycerol, also known as glycerin, is another complex alcohol. It has three alcohol groups attached to propane. It is a clear, syrupy, colorless and odorless liquid with a sweet taste. The flash point is 320°F, and the autoignition temperature is 739°F. Toxicity is low with a mist TLV of 10 mg/m^3 in air. Glycerol is used in the manufacturer of dynamite, pharmaceuticals cosmetics, and food, tobacco and liquors. When the hydrogens are removed from glycerol and three nitro groups are added to the remaining oxygen, nitroglycerin is formed. It is classified by UN/DOT as a Class 1.1 Explosive and is forbidden in transportation unless desensitized. The NFPA 704 designation for nitroglycerin is health—2, flammability—2, reactivity—4 and special—0. It is listed under the UN 4-digit identification number of 0143; when desensitized, it is listed under several UN 4-digit identification numbers depending upon the mixture and make-up of the desensitizer. Orange Guides 127 and 113 are used for desensitized material and 112 for unsensitized material.

Nitroglycerin is a viscous, pale yellow liquid, with a severe explosion risk, and it is highly sensitive to shock and heat. The explosive hazard of nitroglycerin far outweighs the toxicity; however, the toxicity should not be overlooked. It is toxic by ingestion, inhalation and skin absorption. The TLV is 0.05 ppm in air, which is very toxic! Nitroglycerin is a high explosive used in the production of dynamite and rocket propellants, in combating oil well fires and as a vasodilator used as a treatment for those who have heart problems. Structural and molecular formulas of nitroglycerin are shown in Figure 3.69.

Aldehyde Compounds—LF/LR Furfural is an aldehyde, but you wouldn't recognize it by the name, unless you remember that aldehydes may have an "al" ending. It is a "complex hydrocarbon derivative" with three double bonds in the structure. As you might imagine, it may polymerize upon heating! The UN/DOT hazard classification is a Class 3 Flammable Liquid. NFPA has listed a 704 designation as health—3, flammability—2, reactivity—0 and special—0. In the Orange Guide section of the *Emergency Response Guidebook*, the listing is for Guide 132P, where "P" indicates that it may polymerize. The UN four-digit identification number is 1199. It is also a cyclic compound and a combustible liquid with a flash point of 140°F and an autoignition temperature of 797°F. Furfural is a clear liquid when pure, but it turns into reddish brown upon exposure to light and

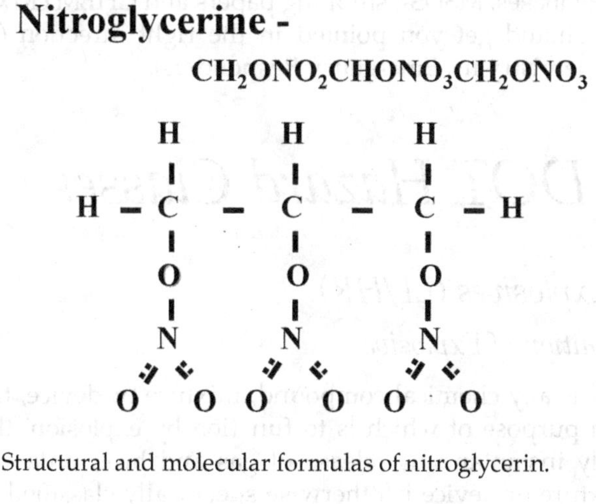

Nitroglycerine -

$CH_2ONO_2CHONO_3CH_2ONO_3$

$$\begin{array}{ccc}
H & H & H \\
| & | & | \\
H-C-C-C-H \\
| & | & | \\
O & O & O \\
| & | & | \\
N & N & N \\
\diagup\diagdown & \diagup\diagdown & \diagup\diagdown \\
O \quad O & O \quad O & O \quad O
\end{array}$$

Figure 3.69 Structural and molecular formulas of nitroglycerin.

air. The odor is similar to that of benzaldehyde. It is absorbed through the skin and is an eye, skin and mucous membrane irritant. The TLV is 2 ppm in air.

Ester Compounds—LF/HR   Some esters are a potential polymerization hazard, whereas others are used as food additives and flavorings. Vanillin, which is present in the flavoring vanilla, would look like a very dangerous chemical looking at the molecular and structural formulas. It is composed of white crystalline needles with a sweet smell. It is a combustible liquid but virtually nontoxic and is used in the manufacture of perfumes, flavoring and pharmaceuticals and as a laboratory reagent. If found in a laboratory, your first instinct might be to treat it as a hazardous material; even though it is really not a hazard, you would err on the side of safety!

Acrylonitrile is a hazardous material that is much worse than it sounds; its real name is vinyl cyanide. Now you may see why they changed the name! Acrylonitrile is a colorless, migratory liquid with a mild odor. The flash point is 32°F and it has rather wide explosive limits of 3%–17% in air. In addition to being a severe fire and explosion risk, it is also toxic. Its primary routes of exposure are by inhalation and skin absorption, and it is a known carcinogen, with a TLV of 2 ppm. The primary use is as a monomer in plastic production and as a grain fumigant.

The important thing to remember is that "applied chemistry" is the rule-of-thumb information used to familiarize responders with basic information about recognizing hazardous chemicals and understand why those materials are hazardous. Being able to pick out elements in a name or formula of a chemical or recognizing a family of compounds helps to determine potential hazards of a compound. Responders still need to thoroughly research information about chemicals from reference sources, computer databases, MSDSs, shipping papers and CHEMTREC. "Applied chemistry" should get you pointed in the right direction (Hazardous Materials Chemistry for Emergency Responders).

Nine DOT Hazard Classes

Class 1 Explosives (LF/HR)
DOT Definition of Explosive

"Explosive" is any chemical compound, mixture or device, the primary or common purpose of which is to function by explosion, that is, with substantially instantaneous release of gas and heat, unless such compound, mixture or device is otherwise specifically classified by the U.S.

Department of Transportation (DOT); see 49 Code of Federal Regulations (CFR) Chapter I. The term "explosives" shall include all materials which are designated as Class 1.1, Class 1.2 and Class 1.3 explosives by the U.S. Department of Transportation, and includes, but is not limited to, dynamite, black powder, pellet powders, initiating explosives, blasting caps, electric blasting caps, safety fuse, fuse lighters, fuse igniters, squibs, cordeau detonant fuse, instantaneous fuse, igniter cord, igniters, small arms ammunition, small arms ammunition primers, smokeless propellants, cartridges for propellant-actuated power devices and cartridges for industrial guns. Commercial explosives are those explosives that are intended to be used in commercial or industrial operations.

Explosives hazard Class 1 is divided into six subclasses: 1.1–1.6 (49 CFR 173.20). Because of their potential danger, subclasses 1.1–1.3 require placards on all highway transportation vehicles regardless of the quantity of explosives being carried (49 CFR 172.504, Table 1). Railroad shipments must always be placarded in all six subclasses regardless of the quantity shipped. Subclasses 1.4–1.6 (49 CFR 172.504, Table 2) fall under the 1,001 lb. rule (49 CFR 172.504 (c)(1), which requires 1,001 lb. or more of the explosives on the vehicle before a placard is required, which means you could have 1,000 lb. or less of a 1.4–1.6 explosive and no placard would be required at all!

Division 1.1 Possessing, detonating or otherwise maximum hazard, such as dynamite, nitroglycerin, picric acid, lead azide, fulminate of mercury, black powder, blasting caps and detonating primers.

Division 1.2 Possessing flammable hazard, such as propellant explosives (including some smokeless propellants), photographic flash powders and some special fireworks.

Division 1.3 includes certain types of manufactured articles that contain Class 1.1 or Class 1.2 explosives, or both, as components but in restricted quantities.

Division 1.4 Consists of explosives that present a minor explosion hazard. The explosive effects are largely confined to the package, and no projection of fragments of appreciable size or range is to be expected. An external fire must not cause virtually instantaneous explosion of almost the entire contents of the package.

Division 1.5 Consists of very insensitive explosives. This division is comprised of substances that have a mass explosion hazard but are so insensitive that there is very little probability of initiation or of transition from burning to detonation under normal conditions of transport. The probability of transition from burning to detonation is greater when large quantities are transported in a vessel.

Division 1.6 Consists of extremely insensitive articles that do not have a mass explosion hazard. This division is comprised of articles that predominately contain extremely insensitive substances and that demonstrate a negligible probability of accidental initiation or propagation. The risk from articles of Division 1.6 is limited to the explosion of a single article.

Forbidden Explosives

"Forbidden explosives" are explosives that are forbidden or not acceptable for transportation by common carriers by rail freight, rail express, highway or water in accordance with the regulations of the DOT, 49 CFR. In addition to the explosives that are approved for transportation by the DOT, there are forbidden explosives that are too unstable or dangerous to be transported. Although these won't likely be transported, they may be found in fixed facilities where they are created and used. The following are some examples of explosives that are forbidden to be offered for transportation: nitrogen triiodide, azides, picrates and metal fulminates.

Other explosives that are forbidden in transportation include explosives that contain a chlorate along with an ammonium salt or an acidic substance including a salt of a weak base and a strong acid; explosives packages that are leaking, damaged, unstable condemned or contain deteriorated propellants; nitroglycerin, diethylene glycol dinitrate or other liquid explosives not authorized; fireworks that combine an explosive and a detonator; fireworks containing yellow or white phosphorus; and toy torpedoes exceeding 0.906 in. outside dimension or containing a mixture of potassium chlorate, black antimony (antimony sulfide) and sulfur, if the weight of the explosive material in the device exceeds 0.01 oz. The Hazardous Materials Table in CFR 49 part 172.101 lists all specific restricted explosives in various modes of transportation and those forbidden from shipment (DOT).

Explosive Families of Compounds (Figure 3.70)

Inorganic Explosives LF/HR

Fulminates, ammonium nitrate and azides are inorganic explosive compounds. Fulminate ions, CNO, are unstable; thus, the salts of fulminates are friction-sensitive explosives. Metals are attached to the CNO ion forming an explosive metal fulminate. Fulminates are among the oldest explosive compounds and have been around since the 1800s. Mercury fulminate was the first inorganic explosive discovered. Mercury II fulminate,

Applied Chemistry and Physics

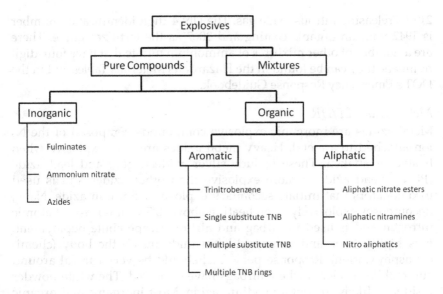

Figure 3.70 Explosive families of compounds.

Hg(ONC)$_2$, is the most common fulminate and is used as the primary explosive in detonators.

A fulminate primary explosive can also be made from silver. Silver fulminate, AgONC, has very little practical value due to its extreme sensitivity to impact, heat, pressure and electricity. The compound becomes progressively sensitive as it is aggregated, even in small amounts; the touch of a falling feather, the impact of a single water droplet or a small static discharge are all capable of explosively detonating an unconfined pile of silver fulminate no larger than a dime and no heavier than a few milligrams. Aggregation of larger quantities is impossible due to the compound's tendency to self-detonate under its own weight.

Silver acetylide (silver carbide), C$_2$Ag$_2$, is also an inorganic explosive that is highly sensitive and cannot be used in detonators. Silver acetylide is a primary explosive. It is a white powder that is sensitive to light. Generally, a chemical explosive must be confined for an explosion to take place. However, silver acetylide maintains a high-energy density and will detonate without confinement. Dry silver acetylide poses an explosion hazard when exposed to heat, shock or friction. When dry, it should not be stored indoors. Because it is light sensitive, it should be stored in a dark room. It should also be stored in an amber bottle.

Ammonium nitrate, NH$_4$NO$_3$, is classified as an oxidizer. It is a colorless or white-to-gray crystal that is soluble in water. It decomposes at

210°C, releasing nitrous oxide gas. The UN 4-digit identification number is 1942 with an organic coating and 2067 as the fertilizer grade. There are a number of other mixtures of ammonium nitrate that have four-digit numbers; they can be found in the Hazardous Materials Tables and in the DOT's Emergency Response Guidebook.

Metal Azides LF/HR

Metal azides are inorganic explosive compounds composed of the N_3 ion attached to a metal. Heavy metal azides are very explosive when heated or shaken. These include silver azide (AgN_3) and lead azide (PbN_3). Lead azide is more explosive than other azides and is used in detonators that initiate secondary explosives. Sodium azide, NaN_3, decomposes explosively upon heating above 275°C. It releases diatomic nitrogen and is used in airbag and airline escape chute deployment. It is highly toxic and behaves like cyanide inside the body (chemical asphyxiation). Response personnel should be very careful around automobile accidents where airbags have deployed. The white powder residue is likely to contain sodium azide. Most inorganic and organic azides are prepared directly or indirectly from sodium azide. Sodium azide is used in the production of metal azide explosive compounds and as a detonator.

Aliphatic Explosive Compounds
(Nitro Hydrocarbon Derivatives) LF/HR

Explosives may also be organized into families; in this case, aliphatic nitro compounds are a hydrocarbon derivative family. Nitro is the one hydrocarbon-derivative family that is classified with an explosive as its primary hazard. However, there are some nitro compounds that have other primary hazards, such as nitrobenzene, which is a poison. This is an exception to the general hazard and, for safety purposes, consider nitros explosive as a group. The nitro group is represented by an atom of nitrogen covalently bonded to two oxygen atoms. The oxygens have a single bond between themselves. This oxygen-to-oxygen single bond is highly unstable and can come apart explosively.

The other bonding spot on the nitrogen is attached to a hydrocarbon or hydrocarbon-derivative backbone of some type. These backbones may include methane and others. A nitro compound is a hydrocarbon with one or more hydrogen atoms removed and replaced by the nitro functional group NO_2. If more than one nitro radical is used, they are represented by the Greek prefix indicating the number: "di-" for two, "tri-" for three and "tetra-" indicating four. When naming compounds from the nitro group, the word "nitro" is used first and the end is the hydrocarbon to which the nitro is attached. Methane has had one hydrogen atom removed, which becomes the methyl radical CH_3 to create a place to attach the nitrogen on the nitro functional group NO_2.

$$CH_3 + NO_2 = CH_3NO_2, \text{nitromethane}$$

Nitromethane, CH_3NO_2, is a colorless liquid that is soluble in water. The specific gravity is 1.13, which is heavier than water. Nitromethane is a dangerous fire and explosion risk, and is shock- and heat-sensitive. It may detonate from nearby explosions. The boiling point is 213°F and the flash point is 95°F. The flammable range only lists a lower explosive limit, which is 7.3% in air; an upper limit has not been established. The ignition temperature is 785°F. Nitromethane may decompose explosively above 599°F if confined, and is a dangerous fire and explosion risk, as well as toxic by ingestion and inhalation. The threshold limit value (TLV) is 100 ppm in air. The UN 4-digit identification number is 1261; the NFPA 704 designation is health—1, flammability—3 and reactivity—4. Nitromethane is used in drag racing to give the fuel in the engine an extra kick to increase speed. It is also used in polymers and rocket fuel.

Another example of a nitro compound is nitroglycerin, which utilizes resonance bonding to allow three oxygen atoms to be attached to one nitrogen while the nitrogen is also attached to a carbon. This resonance bonding is similar to the bonding used for aromatic hydrocarbons. The electrons necessary for bonding oxygen to nitrogen are considered to be in a state of resonance or mesmerism and not associated with a single atom or covalent bond (delocalized electrons).

$$CH_2 + NO_3 + CH + NO_3 + CH_2 + NO_3 = C_3H_5N_3O_9, \text{nitroglycerine}$$

Nitroglycerin, $C_3H_5N_3O_9$, **IUPAC name 1,2,3-trinitroxypropane**, is a pale yellow, viscous liquid. It is slightly soluble in water, with a specific gravity of 1.6, which is heavier than water. It is a severe explosion risk and will explode spontaneously at 424°F. It is much less sensitive to shock when it is frozen. Nitroglycerin freezes at about 55°F. It is highly sensitive to shock and heat, and is toxic by ingestion, inhalation and skin absorption. The TLV is 0.05 ppm in air. Nitroglycerin is forbidden in transportation unless sensitized.

When in solution with alcohol at not more than 1% nitroglycerin, the UN 4-digit identification number is 1204. When in solution with alcohol and more than 1% but not more than 5% nitroglycerin, the UN 4-digit identification number is 3064. The NFPA 704 designation for nitroglycerin is health—2, flammability—2 and reactivity—4. The primary uses are in explosives and dynamite manufacture, in medicine as a vasodilator, in combating oil well fires and as a rocket propellant.

CASE STUDIES

BUTLER, PA, JUNE 9, 1927, NITROGLYCERIN EXPLOSION
By Brian Reed
Butler, PA, Historical Society

It was a seemingly normal Thursday morning on June 9, 1927, in the Lick Hill village just outside of the city of Butler, Pennsylvania. The relative calm of the summer morning was suddenly shattered; however, when a truck carrying nitroglycerin destined for oil well drilling operations exploded while the truck was leaving the main highway and entering a side road; witnesses later reported they saw the truck's tire drop into a large pothole. The explosion killed two men, Frank H. Greer and Richard Coxsan, both of Rouseville and left at least twenty others injured. According to a *Butler Eagle* report, it was "the worst scene of destruction ever seen in Butler County."

The truck was driven by workers of the Bowers Torpedo Company, who were delivering the nitroglycerin to one of their sister companies in the Producers' Torpedo Company. Nitroglycerin is a highly volatile explosive liquid that was used for efficiency in oil well drilling during the early 1900s. The process used a "torpedo," invented in 1865 by Colonel E.A.L. Roberts, in order to blow rocks and debris out of the way of the drill. Explosions during the transport of nitroglycerin were very common—so common, in fact, that many other states had already banned the transportation of the substance.

The explosion left a hole at least eight feet deep and fourteen feet wide in the road and knocked all nearby power lines down. The blast severely damaged 15-20 homes in the area and left nearly twenty people injured. The majority of the people injured in the incident were people who lived in the homes that were destroyed by the blast. The home nearest the blast, the Edward Rettig residence, was completely flattened in the explosion. The blast was so massive that people two miles away in the city of Butler heard and felt it. Village physicians in Lick Hill, quickly overwhelmed by the number of injuries and with no way to call for backup due to the damaged phone systems, were relieved when doctors from Butler who felt the blast came to assist them. A Red Cross Relief Fund was set up and accumulated over $1,400 dollars in donations for the affected families. The workers even asked for donations from all of the sightseers drawn by the blast. Along with personal donations, the Harris Theater in Butler secured 8 separate Vaudeville acts for a benefit show where all of the money raised would be donated to the fund. The event raised nearly $800, and the owner of the Theater even donated $100.

Immediately after the explosion, the call for more restrictions on nitro transportation was set to get answered when city council had a meeting on Monday, June 13. The preexisting ordinance in place was a ban on transportation through the city with a penalty of $500. The new ordinance required all transport vehicles carrying nitroglycerin to submit to a thorough inspection before entering the city, they had to follow a police escort and they were required to have clear markings on their trucks; otherwise, they would have to pay a $1,000 fine. Tragedy can strike anywhere, and on that particular Thursday morning, Butler County felt the force of the destruction. The explosion destroyed the livelihood of the two workers and the nearby families. The village of Lick Hill was left partially flattened, and many people were left homeless following this horrific accident (Butler County Historical Society, used with permission).

CLEVELAND, OK, MARCH 9, 1916, NITROGLYCERIN EXPLOSION

Six hundred quarts of nitro-glycerin exploded near Cleveland, Ok, late today, killing one man and almost causing a panic in Tulsa, thirty-five miles distant, when the concussion shattered windows and rocked frame buildings. Many persons here ran terrified into the streets, fearing an earthquake. The driver, who was hauling the explosive to the Cleveland oil field, was blown to atoms by the accident. The cause of the explosion has not been learned, although it is thought that the nitro-glycerin was detonated by the jar of a rough stretch of road.

Titusville Herald, Pennsylvania, 1916-03-09

BIG HEART, OK, JANUARY 26, 1919, NITROGLYCERIN EXPLOSION

Eight people were killed and more than a score severely injured when a wagon carrying nitroglycerine belonging to the Eastern Torpedo Company exploded in the heart of the residence district here today. The wagon driver and a passenger also on the wagon were blown to atoms. A residence in front of which the explosion occurred was leveled to the ground. Occupants were perhaps fatally wounded and their three-year-old baby boy was killed. Seven other houses in the vicinity were wrecked. The explosion broke every window in the town and shook the ground for hundreds of yards around. All telegraph and telephone communication was destroyed.

> Big Heart has only one doctor. He had a corps of workers attending to the dead and wounded. Pawhuska, Ok, sent physicians and rescue workers in motor cars to the scene. Not all the bodies of the dead and wounded were recovered from the ruins of the houses and casualties may exceed first figures. Only two quarts of nitroglycerin were in the wagon. The cause of the explosion is unknown.
> Lima Daily News, Lima, OH 26 Jan 1919

Aromatic Explosive Compounds

Nitro compounds are also formed when nitro functional groups are added to the benzene and toluene rings and radicals that are formed for hydrocarbon derivatives. The first such compound is the nitrated benzene ring which is called trinitrobenzene (TNB), $C_6H_3(NO_2)_3$. Three nitro functional groups are added to the benzene ring; thus, the prefix "tri" is used in the naming followed by the functional group name "nitro" and ending in "benzene" which is the aromatic before the nitro functional groups were added. Other aromatic explosive compounds are created from the first compound.

$$C_6H_3 + NO_2 + NO_2 + NO_2 = C_6H_3(NO_2)_3, \text{trinitrobenzene}$$

TNB, $C_6H_3(NO_2)_3$, **IUPAC name 1,3,5-trinitrobenzene**, is an aromatic explosive material. It is dangerous and explodes by heat or shock. It is yellow crystal in color. The UN 4-digit identification number is 3367 when wetted with not less than 10% water and 1354 when wetted with not less than 30% water. Trinitrobenzene is a member of the nitro hydrocarbon derivative family.

The common explosive trinitrotoluene (TNT) is created by adding a methyl radical CH_3 to nitro benzene. This compound is formed when three nitro functional groups are attached to the created toluene ring by adding the methyl radical. Because the prefix "tri" is located in front of nitro, there are three nitro groups attached. It is named by the nitro functional group with the prefix "tri" and ending in the name "toluene" because that is the hydrocarbon the nitro functional groups are added to.

$$C_6H_2 + CH_3 + NO_2 + NO_2 + NO_2 = C_6H_2CH_3(NO_2)_3, \text{trinitrotoluene}$$

TNT, $CH_3C_6H_2(NO_2)_3$, **IUPAC name 2-methyl-1,3,5-trinitrobenzene**, is flammable, a dangerous fire risk and a moderate explosion risk. It is light cream to rust in color, and is usually found in 0.5 or 1 lb blocks. It is fairly stable in storage. It will detonate only if vigorously shocked or heated to 450°F; it is toxic by inhalation, ingestion and skin absorption. The TLV is 0.5 mg/m3 of air. The UN 4-digit identification number is 1356 when wetted with not less than 30% water. Other mixtures are listed in the DOT Hazardous Materials Tables with several ID numbers. TNT is one of the common ingredients used in military explosives and is used as a blast effect measurement for other explosives. TNT is a member of the nitro hydrocarbon derivative family. In the following structure, toluene is the backbone for TNT. Three hydrogen atoms are removed from the toluene ring and three nitro functional groups are attached.

Another compound that uses the benzene ring is the explosive trinitrophenol, also known as picric acid. This compound is formed when three nitro functional groups are attached to the benzene ring. Because the prefix "tri" is located in front of nitro, there are three nitro groups attached. It is named by the nitro functional group with the prefix "tri" and ending in the name "phenol" because that is the hydrocarbon functional group that the nitro functional groups are added to. Remember that the radical of benzene is phenol. The ending "ol" is also a clue that the compound is an alcohol. However, though technically an alcohol, phenol has different characteristics from the common alcohol family.

$$C_6H_2 + OH + NO_2 + NO_2 + NO_2 = C_6H_2OH(NO_2)_3, \text{ trinitrophenol}$$

Trinitrophenol (picric acid), $C_6H_2OH(NO_2)_3$, IUPAC name 2,4,6-trinitrophenol, is composed of yellow crystals that are soluble in water. It is a high explosive, is shock- and heat-sensitive and explodes spontaneously at 572°F. Trinitrophenol is reactive with metals or metallic salts, and is toxic by skin absorption. The TLV is 0.1 mg/m³ of air. When shipped in 10%–30% water, it is stable unless the water content drops below 10% or it dries out completely. The UN 4-digit identification number is 1344 when shipped with not less than 10% water. The NFPA 704 designation is health—3, flammability—4 and reactivity—4.

The primary uses are in explosives, matches, electric batteries, etching copper and textile dyeing. Picric acid is often found in chemical laboratories in high schools and colleges, and can be a severe explosion hazard if the moisture content of the container is gone. Picric acid was used by the Japanese during World War II (WWII) as a main charge explosive filler. When in contact with metal, picric acid forms other picrates, which are extremely sensitive to heat, shock, and friction. Great care should be taken when handling WWII souvenirs, because of the possible presence of these picrates.

When the structure of picric acid is compared with that of TNT, the only difference is the fuel that the nitro functional groups are placed on; the number of nitro groups is exactly the same. The explosive power of picric acid is similar to that of TNT. There are other Class 1.1–1.3 materials that are nitro compounds and some Class 1.1–1.3 materials that are made up of other chemicals.

Black powder is a low-order explosive made up of a mixture of potassium or sodium nitrate, charcoal and sulfur in 75%, 15% and 10% proportions, respectively. It has an appearance of a fine powder to dense pellets, which may be black or have a grayish-black color. It is a dangerous fire and explosion risk, is sensitive to heat, and deflagrates rapidly.

Explosive Chemicals Not Considered to Be Explosives

There are a number of other chemicals that are not explosives, but have explosive potential under certain conditions. Oxygen is an oxidizer that causes organic materials to burn explosively. Chlorine is also an oxidizer and may be explosive in contact with organic materials.

Ether is an organic compound that forms explosive peroxides when in contact with air. When a container of ether is opened, oxygen from the air bonds with the single oxygen in each ether molecule and forms an organic peroxide. These peroxides are very unstable and become sensitive to shock, heat and friction. Moving or shaking a container can cause an explosion. Ethers are also very flammable, with wide flammable ranges. Fire is likely to follow an explosion of an ether container. Some examples include ethyl ether, isopropyl ether, methyl *tert*-butyl ether and propylene oxide.

Ethyl ether (diethyl ether), IUPAC name ethoxyethane, is a colorless, volatile, mobile liquid. It is slightly soluble in water, with a specific gravity of 0.7, which is lighter than water. It is a severe fire and explosion risk when exposed to heat or flame. The compound forms explosive peroxides from the oxygen in the air as it ages. Flammable range is wide, from 1.85% to 48% in air. The boiling point is 95°F (35°C), the flash point is –49°F (–45°C) and the ignition temperature is 356°F (180°C). The vapor density is 2.6, which is heavier than air.

Isopropyl ether **(diisopropyl ether)** is highly flammable, with a wide flammable range of 1.4%–21% in air. The boiling point is 156°F (68°C), the flash point is –18°F (–27°C) and the ignition temperature is 830°F (443°C). The vapor density is 3.5, which is heavier than air. In addition to flammability, isopropyl ether is toxic by inhalation and a strong irritant, with a TLV of 250 ppm in air.

Methyl *tert*-butyl ether (MTBE), with molecular formula $(CH_3)_3COCH_3$, is a volatile, flammable and colorless liquid that is sparingly soluble in water. It is used as a gasoline additive. Although the main danger is explosion, MTBE can also form explosive peroxides when in contact with air. It is not as prone to peroxide formation as other ethers, but it can happen after long periods.

Butadiene may also form explosive peroxides when exposed to air.

Potassium metal (K) is a metallic element from family 1 on the Periodic Table of Elements. It can form peroxides and superoxides at room temperature, and may explode violently when handled. Simply cutting a piece of potassium metal with a knife to conduct an experiment could cause an explosion.

Acetone peroxide, $C_9H_{18}O_6$, also called triacetone triperoxide (TATP), is an organic peroxide and a primary high explosive. It takes the form of a white crystalline powder with a distinctive bleach-like odor. It is susceptible to heat, friction and shock. The instability is greatly altered by impurities. It is not easily soluble in water. It is more stable and less sensitive when wet. TATP is an explosive which can be made with liquids that are commonly available from pharmacies and hardware stores. Primary ingredients include hydrogen peroxide (hair bleach), sulfuric acid and acetone. After mixing, the liquid mixture is evaporated, which leaves crystals. TATP can be used as an explosive or as a booster to set off larger charges. It is extremely sensitive and is stored under refrigeration to help maintain stability. Acetone peroxide is also known as "Mother of Satan" because it is highly unstable and liable to detonate with the slightest shock or rise in temperature. Response personnel should be very cautious if any of the precursor chemicals or empty containers are found in or around an occupancy.

Pentaerythritol tetranitrate (PETN), $C_5H_8O_{12}N_4$, a nitrate ester of pentaerythritol, is a white crystalline substance that feels powdery to the touch. In its pure form, PETN melts at 285°F (141.3°C). PETN is most well known as an explosive. Mixed with a plasticizer, it forms a plastic explosive. It's the high explosive of choice because it is stable and safe to handle, but it requires a primary explosive to detonate it. This compound was used by the 2001 shoe bomber, in the 2009 Christmas Day bomb plot and in the 2010 cargo plane bomb plot. It is a shock-sensitive explosive used for demolition, blasting caps and detonating compositions ("Primacord").

Military Explosives

Military explosives are noted for their high shattering power accompanied by rapid detonation velocities. They must be stable, because they are often kept in storage for long periods of time. Because of their intended use, they must detonate dependably after being stored and do so under a variety of conditions. The military explosives used most commonly are TNT, C-3, C-4 and RDX cyclonite. These explosives release large quantities of toxic gases when they explode.

Ammonium picrate, **Dunnite, also called Explosive D, $C_6H_2(NO_2)_3ONH_4$, IUPAC name ammonium 2,4,6-trinitrophenolate**, is a high explosive when dry and flammable when wet. It is composed of yellow crystals that are slightly soluble in water. The UN 4-digit identification number is 1310 for ammonium picrate wetted with not less than 10% water. It is used in pyrotechnics and other explosive compounds. Notice the similarity to picric acid and TNT.

The structure and formula of ammonium picrate do not follow the usual rules of bonding. The ammonium radical has a different bonding configuration, which accounts for the four hydrogen atoms hooked to the nitrogen atom. This hybrid type of bonding involves delocalized electrons. They are electrons that are not associated with any one atom or one covalent bond. Delocalized electrons are contained within an orbital that extends over several adjacent atoms. This is also sometimes referred to as resonance bonding (as in the aromatic hydrocarbon compounds) and in general is beyond the scope of "applied chemistry."

Diazoninitrophenol (DDNP), $C_6H_2N_4O_5$, is a yellowish brown powder. It explodes when shocked or heated to 356°F (180°C); it is dangerous and used as an initiating explosive. It is a primary charge in blasting caps. DDNP is soluble in acetic acid, acetone, concentrated hydrochloric acid and most nonpolar solvents but is insoluble in water. A solution of cold sodium hydroxide may be used to destroy it. DDNP may be desensitized by immersing it in water, as it does not react in water at normal temperature. It is less sensitive to impact but more powerful than mercury fulminate and lead azide. The sensitivity of DDNP to friction is much less than that of mercury fulminate, but it is approximately that of lead azide. DDNP is used with other materials to form priming mixtures, particularly where a high sensitivity to flame or heat is desired. DDNP is often used as an initiating explosive in propellant primer devices and is a substitute for lead styphnate in what are termed "nontoxic" (lead free) priming explosive compositions.

Lead styphnate, $C_6HN_3O_8Pb$, **lead 2,4,6-trinitroresorcinate**, whose name is derived from styphnic acid, is an explosive used as a component in primer and detonator mixtures for less sensitive secondary explosives; is a slurry or wet mass of orange–yellow crystals; must be shipped wet with at least 20% water or water and denatured ethyl alcohol mixture; may explode due to shock, heat, flame or friction if dried; and is used as an initiating explosive. The primary hazard is the blast of an instantaneous explosion and not flying projectiles and fragments. There are two forms of lead styphnate: six-sided monohydrate crystals and small rectangular crystals.

Lead styphnate varies in color from yellow to brown. It is particularly sensitive to fire and the discharge of static electricity. When dry, it can be readily detonated by static discharges from the human body. The longer and narrower the crystals are, the more susceptible lead styphnate is to static electricity. Lead styphnate does not react with metals and is less sensitive to shock and friction than mercury fulminate or lead azide. It is only slightly soluble in water and methyl alcohol, and may be neutralized by a sodium carbonate solution. It is stable in storage, even at elevated temperatures. As with other lead-containing compounds, lead styphnate is inherently toxic to humans if ingested, that is, it can cause heavy metal poisoning.

Nitromannite, $C_6H_8(ONO_2)_6$, **mannitol hexanitrate**, is a powerful explosive. Physically, it is a powdery solid at normal temperature ranges, with a density of 1.6 g/cc. The chemical name is hexanitromannitol, and it is also known by nitromannite, MHN, nitromannitol, nitranitol or mannitrin. It is less stable than nitroglycerin, and it is used in detonators. Mannitol hexanitrate is a secondary explosive formed by the nitration of mannitol, a sugar alcohol. The product is used in medicine as a vasodilator and as an explosive in blasting caps. Its sensitivity is considerably high, particularly at high temperatures >167°F (>75°C) where it is more sensitive than nitroglycerin. It has the highest brisance of any known conventional explosive, even more than nitroglycerin.

RDX, $N(NO_2)CH_2N(NO_2)CH_2N(NO_2)CH_2$, cyclonite, is a high explosive, easily initiated by mercury fulminate and toxic by inhalation and skin contact. The TLV is 1.5 g/m^4 of air. Explosive, which is 1.5 times as powerful as TNT, is an explosive nitroamine widely used in military and industrial applications. It was developed as an explosive that was more powerful than TNT, and it saw its wide use in WWII. RDX is also known as cyclonite. Its chemical name is cyclotrimethylenetrinitramine; name variants include cyclotrimethylene trinitramine and cyclotrimethylene trinitramine. In its pure, synthesized state, RDX is a white, crystalline solid. It is often used in mixtures with other explosives and plasticizers, phlegmatizers or desensitizers. RDX is stable in storage and is considered one of the most powerful and brisant of the military high explosives (Hazardous Materials Chemistry for Emergency Responders).

> *Hazmatology Point: According to the NFPA Fire Protection Handbook, over 90% of explosives in the industrial world are used in mining operations, with the rest used in construction. When responding to fixed or transportation incidents in or around these types of operations, be on the lookout for explosives. When responding to transportation incidents, always consider the possibility of explosives being present. Fire is the principal cause of accidents involving explosive materials. Look for explosive signs, such as placards and labels. Evacuate the area according to the distances listed in the Emergency Response Guidebook orange section. If no other evacuation information is available, a 2,000ft minimum distance should be observed, according to the NFPA Fire Protection Handbook. There is one rule of thumb in responding to incidents where explosives are involved:* DO NOT FIGHT FIRES IF THE FIRE HAS REACHED THE EXPLOSIVE CARGO.

Historic Incidents Involving Common Explosives

From 1841 through WWII, explosives were not in a hazard class because the DOT did not exist. Explosives incidents were by far the most common. Hundreds of civilians and some firefighters were killed in explosions involving dynamite, gun powder and other explosive devices and materials. Most of those incidents involved fixed facilities. There was little or no regulation of the storage, use or locations in which explosives can be stored and used. There was also no restriction on the purchase or use of explosives. For most of the time period, there was no television or radio. When fires would occur, people would gather in crowds to watch the firefighters' work. It was almost a form of entertainment at the time. Unfortunately, many of the explosions occurred after the firemen were on scene and crowds were already there. It is not clear if firemen even knew explosives were involved in many of the incidents.

One of the main chemical explosives listed from Divisions 1.4 to 1.6 is ammonium nitrate, which when mixed with fuel oil (ANFO) becomes a blasting agent. Subjected to confinement or high heat, it may explode but does not readily detonate. Fertilizer grade ammonium nitrate, which is a strong oxidizer above 33.5%, may also explode if it becomes contaminated. Several major incidents have occurred since WWII involving ammonium nitrate and ammonium nitrate fertilizers.

Selected explosives incidents from the 1800s to the mid-1900s			
Type of explosive	Number of incidents	Firefighter fatalities	Civilian fatalities
Nitro glycerine	3	0	23
Dynamite	18	45	100
Black powder	32	0	286
Ammonium nitrate	3[a]	45	412
Munitions	11	0	423
Fireworks factories	7	23	102

[a] Ammonium nitrate incidents during 1947–2013.

CASE STUDIES

SYRACUSE NY AUGUST 30, 1841 THE GREAT GUNPOWDER EXPLOSION

Friday night a fire broke out at Syracuse in Goings Carpenters Shop near the Oswego Canal. It spread with great rapidity and the building was soon enveloped in flames. The village's volunteer fire department companies rushed to the scene. Unknown to all, a gunpowder consignment totaling 650 pounds in 25 kegs, was secretly stored in the Goings Shop. Crowds of citizens flocked to the scene, and soon after a great number had collected. "Powder ! Powder ! There is powder in the building!" But this cry had but a momentary effect on the crowd. The mass moved back a step, then stopped, and in a few moments as firemen were starting their hoses, the twenty-five kegs of powder blew up with one explosion, scattering fragments of the buildings and limbs of human bodies in every direction. As near as could be ascertained, upwards of thirty persons were killed outright, and no less than fifty wounded, some very seriously, and perhaps fatally. From ten to fifteen were so mangled and cut to pieces that it was impossible to recognize them. Every exertion was immediately made to relieve the sufferers.

Everything was done that could have been done under the circumstances. An extra train of cars was run to Auburn for physicians, and our hotel keepers threw open their doors for the reception of the wounded. As to the origin of the fire it is unknown; but it is supposed to have been the work of an incendiary! The fire appeared to have commenced in the top of the building.

> ***Author's Note:*** *This is the first documented hazardous materials incident I could find. Records of this incident are scarce, and I could not find any lists of casualties. However, based on the description of the tragedy, I am certain there must have been firemen killed in the explosion. Syracuse fire department was not established until 1871, so they would not have any record of firefighter deaths.*

SAN FRANCISCO, CA, MAY 22, 1895, NITROGLYCERIN FACTORY EXPLOSION

A terrific report and concussion which was distinctly felt all through the city and towns around the bay for a distance of 40 miles yesterday, was at first believed to have been caused by an earthquake, but proved to be an explosion in the nitro-glycerine and mixing houses of the California Powder Works across the bay. The crew in the glycerine house, four in number, and the foreman of the mixing room were killed, as were also nine Chinese working in the latter department. The explosion occurred in the nitro-glycerine house and was probably caused by a Chinese dropping a can of the explosive. The cause cannot be definitely ascertained, however, as all connected with the building are dead.

There were 200 Chinese in that adjacent mixing room, and at the sound of the explosion all ran. The force of the explosion was tremendous. Huge pieces of wood were thrown into the bay, a distance of half a mile, and nitro-glycerin tanks were hurled a distance of 500 yards. Hands, legs and other parts of the mutilated remains of the dead were scattered along the road or a mile. The nitro-glycerin house first went up, and the mixing storehouse and gun cotton in the premises followed.

The nitro-glycerine house of which not a vestige now remains, was a 8-story frame structure, 1200 feet by 50 feet. It contained 8,000 pounds of nitroglycerin and 2,000 pounds of hercules powder. A remarkable feature of the explosion is that although the storehouse containing 1,000 pounds of hercules powder is completely wrecked its contents are intact. In all 100,000 pounds of explosives went up with a roar and with a sheet of flame.

Fort Wayne News, Fort Wayne, IN, 23 May 1895

SHAMOKEN, PA, NOVEMBER 5, 1888, NITROGLYCERIN EXPLOSION

Intelligence has just been received here of the explosion of a nitro-glycerine tank near Shamokin, Pa., a town twenty-two miles from here at an early hour this morning. Great damage was done to surrounding property, but it is not know whether anyone was killed or injured. The magazine belonged to the Torpedo Company of Delaware. The explosion was terrific and was felt twenty-five miles away. Houses and shanties were shattered in the immediate vicinity. At Sewickley, five miles distant, the shock was so heavy that people ran panic-stricken from their homes. It was reported that a large number of persons had been killed and injured, but it is now believed that no one was hurt, although nothing definite is known.

A Chronicle-Telegram special from Shamokin says the cause of the nitroglycerine explosion is a mystery. It's supposed, however, that a tramp who was seen in the neighborhood, being ignorant of the danger, in some way agitated the stuff and the explosion followed. There were three tons in the magazine, and the concussion shook the very foundations of the houses between Rochester and Pittsburgh. The earth was churned up for a distance of 500 yards. Trees and eighth of a mile away were rent asunder, and many houses were wrecked. The residence of George and William Wilson and Thomas McCoy, situated a half mile away, were completely shattered and the occupants thrown from the windows. Fortunately they were not seriously injured and as far as can be ascertained no one was killed, with the possible exception of the tramp. The loss will be heavy, but cannot now be ascertained.

Coffeyville Weekly Journal, Kansas, 1888-11-08

CAMPBELL COUNTY, TN, DECEMBER 14, 1895, DYNAMITE EXPLOSION

By a premature explosion of powder and dynamite at 3 o'clock Thursday afternoon four men were killed at Lafollette, Campbell County. Four other men are missing and their bodies have not been found. The explosion occurred on the works of the Lafollette Railroad, near Lafollette, where a new road is being projected from Lafollette to Jellico. A large force of men were at work blasting for a cut. Three kegs of powder had been placed in a 16-foot aperture and four sticks of dynamite were also added to the deadly charge.

A fuse was attached to the powder and the men ran from the place to a safe distance to await the result. They waited 14 minutes and the charge failed to go off. The men were then ordered to break into the hole by the foreman to readjust the fuse. They obeyed the order, all being apparently satisfied that the fire had become extinct. Just as they re-entered the hole, however, the terrific explosion occurred, and the earth and stone for many yards around were dislodged and thrown into conglomerate mass with the remains of the unfortunate victims.

At the latest received last night a rescuing force was still at work removing the dirt and rock.

Evening Bulletin, Maysville, Kentucky, 1895-12-14

VESTAL, NY, JUNE 10, 1901, TRAINS DESTROYED BY DYNAMITE EXPLOSION AFTER CRASH

Rescuers and wrecking crews worked until after noon today at Vestal, ten miles west of here, where an explosion of dynamite in a Delaware, Lackawanna and Western freight car wrecked two colliding trains, killed five trainmen, and injured several others. Late this afternoon parts of the last body to be accounted for were found a quarter of a mile away from the scene of the explosion. The dynamite was in a car of freight train No. 61, which was taking water when a double-header wildcat freight train crashed into the stationary train. The impact exploded the dynamite, and both trains were wrecked, and pieces of the locomotives of the wildcat train were found today half a mile distant. The tracks were blocked until after midday. Aside from the damage to railroad property, much minor damage is reported. Nearly every house in the villages of Vestal and Union, which is across the Susquehanna from the wreck, lost more or less of its window glass, while houses and barns near the scene were badly shattered. None of the inmates, however, was injured.

The New York Times, New York, June 10, 1901

SPENCER, NC, OCTOBER 2, 1908, MAGAZINE EXPLOSION

Two lives were lost and 20 or more persons were injured in Spencer by the explosion of a powder storage house in the yards of the Southern Railway company, and most of the buildings nearby were damaged by the shock and the fire which followed. It is feared that other bodies are in the ruins. Fire was discovered in the storage room of the powder house and the Spencer shop fire department rushed to the scene. It is said that a powder magazine contained a half carload of powder, dynamite and other explosives. The fire consumed 20 or more cars and other material. Every building in Spencer and for many miles around was badly shaken up and damaged. Of the Southern's buildings, the blacksmith shop is the most seriously damaged. The large new machine shop, 200×600 feet, was also badly demolished, all windows being torn out and a number of columns torn down.

Coshocton Daily, Age, Ohio, 1908-10-02

TEXAS CITY, TX, APRIL 16, 1947

The SS Grand Camp was at the port taking on a load of ammonium nitrate fertilizer to be shipped to Europe as part of the rebuilding process following WWII. Around 8:10 a.m., fire was reported deep in cargo hold number 4 of the Grand Camp. The captain of the ship ordered the hold sealed and steam injected into the burning hold (the heat and confinement necessary for an explosion!). This confinement and injection of steam is likely to have resulted in the elevation of the temperature of the ammonium nitrate and the explosion that occurred. Texas City's volunteer fire department responded to the initial report of fire with their four fire engines led by Chief Henry J. Baumgartner.

At approximately 9:12 a.m., an explosion occurred within the hold of the Grand Camp. Instantly, all 27 members of the Texas City Volunteer Fire Department at the scene were killed, and some bodies were disintegrated by the heat and blast pressure of the explosion. All that remained of their fire engines were piles of twisted metal. Texas City lost all but one of their firefighters and all of their apparatus in the explosion (*Firehouse Magazine*).

MARSHALL'S CREEK, PA, JUNE 26, 1964

Fire in a cargo truck, thought to be caused by spontaneous heating following the blowout of a tire, quickly spread to the cargo compartment of a 28 ft tractor-trailer truck. The driver had disconnected the trailer off the roadway and had driven the tractor several miles to

Figure 3.71 As the firefighters approached the trailer, a detonation occurred. (Courtesy of Marshalls Creek, PA, Volunteer Fire Department.)

a service station. While the driver was gone, the tires ignited. The truck was carrying 4,000 lb of 60% standard gelatin dynamite in boxes and 26,000 lb of nitro carbo nitrate blasting agent in 50 lb bags. The compound is a mixture of 850 lb of ammonium nitrate fertilizer to 7 gal of No. 2 diesel fuel. Another passing tractor-trailer driver reported the fire to the Marshall's Creek Fire Department.

The driver reported that there were no markings on the trailer. Three fire engines responded, and an attack line was pulled to fight the blaze. As the firefighters approached the trailer, a detonation occurred (Figure 3.71). The fire had reached the explosive cargo, and the resulting explosion killed six people, including three firefighters, the truck driver who reported the fire and two bystanders. Property damage was over $600,000, including all three of the fire engines (*Firehouse Magazine*).

> **Hazmatology Point:** *Marshall's Creek incident was one of the key incidents that led to the U.S. Department of Transportation placard and label system and Emergency Response Guidebook.*

RICHMOND, IN, APRIL 6, 1968

At 1:47 p.m., leaking natural gas from a transmission line(s) under a building housing a firearms store was ignited by gun fire in the firing range. A secondary explosion of up to a ton of smokeless powder,

Figure 3.72 A secondary explosion of up to a ton of smokeless powder, along with some black powder that was stored in the basement, killed 41 persons, injured another 150 and caused over $2 million in property damage. (Courtesy Richmond IN Fire Department.)

along with some black powder that was stored in the basement, killed 41 persons, injured another 150 and caused over $2 million in property damage (Figure 3.72). Three buildings were destroyed by the explosion and five others damaged (*Firehouse Magazine*).

WACO, GEORGIA, JUNE 5, 1971

An automobile collided with a truck carrying 25,414 lb of explosives, resulting in a fire and the explosion of the dynamite cargo. The fire started as a result of gasoline and diesel fuels spilling from the vehicles and then igniting. Heat transfer from the fires caused the nitroglycerin-based dynamite to detonate, killing two firefighters, a tow truck driver and two bystanders; injuring another 33 persons and causing over $1 million in property damage.

KANSAS CITY, MO, NOVEMBER 29, 1988

Thirty-one years ago, on November 29, 1988, at approximately 03:40 hours, the Kansas City, MO, Fire Department received a call for a fire at a highway construction site (Figure 3.73). There were several explosions following the arrival of the fire department. It was reported by the Kansas City Fire Department that the first explosion

Figure 3.73 On November 29, 1988, at approximately 03:40 hours, the Kansas City, MO, Fire Department received a call for a fire at a highway construction site. (Courtesy Kansas City, MO, Fire Department.)

involved a split load of materials in a trailer/magazine. One compartment had approximately 3,500 lb of ammonium nitrate/fuel oil mixture (ANFO). The rest of the contents were approximately 17,000 lb of ammonium nitrate/fuel oil mixture with 5% aluminum pellets. In the second trailer/magazine, there were approximately one thousand 30 lb "socks" of ammonium nitrate/fuel oil mixture with 5% aluminum pellets. Pumper 30 was dispatched and arrived on scene at 03:52 a.m. At 04:08 a.m., 22 min after Pumper 41 arrived and approximately 16 min after Pumper 30 arrived, the magazine exploded killing all six firefighters assigned to Pumper 41 and Pumper 30.

Bucks County Courier Times Doylestown Pennsylvania 1971-06-05.

NEW YORK CITY, NY, FEBRUARY 26, 1993

On February 26, 1993, at 12:18 p.m., international terrorism struck the United States. The World Trade Center in New York City was bombed by Islamic fundamentalists under suspected mastermind Ramzi Yousef (Figure 3.74). The World Trade Center boasts two high-rise towers. Each tower was 110 stories tall, with several other buildings completing the complex. When fully occupied, there are over 150,000 people in the buildings. Noontime is the busiest period of day at the complex and would be the ideal time to launch a terrorist attack to produce a large number of deaths and injuries. Components of the World Trade Center bomb included pellets and bottled hydrogen. The van that Yousef used had four 20 ft (6 m) long fuses, all

Figure 3.74 The World Trade Center in New York City was bombed by Islamic fundamentalists under suspected mastermind Ramzi Yousef. (Courtesy Fire Department New York.)

covered in surgical tubing. He calculated that the fuse would trigger the bomb in twelve minutes after he would use a cheap cigarette lighter to light the fuse. The materials to build the bomb cost some 300 explosive effects from the 1,300-lb bomb caused a crater 180 ft deep, 100 ft long, and 200 ft wide in the underground parking garage. The crater was six levels deep. A rented truck was used to transport the bomb into the parking garage. Six people died and 1,042 were injured (History.com).

OKLAHOMA CITY BOMBING, OK, APRIL 19, 1995

The terrorist bombing at the Oklahoma City Federal Building involved a homemade mixture of ammonium nitrate fertilizer and fuel oil (Figure 3.75). As a result of the terror attack, 168 people, many of them children died and another 600 were injured. No emergency responders were killed at either incident since the explosions occurred before their arrival, but that may not always be the case. Over 800 buildings sustained some type of damage from ground shock and blast pressure. Of the buildings damaged, 50 would have to be demolished. Windows were broken as far as 2 miles from the blast site, and the blast was heard 50 miles away. It registered 3.5 on the open-ended Richter scale in Denver, Colorado (*Firehouse Magazine*).

Applied Chemistry and Physics

Figure 3.75 The terrorist bombing at the Oklahoma City Federal Building involved a homemade mixture of ammonium nitrate fertilizer and fuel oil.

WEST, TX, APRIL 17, 2013

On April 17, 2013, a fire and subsequent explosion occurred at the West Fertilizer Company in West, Texas. Firefighters from the West Volunteer Fire Department were fighting a fire at the facility when the explosion occurred. Ammonium nitrate was located in a bin inside a seed and fertilizer building on the property. The explosion registered 2.1 on a seismograph reading from Hockley, Texas, 142 miles away (Figure 3.76). Fifteen people, mostly emergency responders,

Figure 3.76 The explosion registered 2.1 on a seismograph reading from Hockley, Texas, 142 miles away.

were killed, over 200 were injured and 150 buildings sustained damage. Investigators confirmed that ammonium nitrate was the source of the explosion. According to the U.S. Environmental Protection Agency (EPA), there was a report of 240 tons of ammonium nitrate on the site in 2012 (*Firehouse Magazine*).

Hazmatology Point: *When responding to fixed or transportation incidents in or around construction sites, mining operations or facilities that retail agricultural fertilizers, be on the lookout for explosives or chemical oxidizers such as ammonium nitrate or commercial grade ANFO. When responding to transportation incidents, always consider the possibility of explosives or oxidizers being present. Fire is the principal cause of accidents involving explosive materials. Look for explosive signs, such as placards and labels. Evacuate the area according to the distances listed in the DOT Emergency Response Guidebook orange section. If no other evacuation information is available, a 2,000 ft minimum distance should be observed, according to the NFPA Fire Protection Handbook. There is one rule of thumb in responding to incidents where explosives or chemical oxidizers are involved:* DO NOT FIGHT FIRES IF THE FIRE HAS REACHED THE EXPLOSIVE STORAGE AREA OR CARGO. THE SAME APPLIES TO CHEMICAL OXIDIZERS SUCH AS AMMONIUM NITRATE.

Homeland Security Monitoring of Ammonium Nitrate

Since the bombing of the Murrah Federal Building in Oklahoma City, the Department of Homeland Security does monitor the storage and sale of bulk ammonium nitrate fertilizer used for agricultural purposes.

Author's Note: *Following the West, TX, ammonium nitrate explosion, I was doing some research for an article in* Firehouse Magazine. *I wanted to find out where ammonium nitrate was being stored and sold in Nebraska. A farmer friend of mine told me to go to the farmer's coop where he purchased ammonium nitrate for his pasture. When I went there, they said they did not sell it. Other facilities were contacted, and none of them would divulge information about ammonium nitrate. It appears the Homeland Security Regulations are working.*

This oversite has led to the reduction of ammonium nitrate used by farmers and stored for sale by retailers across the country. It has also caused the retailers who still handle the product to be very careful about information regarding its presence and can only be sold to licensed farmers.

According to the Nebraska Agribusiness Association, there are 31 sites in Nebraska that still store and sell ammonium nitrate fertilizer. Responders in agricultural communities should be aware of the types of fertilizers stored at local facilities. Ammonium nitrate is not one of the extremely hazardous substances covered by the Emergency Planning and Community Right-to-Know Act (EPCRA). However, fire departments have the right to information involving chemicals at a facility for the purposes of pre-planning even if the chemicals are not regulated under the act. The State Emergency Response Commission (SERC) in each state and the Local Emergency Planning Committee (LEPC) may also be of assistance.

Chemical Notebook

Ammonium nitrate, NH_4NO_3, is a strong inorganic oxidizer that can be an explosive all by itself under certain conditions. It is primarily used as an agricultural fertilizer for its nitrogen content. Ammonium nitrate is the primary component of ANFO, a commercially available explosive as well. Response personnel should deal with ammonium nitrate incidents with a great deal of caution. Ammonium nitrate is a colorless or white-to-gray crystal that is soluble in water. It decomposes at 210°C (392°F), releasing nitrous oxide gas and ammonia. Ammonium nitrate itself does not burn, but as an oxidizer supports and enhances combustion. When in contact with other combustible materials, the fire hazard is increased.

A fire involving ammonium nitrate in an enclosed space can lead to an explosion. Because it is an oxidizer, a fire involving ammonium nitrate can occur in the absence of atmospheric oxygen. Ammonium nitrate may explode when exposed to strong shock or high temperatures under confinement. Contaminants may increase the explosion hazard of ammonium nitrate. Organic materials, such as chlorides, and some metals, such as chromium, copper, cobalt and nickel, can make explosions involving ammonium nitrate more energetic. The NFPA 704 designation hazards for ammonium nitrate health—1, fire—0, reactivity—3 and special information—OX for oxidizer. The UN 4-digit identification number is 1942 with an organic coating and 2067 as the fertilizer grade. There are a number of other mixtures of ammonium nitrate that have four-digit numbers; they can be found in the DOT's Hazardous Materials Tables and in the DOT's Emergency Response Guidebook.

Categories of Chemical Explosions

Physical Explosion

In a physical explosion the high-pressure gas is produced by mechanical means, that is to say, even if chemicals are present in the container, they are not affected chemically by the explosion. In a chemical explosion the high-pressure gas is generated by the chemical reaction that takes place. The two major explosion types can be further divided into four sub types that result in the release of high pressure gas. The *first* type of explosion occurs as the result of the physical overpressurization of a container, causing the container to burst, such as the case of a children's balloon bursting when too much air is placed in the balloon.

Physical/Chemical Explosion

The *second* type of explosion occurs via physical or chemical means as in the case of a hot water heater or boiler explosion. The water inside the heater turns to steam when overheated, which results in a pressure increase inside the container, and the container fails at its weakest point. This is by far the most common type of accidental explosion that takes place.

Chemical Explosion

The *third* type of explosion involves a chemical reaction: the combustion of a gas mixture. Many of the explosive materials that are regulated and allowed to be offered for transportation by the DOT would involve this type of explosion should an accident occur. Therefore, most of our discussion of explosions and explosive materials in this hazard class will involve materials that explode by chemical means, with the high pressure being created by a chemical reaction. There are other types of explosions that produce high-pressure gases, some of which will be discussed in later hazard classes.

Explosives Tetrahedron

Just like the fire tetrahedron, there are several components (Figure 3.77) that must be present for a chemical explosive material like ammonium nitrate to explode. Since WWII has been the most common cause of

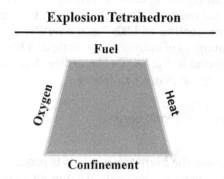

Figure 3.77 Explosive tetrahedron.

firefighter and civilian deaths from explosives. An explosion involving a chemical explosive is really a fast-moving fire. In simple terms, an explosive that functions via chemical reaction creates a rapidly burning fire that is made possible by the presence of a chemical oxidizer. You must have a fuel, oxygen, but in the case of an explosive material oxygen from the air is not enough, so you must have a chemical oxidizer. You must have heat or something to initiate the explosion. Finally, there must be something to confine the explosive materials in order for an explosion to occur. Without confinement, the materials will just burn off. Confinement can be pretty much anything that will get the job done including the material itself. If you pile enough explosive material together the sheer weight of the material itself can provide the confinement for the explosion to take place.

Types of Chemical Explosions

Detonation
Detonation is an instantaneous decomposition of the explosive with the release of high heat and pressure shock waves. Blast pressures can be as much as 700 tons/in.^{2}. Pressure and heat waves travel away from the center of the blast equally in all directions. The reaction occurs at supersonic speed or faster than the speed of sound, which is 1,250 ft/s, above 3,300 ft/s up to 29,900 ft/s or over 20,300 miles/h!

Often the terms "explosion" and "detonation" are used interchangeably, which is incorrect. An explosion may very well be a detonation; however, an explosion that is not a detonation can occur.

Deflagration
Deflagration is the burning of fuel accelerated by the expansion of gases under the pressure of containment, which causes the containing vessel to break apart explosively. In contrast to detonation, materials that deflagrate do so slower than the speed of sound. Under the right conditions, a material that normally deflagrates can detonate. The term "detonate" refers to an instant, violent explosion that results when shock waves pass through molecules and displace them at supersonic speed.

> *Hazmatology Point: In terms of dangers to emergency responders, the only real difference between detonation and deflagration is perception of the danger. When a detonation occurs, it happens faster than the speed of sound. Deflagration occurs at less than the speed of sound. If you are caught in a detonation facing away from the explosion, you will not hear the blast or the blast wave that kills you. If you are caught in a deflagration facing away from the explosion, you will hear the explosion just before the blast wave kills you. Your only defense against incidents that involve explosives is the knowledge that these incidents are losers.*

Awareness of the danger and being a safe distance away are your only protection against injury or death. Do not fight fires involving explosive materials!

Yield vs. Order

The yield of an explosive is associated with the rate or speed in which the explosion occurs. This is an indication of whether an explosive will detonate or deflagrate. A high-yield explosive detonates, and the blast pressure shatters materials that it contacts. Examples of high-yield explosives are dynamite, TNT, nitroglycerin, detonating cord, C-3, C-4 and explosive bombs. A low-yield explosive deflagrates and is used to push and shove materials. Examples of low-yield explosives are black powder and commercial ammonium nitrate. Deflagrating materials are often used to move rocks in road construction, quarries and mining.

Order has to do with the extent and rate of a detonation. A high-order detonation is one in which all of the explosive material is consumed in the explosion, and the explosion occurs at the proper rate. The proper rate in this case would be supersonic. Thus, a low-order explosion would occur as an incomplete detonation or at less than the desired rate. Yield involves the specific explosive material that is used, and order indicates the way in which the explosive detonates. The hazards to emergency responders are obvious. If an explosion is low order, not all of the explosive material is consumed, and therefore, the remaining material presents a hazard. Whether high or low yield, high or low order, all explosives should be treated as high-yield, high-order Class 1.1 explosives. There is no way for responders to know in the field. Do not mess with explosives, call in the experts and stay a safe distance from the incident.

Explosive Effects

Phases of Explosions

There are primary and secondary effects of explosions. Primary effects are blast pressure, thermal wave and fragmentation. Blast pressure has two phases: the positive and the negative. In the positive phase, the high-pressure gas, heat wave and any projectiles travel outward.

Blast pressure, is generally the most destructive element of an explosion. If the explosion is a detonation, the waves travel equally in all directions away from the center of the explosion. The negative phase occurs right after the positive phase stops. A partial vacuum is created near the center of the explosion by all of the outward movement of air from the blast pressure. During the negative phase, the debris, smoke and gases produced by the blast are drawn back toward the center of the explosion origin, then rise in a thermal column vertically into the air, and eventually carried downwind by the air currents. The negative phase may last up to three times as long as the positive phase.

If the explosion is a result of an exothermic (heat-producing) chemical reaction, the shock wave may be preceded by a high-temperature thermal wave that can ignite combustible materials. Some explosives are designed to disperse projectiles when the explosion occurs, for example, antipersonnel munitions and hand grenades. Other explosives may be in metal or plastic containers to provide the necessary confinement for an explosion to occur. These containers may become projectiles when the explosion occurs. Projectiles from an explosion travel out equally in all directions from the blast center just as the blast pressure does. In order for the waves or projectiles coming off the explosion to travel equally in all directions, the velocity (speed) of the release of the high-pressure gases must be supersonic or faster than the speed of sound. This occurs only in a detonation, not in a deflagration.

Fragmentation may come from the container that holds the explosive material and from materials in close proximity. Objects in the path of the explosion are broken into small parts by the force of the blast pressure, creating fragments. These fragments may have jagged or sharp edges. The fragments will travel away from the blast center at high speeds, greater than 2,700 ft/s, faster than a speeding bullet!

Thermal hazards are the generation of heat by the explosion. The amount of heat produced will depend on the type of material that is involved. There is a flash and a fireball associated with almost any chemical explosion. The more rapid an explosion, the greater the effects produced by the heat. Although the total heat produced may be similar in each explosion, a detonation will produce the most heat over a larger area, because of the speed of the explosion.

Secondary effects of an explosion are shock-wave modification, fire and shock-wave transfer. There are three ways that a shock wave can be modified: it may be reflected, focused or shielded. Reflection refers to the shock wave striking a solid surface and bouncing off. When a shock wave strikes a concave (curved) surface, the force of the shock wave is focused, or concentrated, on an object or small area once it bounces off the concave surface. This effect is similar to the principle behind satellite dishes. When a signal reaches a satellite dish from the satellite in space, the signal is focused on the electronic sensor protruding out of the front of the satellite dish. Shielding simply means that the shock wave encounters an object too substantial to be damaged by the wave, so the shock wave goes around the object or is absorbed by it. The area immediately behind the object provides a place of shelter from the shock wave. Fire and shock-wave transfer involves the transfer of the shock-wave energy and fire to other objects, causing fires and destruction.

Overpressure Damage to Property

0.5 – 1 psi	Window glass breakage
1-2 psi	Buckling of corrugated steel
2-3 psi	Shattering of concrete
3-4 psi	Steel panel building collapse
5 psi	Utility poles snap
7 psi	Overturn rail cars
7-8 psi	Brick wall failure

Figure 3.78 An overpressure of just 0.5–1.00 psi can cause windows to break and knock down personnel.

Overpressure

An overpressure of just 0.5–1.00 psi can cause windows to break and knock down personnel (Figure 3.78). At 5 psi eardrums can be ruptured and wooden utility poles can be snapped in two. Ninety-nine percent of people exposed to overpressures of 65 psi would die. If the explosion is a result of an exothermic (heat-producing) chemical reaction, the shock wave may be preceded by a high-temperature thermal wave which can ignite combustible materials.

There are also materials that emit projectiles out of the blast center. The shock wave, heat wave or projectiles may or may not be sent out equally in all directions from the center of the blast. In order for the waves coming off the explosion to travel equally in all directions, the velocity (speed) of the release of the high-pressure gases must be sonic or faster than the speed of sound.

Dust Explosion

This third subgroup of potentially explosive materials involves a chemical reaction type of explosion, sometimes referred to as combustible dusts (Figure 3.79). Dust explosions that have been occurring since the mid-1800s have been documented. Instances have increased and continue to the present day. A call to action has been requested by the U.S. Chemical

Explosive Dusts

Coal	Crude Rubber	Peanut Hulls
Cork	Titanium	Aluminum
Flour	Cornstarch	Soy Protein
Silicon	Walnut Shell	Pea Flour
Sugar	Zirconium	Magnesium

Figure 3.79 Explosive dusts.

Safety Board (CSB) for the solutions for reducing dust explosions. From 2006 to 2017, 105 incidents occurred in the United States involving combustible dusts. Fifty-nine fatalities resulted along with 303 injuries. Those facilities that had more than ten dust explosions included metal industries, lumber and food products facilities, and other industrial facilities.

Combustible dusts are, in many cases, ordinary combustible materials or other chemicals that, because of their physical size, have an increased surface area (Figure 3.80). This increased surface area exposes more of the particles to oxygen when they are suspended in air. For example, a solid block of wood has a surface area of 1.5 ft². When that block of wood is turned into sawdust, the surface area increases to several thousand square feet. Much more of the surface area is exposed to oxygen in the air. When these dusts are suspended in air, they can become explosive if an ignition source is present.

There are three phases in a dust explosion: initiation, primary explosion and secondary explosion (Figure 3.81). Initiation occurs when an ignition source contacts a combustible dust that has been suspended in air. This creates the primary explosion, which shakes more dust loose from the confined area and suspends it in air. The secondary explosion then occurs, which is usually the larger of the two explosions because there is more fuel present. Explosive dusts include, but are not limited to, the following dusts: grain, saw, coal, cork, magnesium, zirconium, crude rubber, sugar, cornstarch, pea flour, walnut shells, peanut hulls, aluminum, flour, titanium and silicone.

One of the major facilities where dust explosions occur is grain elevators; explosions occur when grain dust is suspended in air in the presence of an ignition source. The primary danger area where the explosion is likely to occur within most elevators is the "leg," or the inclined conveyor, the mechanism within the elevator that moves the grain from the entry point to the storage point. In older elevators, explosions occur in the head house on top of the elevator. For a dust explosion to occur, five factors

Surface Area Comparison

Solid   **Dust**
6 Inch Cube   **Identical Volume**

Surface Area = 1.5 Square Feet

Surface Area = Several Thousand Square Feet

Figure 3.80 The finer the particles of dust, the greater the surface area and the more likely a dust explosion may occur.

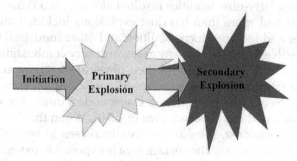

Figure 3.81 There are three phases in a dust explosion: initiation, primary explosion, and secondary.

must be present: an ignition source, a fuel (the dust), oxygen, a mixing of the dust and the oxygen and confinement. The explosion will not occur unless the dust is suspended in air within an enclosure at a concentration that is above its lower explosive limit (Figure 3.82).

Hazmatology Point: During the 1980s, the State of Nebraska led the nation in deaths from grain elevator explosions. At the request of the governor to do something about the explosions, Nebraska's State Fire Marshal implemented a grain elevator inspection program. Elevators were inspected at least once a year. Following several years of inspections and repair orders, the death rate from elevator explosions were greatly reduced. As proven by the Nebraska program, grain elevator dust explosions can be prevented.

Combustible dusts may be present in many different types of facilities. Common places for combustible dusts to be found are in grain elevators, flour mills, woodworking shops and dry-bulk transport trucks.

Explosive Properties Dust

Type of Dust	Ign Temp	LEL
Wheat/corn/oats	806° F	55
Wheat Flour	380° F	50
Cornstarch	734° F	40
Rice	824° F	50
Soy Flour	1004° F	60

Figure 3.82 Dust explosion will not occur unless the dust is suspended in air within an enclosure at a concentration that is above its lower explosive limit.

Dusts in facilities have caused many explosions over the years that have killed and injured employees as well as firemen. An explosion occurred in a facility in Anne Arundel County, Maryland, that had many hazardous materials on-site. At first, it was thought that one of the chemicals had exploded. The fire department and the hazmat team were called to the scene. Investigation revealed that the explosion occurred in a dust collection system; it was a combustible dust explosion. Dust explosions can be prevented by proper housekeeping and maintenance practices at facilities where these types of dusts are present (Hazardous Materials Chemistry for Emergency Responders).

Historic Dust Explosion Incidents

Chicago, IL August 5, 1897 Grain Elevator Explosion

Seven and probably eight lives were lost in an explosion which took place this evening during a fire in the northwestern grain elevator at Cook and West Water Streets. Three of the dead are firemen. The body of another fireman is thought to be buried in the ruins of the elevator; and three people were blown into the Chicago River. Nine firemen were injured. From the force with which the explosion swept the spot on which they were standing it is certain that they must have been instantly killed.

Titusville Morning Herald Pennsylvania 1897-08-06.

Milwaukee, WI April 22, 1926 Sawdust Explosion

Three firemen were killed and 14 injured in a sawdust explosion at the Marsh Woodworks Company. Fire earlier in the day had been brought under control. Firemen detailed to clean up were at work when the explosion occurred.

Nevada State Journal, Reno, NV 22 Apr 1926.

Corpus Christi, TX April 7, 1981 Grain Elevator Explosion

An explosion and fires ripped the top off several silos at the Corpus Christi Public Grain Elevator at the city's seaport, on Texas' middle Gulf coast (Figure 3.83). This was the worst explosion of its kind in Texas history. Nine people were killed and thirty were injured. An emergency medical technician, who asked not to be identified, said there were bodies lying all over the ground. Flames ignited by the blast and fanned by 35 mph winds sent thick, black clouds of smoke billowing 500 feet into the air and troubled firefighters who warned bystanders about the possibility of additional explosions. The experts said that there were 12 explosions that ripped through the grain elevator at 1,500 feet per second.

Fifty firefighters kept watch on the smoldering fire overnight to prevent any other explosions, which was caused by an accumulation of grain dust, authorities said. The fire still was burning early the next day,

Figure 3.83 This was the worst explosion of its kind in Texas history. Nine people were killed and thirty were injured.

although firefighters said they had isolated the flames to one section of the silos. Acting Fire Chief E. E. Irwin said "the fire would be allowed to burn itself out".

Corpus Christi Caller Times 1981-04-06.

Bellwood, NE April 7, 1981 Grain Elevator Explosion

On the same day as the explosion that occurred in Corpus Christi, TX, another grain elevator exploded in Bellwood, Nebraska killing two and injuring one and causing severe damage. 1981 saw 21 grain elevator explosions in 12 different states, 5 of those occurring in the State of Nebraska. All of the deaths occurred in Nebraska and Texas, a total of 13 deaths and 62 injuries. The injuries occurred in 9 states, with Nebraska and Texas leading the way with 7 and 33 respectively.

Columbus Times, Columbus, NE, April 8, 1981.

Nuclear Explosions

The fourth subgroup of explosion is thermonuclear, which is the result of a tactical decision, a weapon that malfunctions or an act of terrorism. There are two types of nuclear explosions: air burst and ground burst. Both are engaged as a result of a tactical objective.

Air burst is designed to knock out all electronic equipment, disrupting communications and computer usage. This type of explosion does not create fallout, because it does not reach the ground and does not suck up debris in the negative phase of the explosion.

Ground burst is designed for mass destruction of everything it contacts. Initially, during the positive phase of the ground blast, there is a thermal wave that is released first, followed by a shock wave. During the negative phase of the explosion, the debris from the explosion is drawn into the cloud, travels downwind and then falls back to the ground. The debris, while in the cloud, is contaminated with radioactive particles, and radioactive fallout is created. With the onset of terrorism that has occurred over the past several years, another type of nuclear bomb called the dirty bomb has become a concern for emergency responders. While the dirty bomb is not technically a nuclear explosion, it does create radioactive contamination similar to a ground burst. A dirty bomb uses conventional explosive materials to disseminate radioactive materials and cause contamination of areas where they are released. This topic will be discussed further in Hazard Class 7 Radioactive Materials (Hazardous Materials Chemistry for Emergency Responders).

Hazard Class 2 Compressed Gases (LF/HR)

The compressed gas class is divided into three Divisions: Division 2.1 materials are flammable compressed gases LF/HR, Division 2.2 materials are nonflammable compressed gases HF/LR and Division 2.3 materials are poison compressed gases LF/HR. Each of the compressed gas categories presents its own special hazards in addition to the hazard of being under pressure in a specially designed and regulated pressure container. Pressures range from atmospheric 5 to 6,000 psi. The higher the pressure, the more substantial the container must be to hold the pressure. The higher the pressure, the greater the danger when the pressure is released or the container fails.

Gases in this class may also be liquefied in order to ship larger quantities more economically. Incidents are more frequent from gases that have been liquefied then just compressed gases in containers. Because of the pressure of the compressed gases, containers are small. As larger containers would be too heavy and bulky in order to hold high pressures, only the small containers are practical.

Division 2.1 Flammable Gases LF/HR

Class 2.1 materials are flammable. The U.S. Department of Transportation (DOT) defines a flammable gas as "any material which is a gas at 20°C (68°F) or less and 14.7 psi of pressure or above, which is ignitable when in a mixture of 13 percent or less by volume with air or has a flammable range with air of at least 12 percent regardless of the lower limit." The DOT has established criteria for a material to be classified as a flammable gas. First of all, its lower explosive limit must be below 13%. If it is then it is classified as a flammable gas. Some materials have wide flammable ranges which make them much more dangerous than materials with narrow ranges. So the DOT says that if a material has a flammable range greater than 12 percentage points regardless of what the lower limit is, it is classified as a flammable gas. Examples of 2.1 flammable gases are hydrogen, propane and butane.

Division 2.2 Nonflammable Gases LF/LR

Division 2.2 includes materials that are nonflammable and nonpoisonous compressed gases. The materials in this class can be compressed gases, liquefied gases, pressurized cryogenic gases and gases in solution. These materials are nonflammable, but that does not mean the containers cannot (BLEVE) under flame impingement conditions on the container from a fire involving other materials, or from a damaged or weakened container.

The DOT definition of a nonflammable gas is "a material that exerts in the packaging an absolute pressure of 41 psi or greater at 20°C, and does not meet the definition of Division 2.1 or 2.3." If the pressure is less than 41 psi, they do not belong in this category.

There are also some compounds that we will scientifically call the "foolers" in this category. Anhydrous ammonia is regulated by the DOT as a nonflammable compressed gas. United States of America is the only country in the world that placards anhydrous ammonia this way. Everywhere else, it is placarded as a poison gas and corrosive. Not to mention that it is also flammable under certain conditions. If it is inside a building or in a confined space, it may very well be within its flammable range and burn if an ignition source is present. In Shreveport, Louisiana, one firefighter was killed and one badly burned in a fire involving anhydrous ammonia. A leak developed inside a cold storage plant. The firefighters donned Level A chemical protective clothing and went inside to try to stop the leak. Something caused a spark and the anhydrous ammonia caught fire.

If anhydrous ammonia met the DOT definition of a flammable gas, which is a LEL of less than 13% and a flammable range of greater than 12% points, it would be placarded as such. It does not, however, meet the DOT definition of a flammable gas. It has a LEL of 16% and a range of 16%–25%, which is 10 percentage points.

Cryogenic liquids do not have a hazard class of their own. They are included in Class 2 divisions depending on the hazard of the cryogenic liquid. Some cryogenic liquids do not fit into any of the divisions. These are basically inert materials as far as the DOT system goes. However, cryogenic liquids as a group have some dangerous hazards to responders and the public if they are released from their containers. Personally, I have reached out to the DOT about these issues and the issues with anhydrous ammonia within the last year. It will be interesting to see if any changes are made to the 2020 *Emergency Response Guidebook* (ERG).

Division 2.3 Poison Gases LF/HR

Subclass 2.3 materials are poison gases. Poisons are among the most dangerous hazardous materials for emergency responders. They are also gases, which is the most difficult physical state for responders to deal with. Primarily they are an inhalation hazard, though some may also be absorbed through the skin. The DOT definition of poison gas is "a material that is a gas at 20°C or less at 14.7 psi and is so toxic to humans as to pose a hazard to health during transportation, or in the absence of adequate data on human toxicity, is presumed to be toxic to humans because when tested on laboratory animals it has an LC50 value of not more than 5000 ml/m^3."

These materials are considered so toxic that when transported the vehicle must be placarded regardless of the quantity. The potential exists for Division 2.3 materials to affect large populations by creating gas clouds that are toxic. Examples of some 2.3 poison gases are fluorine, chlorine, carbon monoxide, hydrogen sulfide and chloropicrin–methyl bromide mixture. Poison gases may be encountered as gases, liquefied gases or cryogenics. The placard will indicate poison gas, but it will not tell you that the material has been liquefied or turned into a cryogenic liquid. The container type will help determine the physical state of the materials.

Toxicological Terms

Toxicology is the study of toxic or poisonous substances and their effects on the body. It relates to the physiological effect, source, symptoms and remedial measures for the materials.

Asphyxiates displace the oxygen in the air so that there is not enough to sustain life without respiratory protection.

Corrosives can be fuming acids that produce corrosive and toxic gases, such as nitric and sulfuric. Some acids are made by dissolving a gas in water. Fuming acid gases can react with moisture in the lungs to reproduce the acid inside the lungs.

Carcinogens are materials that cause cancer. This information has been obtained by studying populations exposed to materials for long periods, usually in the workplace. The information can also be obtained through tests on laboratory animals. This does not mean that the material will actually cause cancer in humans, but it may be a good indication. Examples of known carcinogens include radon and vinyl chloride.

Mutagens cause mutations or changes in genetic material. Changes may affect the current generation that was exposed or future generations. Examples include ethylene oxide and emissions from the combustion of crude glycerin and diesel fuel.

Teratogens cause one-time birth defects when the pregnant woman is exposed or the father is exposed and passes the effect on to the mother. These types of materials do not permanently damage the reproductive system. Normal children can be produced as long as repeated exposures do not occur. Examples some inhalation anesthetics are suspected and carbon monoxide.

Irritants are materials that cause irritation to the respiratory system. Examples are tear gas and some disinfectants.

Several factors influence toxic effects. One is the concentration of the toxic material. Concentrations are expressed in terms of parts per million or billion, percentage or milligrams per kilogram. The higher the concentration of the toxicant, the more serious the effects will be. Types of exposure include:

- **Acute.** It is a one-time exposure. Depending on the concentration and duration of exposure, there may or may not be toxic effects.
- **Subacute.** It involves multiple exposures with a period of time between exposures.
- **Chronic.** It involves multiple exposures. An emergency responder can have an acute exposure to a toxic material at the scene of a hazmat incident. A single exposure may not produce a problem, whereas multiple exposures may cause damage. It is important to monitor exposures for personnel to determine any illness several days after the incident. An example of chronic exposures to responders is by continually wearing contaminated bunker gear. Toxic materials are forced into the fabric from heat and smoke. Wearing contaminated bunker gives you a chronic exposure to those toxic materials. Some fire gases are so toxic we would wear Level A chemical suits if they were released at a hazmat incident.

Target organs. It takes some toxic materials several hours to several days to reach the susceptible target organs and cause damage. The response of the body to a hazardous material is based largely on concentration, length of exposure, type of exposure, route of exposure, susceptible target organ and other health-related variables such as age, sex, physical condition and size. This is referred to as dose/response. The type of response depends on which part of the body is affected by a particular toxic material. This is referred to as the "target organ." The target organ should not be confused with routes of exposure. A pesticide may enter the body through inhalation, but may have an effect on the central nervous system.

Routes of exposure. There are four routes in which toxic materials can enter the body and cause damage: inhalation, absorption, ingestion and injection. Inhalation involves material entering the body through the respiratory system and that's where damage usually occurs. In absorption, material enters the body through the skin, eyes or some other tissue. Damage may occur to the point of contact or the material may travel to the susceptible target organ and cause harm there. Ingestion occurs when the material enters through the mouth and travels to the susceptible target organ. Absorption can also occur after ingestion. Injection involves a sharp object that is contaminated with a toxic material breaking the skin or the material entering through an open wound in the skin.

Exposure rates involve measurement of workplace exposures of hazardous materials based on tests conducted on animals. The results are then projected to humans based on the weight ratios. The emergency responder's workplace is the incident scene, and the concentrations of toxic materials may be much higher than the values indicated through the following terms:

Threshold limit value (TLV), a time-weighted average (TWA), is the maximum amount of acceptable exposure for 8h a day 40h a week without ill effects.

TLV-STEL (short-term exposure limit) is the maximum concentration averaged over a 15-min period to which healthy adults can be safely exposed for up to 15 min. Exposure should not occur more than four times per day with at least 1 h between exposures.

TLV-ceiling is the maximum concentration to which a healthy adult can be exposed without injury.

PEL (permissible exposure limit) is the maximum concentration averaged over 8h to which 95% of healthy adults can be repeatedly exposed for 8h a day 40h a week.

IDLH (immediately dangerous to life and health) is the maximum amount of a toxic material that a healthy adult can be exposed to for up to 15 min and escape without irreversible health effects.

Not all people are affected by toxic materials in the same way. Variables include age, sex, weight, general health, body chemistry and physical condition. These can all affect the way individuals respond to a toxic material. Older people and younger children may be affected by a toxic material, whereas other adult persons are not. Some materials affect females, but not males, and some will affect both. Females who are pregnant may be affected by a toxic material that will cause damage to the fetus, and a male may not be affected.

Toxicity is based on dose and size or weight of the test animal. This is then projected to humans based on the weight ratio of the dose, the weight of the animal and the weight and health of the human. Several protective measures may minimize the effect of toxic materials. Antidotes are available for a small number of toxic materials, but they must be administered immediately after exposure. Your body has the ability to filter out some toxic materials through the normal process of eliminating wastes.

Protect yourself from toxic materials by wearing protective clothing and avoiding contact with toxic materials. Wearing self-contained breathing apparatuses (SCBAs) is the single most important protective measure you can do for protection against gases that you cannot see. If there is the slightest chance that you are dealing with a toxic gas, wear the SCBA until the area can be monitored to determine when it is safe. As mentioned earlier, fires produce toxic gases. So, we wear SCBA; however, the most dangerous time of a fire in terms of toxic gases is during overhaul when those gases are still present even though we cannot see them. SCBA should be worn until monitoring is done to determine that it is safe to take them off. Practice contamination prevention. Establish zones, deny entry and provide protection to responders and the public.

Common Class 2 Hazardous Materials

Liquefied Compressed Gases

The liquefaction of gases presents a hazard in itself. Liquefied gases have large liquid-to-gas expansion ratios. That is to say, a very small amount of a liquid leaking from a container can form a very large gas cloud. This increases the danger of flammability if an ignition source is present and asphyxiation or toxicity when vapor clouds form. Some liquefied gases such as propane and butane are ambient temperature liquids. Cryogenic liquefied gases, on the other hand, are very cold liquids. This difference in temperatures can cause problems for responders unless handled properly.

Liquefied Petroleum Gases—LF/HR

Liquefied petroleum gases (LPGs) include several similar mixtures that contain mostly propane and butane with small amounts of propylene and butylenes. A mixture does not change the chemical makeup of any of the components. Mixtures do, however, change the physical characteristics of the individual components to something between the highest and the lowest of the components physical characteristics. So the mixture has its own physical characteristics. Since propane is the primary LPG that is used in the United States, propane and LPG are often used synonymously.

Chemical Notebook **Propane (C_3H_8)** has a boiling point of −42°F and a vapor pressure of 858 at 21°C: kPa. Flammable range is 2.1–10.1 and its ignition temperature is 851°F. Propane expands from 1 gallon of the liquid to over 270 gallons of gas and is heavier than air (1.53). This means that a very small spill or leak can produce a large gas or vapor cloud. Both liquefied compressed and cryogenic gases can also produce a significant asphyxiation hazard in addition to any other hazards they might have. We are usually very aware of the flammability hazards of propane and other flammable gases, but we may overlook the asphyxiation hazard.

Liquefied gases are gases that have been liquefied by bringing the gas to its critical temperature and pressure. At this point, the gas turns into a liquid. Propane, for example, has a critical temperature of 206°F and a critical pressure of 617 psi. At that temperature and pressure, the propane gas becomes a liquid.

Butane (C_4H_{10}) has a boiling point of −0.4°F, a flammable range of 1.86–8.41, an ignition temperature of 761°F and an expansion ratio of 1 gallon of liquid to over 235 gallons of butane gas. It is heavier than air (2.00) and has a critical temperature of 306°F and a critical pressure of 555 psi. At that temperature and pressure, butane becomes a liquid.

Isobutane (i-C_4H_{10}) has a boiling point of −11.75°F, a flammable range of 1.80–8.44, an ignition temperature of 761°F and a vapor pressure of 215 at 21°C: kPa. Butane has an expansion ratio of 1 gallon of liquid to 234 gallons of gas, is heavier than air (2.07) and has a critical temperature of 134.6°C and a critical pressure of 3,650 kPa.

Propylene (C_3H_6) has a flammable range of 2–11.1, an ignition temperature of 856, a boiling point of −53.9°F and a vapor pressure 146 at 0°C: kPa. It has a critical temperature of 197.5 and a critical pressure of 666.3.

Butylene (C_4H_8) has a flammable range of 1.98–9.65, an ignition temperature of 680, a boiling point of 21°F, a vapor density of 1.96 and a vapor pressure 1,939 at 21°C: kPa. It has a critical temperature of 40.2°F and a critical pressure of 146.4.

These gases are all hydrocarbons. You can see the differences in physical properties such as boiling points, ignition temperatures and vapor pressures based upon the compound. Flammable ranges are all similar because as a family the hydrocarbons have narrow flammable ranges. Vapor densities are also similar as most hydrocarbon gases except methane and ethane are lighter than air. Isobutane is branched in its structure. Branching has the effect of lowering the boiling point of the liquid, which allows it to be used in colder climates than butane.

Once liquefied, the gas is kept in the liquid state by pressurizing the tank to keep a constant artificial atmosphere pressing down on the liquid to prevent it from returning to the gas state. Liquids in these tanks are at atmospheric temperature. Irrespective of the temperature that is outside the tank, the liquid inside will be nearly the same. Liquefied gases exist in the tanks well above their boiling points. The only thing keeping the material from boiling and turning back to a gas is the pressure in the container.

These tanks are usually constructed of steel with working pressures generally between 100 and 500 psi. Propane has a working pressure of 250 psi, whereas anhydrous ammonia has a pressure of 265 psi. These tanks are never filled to the top. There is usually a 20% vapor space allowed above the liquid level. Usually, these tanks are not insulated but rather painted with white or aluminum paint to reflect radiant heat. When a pressure container such as the 331, with a boiling liquid inside, is exposed to flame on the vapor space, the container will quickly fail and produce a BLEVE, which is a boiling liquid expanding vapor explosion.

During a BLEVE, when the container opens up, all of the liquid inside instantly turns into a gas because it is existing above its boiling point. If there is an ignition source present, a fire ball may also be created. According to the National Fire Protection Association, BLEVEs occur within 8–30 min after flame impingement starts. with an average time of 15 min or less. If flame impingement occurs on the liquid level of the tank, the liquid in the tank will absorb the heat and protect the container. The heat from the fire will cause an increase in the pressure in the tank, and it will be vented through the relief valve.

If the relief valve cannot relieve the pressure as fast as it builds up, the container may still fail. Relief valves are only designed to relieve pressures created by increases in ambient temperature, not flame impingement. Excess flow valves are installed at product discharge openings. They operate in the event of a failure in discharge hoses or piping. Venting systems are either mechanical pressure relief or frangible disks. Valve protection during a rollover is much the same as that for other MC series tanks.

Class 2 Multiple Hazard Bad Actors

Chemical Notebook

Chlorine (Cl_2), LF/HR Because they are in such wide use, the hazards of common chemicals sometimes are taken for granted. Complacency can set in, and improper procedures may be used by those who work with the chemicals regularly and by emergency responders who deal with the materials during a release, resulting in injury and death. One of these chemicals is chlorine found in most communities in the United States as a gas or in compound with other chemicals that can release the chlorine when in contact with water or other chemicals. It is generally transported and stored as a liquefied compressed gas and is found in 100 to 150 lb cylinders, 1 ton containers and railroad cars.

Chlorine (elemental symbol: Cl) is a nonmetallic element, a member of the halogen family of elements with an atomic number of 17 on the periodic table. Other halogens include fluorine, bromine and iodine. Chlorine was discovered in 1774 by Carl Scheele, who also discovered oxygen and several other important compounds. Scheele called his discovery "dephlogisticated marine acid." Chlorine has an atomic weight of 35.453 and is a greenish-yellow diatomic gas with a pungent irritating odor, but it does not exist freely as a gas in nature. Diatomic gases are elements that do not exist as a single molecule, in this case Cl, but rather as the diatomic molecule Cl_2. Other elements that are diatomic are hydrogen, nitrogen, bromine, iodine, fluorine and oxygen. (Oxygen is often referred to as O_2 because it is a diatomic element.)

The primary source of chlorine is in the minerals halite (rock salt), sylvite and carnallite, and from the chloride ion (sodium chloride) in seawater. It can be liquefied for more economical shipping, storage and use.

Chlorine is toxic by inhalation (1 ppm in air), nonflammable, nonexplosive and a strong oxidizer (stronger than oxygen). Because chlorine is a strong oxidizer, it will support combustion even though it is nonflammable. Chlorine has a National Institute for Occupational Safety and Health (NIOSH) IDLH rating of 10 ppm and an exposure limit TWA of 1 ppm. The Occupational Safety and Health Administration (OSHA) ceiling for chlorine is 1 ppm. The maximum airborne concentration is 3 ppm. This is the amount to which a person could be exposed for up to 1 h without experiencing or developing irreversible or other serious health effects or symptoms that could impair the ability to take protective action.

Chlorine gas irritates the mucus membranes, and the liquid burns the skin or causes irritation to the skin and may cause burning pain, inflammation and blisters. Tissue in contact with cryogenic liquid chlorine can cause frostbite injury. Chlorine's odor threshold is about 3.5 ppm, although some report that odor can be detected below the 1 ppm OSHA ceiling and TWA. Short-term exposure to low concentrations of chlorine (1–10 ppm) can result in sore throat, coughing and eye and skin irritation. After a few breaths at 1,000 ppm, chlorine can be fatal. Exposures to chlorine should not exceed 0.5 ppm (an 8 h TWA over a 40 h week).

Chlorine is not known to cause cancer. Reproductive and developmental effects are not known or documented. It has a boiling point of 29°F, a freezing point of –150°F, a gas density of 2.5 (making it heavier than air), a specific gravity of 1.56 (heavier than water) and a vapor pressure of 5,168 mmHg at 68°F. The vapor pressure of chlorine is 53.1 psi at 32°F and 112.95 psi at 77°F.

Chlorine is slightly soluble in water and reacts with a variety of other chemicals, including aluminum, arsenic, gold, mercury, selenium, tellurium, tin and titanium. Carbon steel ignites near 483°F in contact with chlorine. It reacts with many organic materials creating violent or explosive results. Chlorine also reacts violently with acetylene, ether, turpentine, ammonia, fuel gas, hydrogen and finely divided materials. It is placarded and labeled as a Division 2.3 poison gas in transportation and OSHA-mandated fixed storage. Nonbulk containers will also have the corrosive label displayed. Chlorine has a UN four-digit identification number of 1017 and a National Fire Protection Association (NFPA) 704 designation of toxicity—3, flammability—0, reactivity—0 and special information oxy (oxidizer).

Chlorine was used during World War I as a choking (pulmonary) agent. One of the primary uses of chlorine around the world is in the chlorination of drinking water and treatment of sewage. It is also widely used in swimming pools. Chlorine is used in the production of paper products as a bleach and in dyestuffs, textiles, petroleum products, medicines, antiseptics, insecticides, food, solvents, paints, plastics and many other consumer products. Exposure to chlorine can cause various signs and symptoms depending on the amount and length of time exposed. There is no available antidote for chlorine exposure. Effects can be treated, and most people exposed that survive acute exposure will likely recover with little if any side effects. Listed below are potential symptoms:

- Coughing
- Chest tightness
- Burning sensation in the nose, throat and eyes
- Watery eyes (contact with liquid can cause blindness)
- Blurred vision
- Nausea and vomiting
- Burning pain, redness and blisters on the skin if exposed to gas and skin injury, similar to frostbite if exposed to liquid (cryogenic) chlorine
- Difficulty breathing or shortness of breath
- Fluid in the lungs

Exposure to low concentrations (1–10 ppm) is likely to result in eye and nasal irritation, sore throat and coughing. Higher concentrations (greater than 15 ppm) are likely to result in rapid onset of respiratory distress with airway constriction and accumulation of fluid in the lungs (pulmonary edema). Additional symptoms may include rapid breathing, blue discoloration of the skin, wheezing, rales or hemoptysis. Pulmonary injury may progress over several hours and lung collapse can occur. It is estimated that the lowest lethal concentration for a 30 min exposure is 430 ppm. While these symptoms can also be present with exposure to other inhalation hazards, investigation of the site and circumstances should clear up the chemical involved in most cases.

Chlorine usually does not just appear; it has a distinctive color and odor. Examinations of containers and reports of witnesses can be helpful in positive identification. There are usually not any long-term health effects from sudden exposures to chlorine vapor for those who survive. Complications such as pneumonia during treatment can occur. Chronic bronchitis can also develop in people who contract pneumonia.

Although it is a gas, chlorine can cause irritation and burns in contact with the skin. Therefore, firefighter turnouts are not appropriate for chlorine exposures inside the "hot zone" of a hazardous materials incident. In the past, firefighters were known to wear firefighter turnouts with petroleum jelly covering the exposed skin. Chlorine is a poison gas and requires SCBA and full Level A chemical protective clothing for anyone knowingly going into an atmosphere where chlorine is present. OSHA allows Level B protection for unknown atmospheres, which could include chlorine, but as soon as it is known that chlorine is present, protection should be changed to Level A.

Generally, gases do not present a serious contamination concern, because it is unlikely they will stay on chemical protective clothing. When exposed to chlorine gas, responders will need to go through a minimal decontamination reduction corridor. Liquid exposure to chlorine or compounds of chlorine may require a more extensive decontamination effort. Victims will require decontamination quickly to reduce damage to skin and eyes. Emergency decontamination would be appropriate by first responders if done from a safe distance, avoiding vapor and run-off. Exposure of victims to gas will result in minimal contamination. Removing clothing can limit the exposure to liquid chlorine and any gas that may be trapped in the victims' clothing (*Firehouse Magazine*).

When released from a container, chlorine is most concentrated at the point of the release. As with many gases and vapors, the concentration diminishes the farther away from the source you get. Evacuation and isolation distances found in the DOT ERG are based on computer modeling of chlorine releases. Isolation (hot zone) for small spills (those from a small container or a small leak from a large container) is 100 ft. From a large container (several small containers or a large leak from a large container), the recommended isolation distance is 200 ft.

Evacuation distances are categorized into day and night spills. This is because the environment tends to be more stable at night, meaning a cloud will stay together longer and travel farther before dissipating. The evacuation distance for small day or night spills is one-tenth of a mile. The evacuation distance is three-tenths of a mile for a large day spill and seven-tenths of a mile for a large night spill. Several factors influence the amount of time a cloud of gas will stay together, including temperature, humidity, and wind direction and velocity. Chlorine dissipates best in warm, windy weather. It is a common industrial chemical in the top 10 produced chemicals annually. It is used and transported to and through almost any community in the United States, and releases do occur (DOT, ERG).

Historic Chlorine Incidents

World War I Usage

On April 22, 1915, the German army released a large cloud of chlorine at Ypres, France, resulting in the deaths of 5,000 Allied soldiers and the injury of 10,000 more. It could also be a potential weapon for terrorists because of its common use and availability. As a result of the military use of chlorine, much data is available about human exposure and the expected effects, both long and short term. Release of chlorine from containers as an act of terrorism could be very effective in killing hundreds if not thousands of people (*Firehouse Magazine*).

Henderson, NV, 1991

A massive leak of liquefied chlorine gas occurred on May 6, 1991, in Henderson, NV, in a plant that produces chlorine gas from sodium chloride. More than 200 people sought aid at local hospitals for respiratory distress caused by inhalation of the chlorine, with 30 admitted for treatment. Several first responders and the battalion chief in charge were overcome by chlorine at the main entrance of the plant when they responded. Over 700 people were evacuated and taken to shelters with 2,000 to 7,000 others taken elsewhere. The alarm was delayed when plant employees thought they could handle the release internally. Fire department response was the result of several reports from the public of strong odor in the area. The release was caused by failure of pipes corroded by leaking acid from a heat exchanger that ate through the pipes, resulting in the release of more than 70 tons of chlorine (NASA).

Sun Bay, FL, 1998

Another incident occurred on Oct. 3, 1998, in Sun Bay South, FL. This was the sixth release from the same facility in three years. Chlorine is pumped from railroad cars into the facility and five of the six releases occurred during off-loading operations. The most recent incident was caused by a cap that burst. A dozen people experienced difficulty breathing and one employee experienced a burned trachea and other injuries.

St. Louis, MO, 2002

In St. Louis, MO, in August 2002, a chlorine release caused the injury of 63 people, including workers and nearby residents, during the off-loading of a chlorine railroad car. Once again, an automatic shutdown system failed to operate. Approximately 48,000 pounds of chlorine was released.

Atlanta, GA

In Atlanta a small-pressurized cylinder fell from a truck in the garage of the Hilton Hotel. The resulting leak of chlorine, a 2.3 poison gas, sent thirty-three people to the hospital including six firefighters and four police officers. Chlorine is one of the most common poison gases transported and stored. In addition to being a poison, chlorine is also a very strong oxidizer. Chlorine will behave much the same way as oxygen in accelerating combustion during a fire. Chlorine is also corrosive. It may be encountered in 150 lb. cylinders, 1-ton containers, and in tank truck and rail tank car quantities. Because chlorine is so common its hazards are sometimes taken for granted. There was a time when firefighters handled chlorine leaks with turnouts and SCBAs. That is no longer an acceptable practice. Poison gases pose a large threat to not only the public but also to emergency responders and the proper protective clothing and SCBA must be worn for protection. Chlorine is used as a swimming pool chlorinator, a water treatment chemical and has many other industrial uses.

North Carolina, 2003

Firefighters and hazmat personnel responded to a leak at a water distribution plant in North Carolina on June 13, 2003. Chlorine alarms in a pump house alerted personnel to the leak. No one was injured as hazmat personnel entered the facility and found a manifold connected to 150-pound cylinders of chlorine had malfunctioned. Responders were able to shut off valves on the cylinders stopping the leak. Pre-planning and following proper procedures led to a successful outcome.

Glendale, AZ, 2003

On Nov. 17, 2003, in Glendale, AZ, a leak occurred as chlorine was being loaded into a railroad car. The incident forced the evacuation of the surrounding area for hours. Fourteen people were treated for symptoms such as nausea, throat irritation and headaches. A preliminary investigation indicated that safety devices apparently failed. A full investigation has been initiated by the U.S. Chemical Safety Board (*Firehouse Magazine*).

Cryogenic Liquids LF/HR

Cryogenic liquefied gases are very cold liquids. The DOT definition of a cryogenic liquid, sometimes referred to as "refrigerated liquids," is any liquid with a boiling point below –130°F. Other sources list boiling points from –100°F to –200°F. If cryogenic liquids are shipped above 41 psi and have no other hazard, they are considered a compressed gas and would be placarded as a nonflammable compressed gas. Cryogenics may carry other placards such as flammable gas, poison gas or oxidizer. If cryogenics do not have any other placardable hazard, they are not considered a hazardous material by the DOT.

Materials listed under Hazard Class 2, which are shipped as liquefied gases, such as cryogenics, exhibit other hazards not indicated by the placard. Cryogenics, sometimes called refrigerated liquids, have boiling points of –130°F below zero or greater. Therefore, all cryogenic liquids are above their boiling points at ambient temperatures. Liquid helium has a boiling point of –452°F below zero, which is the coldest material known. It is also the only material on earth that never exists as a solid, only as a cryogenic liquid and as a gas.

Unlike propane and other liquefied gases, gases that are liquefied into cryogenics are liquefied through a process of alternating pressurization, cooling and ultimate decompression. Therefore, they do not require pressure to keep them in the liquid state. However, if they will be in a container for a long period of time, they are pressurized to keep them liquefied as long as possible. Nonpressurized cryogenics are kept cold by the temperature of the liquid and the insulation around the tanks. The cryogenic liquefaction process begins when gases are placed into a large processing container. They are pressurized to 1,500 psi. The process of pressurizing a gas causes an increase in heat. The molecules within a container move faster causing more collisions with each other and the walls of the container. As the molecules collide, heat is generated.

For example, the top of an SCBA bottle or an oxygen bottle becomes hot as it is being filled. This is a result of the molecules colliding in the bottles. Once the pressure of 1,500 psi is reached, the material is cooled to 32°F by using ice water. Once cooled, the pressure is once again increased, this time up to 2,000 psi, again accompanied by an increase in temperature. The material is then cooled to –40°F with liquid ammonia. Once the material is cooled, all of the pressure is released at once, and the resulting heat decrease turns the gases into cryogenic liquids. During the decompression process, the heat present within the container decreases as the pressure rapidly decreases.

Many of the gases found on the periodic table of chemical elements are extracted from the air and turned into cryogenic liquids. These include neon, argon, krypton, xenon, oxygen and nitrogen. All of the gases except oxygen are considered inert, that is to say, they are nontoxic, nonflammable and nonreactive. To extract these materials from the air, the air is first turned into a cryogenic liquid by the process mentioned previously. Then the liquid air is processed through a type of distillation tower where each component gas is extracted off as it reaches its own boiling point. Once extracted, the gases are then liquefied by the same process of alternate pressurization and decompression. Some other common materials that are made into cryogenics include flammable methane (LNG), hydrogen and oxidizers such as oxygen, fluorine and nitric oxide (*Firehouse Magazine*).

Chemical Notebook

Helium, He, LF/LR Helium is a gaseous, nonmetallic element of the noble gas family, family eight on the periodic table. It is a colorless, odorless and tasteless gas. It is nonflammable, nontoxic and nonreactive. It has a boiling point of –452°F. It is slightly soluble in water. Even though it is an inert gas, it can still displace oxygen and cause asphyxiation. The vapor density is 0.1785, which is lighter than air. Helium is derived from natural gas by liquefaction of all other components. It has a UN four-digit identification number of 1046 for the compressed gas and 1963 for the cryogenic liquid.

Neon, Ne, LF/LR Neon is a gaseous, nonmetallic element of the noble gas family. It is a colorless, odorless and tasteless gas. Neon is present in the earth's atmosphere at 0.0012% of normal air. It is nonflammable, nontoxic and nonreactive. It does not form chemical compounds with any other chemicals. It is, however, an asphyxiant gas and displaces oxygen in the air. The boiling point of neon is −410°F. It is slightly soluble in water. Neon has a vapor density of 0.6964, which is lighter than air. The UN 4-digit identification number is 1065 for the compressed gas and 1913 for the cryogenic liquid. Its primary uses are in luminescent electric tubes and photoelectric bulbs. It is also used in high-voltage indicators, lasers (liquid) and cryogenic research.

Argon, Ar, LF/LR Argon is a gaseous, nonmetallic element of family eight. It is present in the earth's atmosphere to 0.94% by volume. It is a colorless, odorless and tasteless gas. It does not combine with any other chemicals to form any compounds. The boiling point is −302°F. It is slightly soluble in water. The vapor density is 1.38 so it is heavier than air. The UN 4-digit identification number is 1006 for the compressed gas and 1951 for the cryogenic liquid. It is used as an inert shield in arc welding electric and specialized light bulbs (neon, fluorescent, and sodium vapor), in Geiger counter tubes and in lasers.

Krypton, Kr, LF/LR Krypton is a gaseous, nonmetallic element of family eight. It is present in the earth's atmosphere to 0.000108% by volume. It is a colorless and odorless gas. It is nonflammable, nontoxic and nonreactive. It is, however, an asphyxiant gas and can displace oxygen in the air. At cryogenic temperatures, krypton exists as a white, crystalline substance with a melting point of 116°K. The boiling point of krypton is −243°F. Krypton is known to combine with fluorine at liquid nitrogen temperature by means of electric discharges or ionizing radiation to make KrF_2 or KrF_4. These materials decompose at room temperature. Krypton is slightly water soluble. The vapor density is 2.818, which is heavier than air. The UN 4-digit identification number is 1056 for the compressed gas and 1970 for the cryogenic liquid. It is used in incandescent bulbs, fluorescent light tubes, lasers and high-speed photography.

Xenon, Xe, LF/LR Xenon is a gaseous, nonmetallic element of family eight. It is a colorless and odorless gas or liquid. It is a gas at standard temperatures and pressures. It is nonflammable and nontoxic, but it is an asphyxiant and displaces oxygen in the air. The boiling point is −162°F. The vapor density is 05.987, which is heavier than air. It is chemically unreactive; however, it is not completely inert. The UN 4-digit identification number is 2036 for the compressed gas and 2591 for the cryogenic liquid. Xenon is used in luminescent tubes, flash lamps in photography and lasers and as an anesthesia.

Xenon Compounds LF/LR xenon combines with fluorine by mixing the gases, and the mixture is heated in a nickel vessel to 400°C, and cooled. The resulting compound is xenon tetrafluoride, XeF_4. The resulting product is composed of large colorless crystals. Compounds of xenon difluoride, XeF_2, and hexafluoride, XeF_6, can also be formed in a similar manner. The hexafluoride compound melts to a yellow liquid at 122°F and boils at 168°F. Xenon and fluorine compounds also combine with oxygen to form oxytetrafluoride, $XeOF_4$, which is a volatile liquid at room temperature. These compounds with fluorine must be protected from moisture to prevent the formation of xenon trioxide, XeO_3, which is a dangerous explosive when dried out. The solution of xenon trioxide is a stable weak acid that is a strong oxidizing agent.

Expansion Ratios

Cryogenic liquids have very large expansion ratios, some as much as 900 or more to 1; they can form massive vapor clouds (Figure 3.84). These vapor clouds can obscure vision of the source of the leak making the location of the source difficult to find. Vapor clouds from cryogenic liquids can travel great distances and require evacuation of the public. The visible cloud is not the total extent of the hazard. The warmer air on the outer edge of the vapor cloud causes the gas to become invisible. It is then possible to be in an oxygen-enriched atmosphere, a flammable atmosphere or a gas that can cause asphyxiation and the gases will not be visible. Because of the large liquid-to-gas expansion ratios, the other hazards of cryogenic liquids are magnified many times. If the cryogenic liquid is flammable or toxic, these hazards are intensified because of the potential of large gas cloud production from a very small amount of liquid. As the size of the leak increases, so does the size of the vapor cloud. This means that 1 gallon of a cryogenic liquid can produce as much as 900 gallons of a gas.

Cryogenic Expansion Ratios

Fluorine	981 to 1
Oxygen	862 to 1
Argon	841 to 1
Hydrogen	840 to 1
Helium	754 to 1
Nitrogen	697 to 1
LNG	637 to 1

Figure 3.84 Cryogenic liquids have very large expansion ratios.

Because of the large expansion ratios, cryogenic liquids can turn into gases and displace oxygen in the air which can harm responders by simple asphyxiation. Simple asphyxiation is not poisoning, but rather just not enough oxygen to breathe. The atmosphere contains about 21% oxygen. When the oxygen in the lungs, and ultimately in the blood, is reduced, unoxygenated blood reaches the brain and the brain shuts down. It may only be a few seconds between the first breath and collapse. Response personnel should always put SCBA on prior to entering a confined space or other area where asphyxiant gases may be present.

Hazards to Responders

Being very cold, cryogenic liquids can cause frostbite and solidification of body parts. When the parts thaw out, the tissue is irreparably damaged. Touching uninsulated piping and valves of cryogenic liquid containers can cause skin to stick to the metal, much like when kids put their tongues on an ice cube tray or fence post in the winter and it gets stuck. There isn't any protective clothing or equipment that can be worn to protect the body from the effects of contact with a cryogenic liquid. Anything that the cryogenic liquid contacts will become solidified and brittle. Gloves can provide protection from skin contacting piping and valves. Body parts that have come in contact with cryogenics should be treated like frostbite. Parts should be flushed with water that is cool to luke warm to limit additional tissue damage.

Chemical Notebook

Nitrogen, N2, LF/LR Nitrogen is a gaseous, nonmetallic element. It is a colorless, odorless and tasteless gas that makes up 78% of the air that is breathed. The boiling point of nitrogen is –320°F. It is slightly soluble in water. Nitrogen does not burn and is nontoxic. It may, however, displace oxygen and be an asphyxiant gas. The vapor density of nitrogen is 0.96737, which makes it slightly lighter than air. The UN 4-digit identification number is 1066 for the compressed gas and 1977 for the cryogenic liquid. As a cryogenic liquid, the NFPA 704 designation is health—3, flammability—0 and reactivity—0. Nitrogen is used in the production of ammonia, cyanides and explosives, as an inert purging agent and as a component in fertilizers. It is usually shipped in insulated containers, insulated MC 338 tank trucks and tank cars.

Oxygen, O2, LF/HR Like nitrogen, oxygen is a nonmetallic elemental gas. Oxygen makes up approximately 21% of the air breathed. The boiling point of oxygen is –297°F. It is nonflammable but supports combustion. Oxygen can explode when exposed to heat or organic materials. The vapor density of oxygen is 1.105, which makes it slightly heavier than air. Oxygen is incompatible with oils, grease, hydrogen, flammable liquids, solids and gases. The UN 4-digit identification number for oxygen is 1072 for the compressed gas and 1073 for the cryogenic liquid. The NFPA 704 designation for liquid oxygen is health—3, flammability—0 and reactivity—0. Liquid oxygen is shipped in Dewier flasks and MC 338 tank trucks. It may also be encountered in cryogenic rail cars.

Carbon Dioxide, CO_2, LF/LR   Carbon dioxide is a colorless and odorless gas. It can also be a solid (dry ice), which will undergo sublimation and turn back into carbon dioxide gas, or a cryogenic liquid. It is miscible with water. It is not flammable or toxic but can be an asphyxiant gas and displace oxygen. In 1993 two workers were killed on board a cargo ship when a carbon dioxide fire extinguishing system discharged. The oxygen in the area was displaced by the carbon dioxide and the men were asphyxiated. Carbon dioxide has a vapor density of 1.53, which is heavier than air. It may be shipped as a cryogenic or liquefied compressed gas. Frozen (solid) carbon dioxide gas is known as dry ice. It has a UN 4-digit number of 2187 for the cryogenic and 1013 for the compressed gas. The NFPA 704 designation is health—3, flammability—0 and reactivity—0. It is used primarily in carbonated beverages and fire extinguishing systems.

When responding to fires involving liquefied gas tanks, firefighters often apply water to cool the tanks. Cryogenic liquids are already colder than water at any temperature, and water will act as a superheated material causing the cryogenic to heat up and vaporize faster. This will cause pressure to build up inside the tank, and the tank may fail violently. Care should be taken when applying water to "cool" cryogenic containers.

Hydrogen, H_2, LF/HR   Hydrogen is a clear, odorless and tasteless gas. It is lighter than air, and in a leak it rises up and dissipates in the air rather than spilling on the ground or surfaces. Hydrogen was used at one time in the early 1900s for dirigibles (airships) because it is significantly lighter than air. Hydrogen was used because of its high lifting capacity, availability and low cost. However, it is highly flammable as was demonstrated when the *Hindenburg* caught fire on May 6, 1937, in New Jersey while attempting to dock with its mooring mast at Naval Air Station Lakehurst and was destroyed. On board were 97 souls, 36 passengers and 61 crewmen. There were 36 fatalities, 13 passengers, 22 crewmen and one working on the ground. Helium gas has the same attributes as hydrogen; however, it is not flammable. Another drawback is that helium is rare and very expensive. Most airships built since the 1960s have used helium, and some used hot air.

Besides being lighter than air and extremely flammable, it also burns with an almost invisible flame. This can be a danger and a challenge for

Figure 3.85 Fire departments that have a known hazard of hydrogen in their jurisdictions carry a straw broom; as they approach a potential hydrogen fire, the broom will ignite before the firefighters reach the flames.

emergency responders. Fire departments that have a known hazard of hydrogen in their jurisdictions carry a straw broom; as they approach a potential hydrogen fire, the broom will ignite before the firefighters reach the flames (Figure 3.85). Hydrogen is used by industries for treating metals, petroleum refining process and the production of chemicals. It also has been used as a fuel in the space program.

Hydrogen is sometimes shipped and stored as a cryogenic liquid and has a boiling point of –423°F. There are very significant hazards to firefighters responding to a leak. First of all, they are very cold and have a wide liquid-to-vapor expansion ratio of 1 gallon of liquid turns into 848 gallons of gas. Hydrogen is not toxic but an asphyxiation hazard. Hydrogen has a very wide flammable range, a flashpoint of –423°F and an ignition temperature of 1,085°F. While hydrogen will burn almost any time and any place, great care should be taken to keep ignition sources away. Comparing other fuel ignition temperatures between methane at 1,003°F, propane at 914°F, ethanol at 369°F and gasoline at 450°F–900°F, shows a wide variance depending on the brand and additives (Hazardous Materials Chemistry for Emergency Responders).

Anhydrous Ammonia—HF/HR Anhydrous ammonia (NH_3) is one of the leading sources of nitrogen fertilizer applied to agricultural crops in the United States. It is used as a fertilizer because by molecular weight, it is 82% nitrogen, readily available and easy to apply. There are also disadvantages and potential dangers involved in handling and use of anhydrous ammonia. Ammonia gas is compressed into a liquid for storage and use. It is stored and handled under pressure, which requires the use of specially designed and well-maintained equipment. At 60°F ammonia has a vapor pressure of 93 psi, whereas at 100°F the vapor pressure is 200 psi. As the temperature of the ammonia rises, so does the vapor pressure. Liquid anhydrous is a very cold liquid, with a boiling point of –28°F, but a high-pressure flashing liquid release can produce temperatures of –60°F or lower, and can cause serious thermal burns very quickly.

Liquid ammonia when released quickly returns to the gas state at the expansion rate of 850 gallons of ammonia gas for every 1 gallon of liquid. Additionally, employees of suppliers, end users and emergency responders need to be adequately educated about the procedures and personal protective equipment (PPE) necessary to safely handle and work around anhydrous ammonia. Because anhydrous ammonia is so common in many parts of the country, complacency may develop during dispensing, transportation and use of anhydrous ammonia. Ammonia is primarily transported in rural areas from the distributor to the farm in nurse/applicator tanks usually pulled behind farm pickup trucks (Figure 3.86). The liquid capacity of these tanks ranges from 1,000 to 3,000 gallons, and it is estimated that there are more than 200,000 of these tanks in use nationwide.

During filling of the tanks, they are required to be no more than 85% full of ammonia. Some nurse tank trailers are equipped with two tanks on the same trailer. State law controls the number of trailers that can be towed at one time over the public roads and the speed of travel. The number of nurse tanks allowed to be towed ranges from one to three depending on the state law. Ammonia nurse tanks are required to have a minimum of 5

Figure 3.86 Ammonia is primarily transported in rural areas from the distributor to the farm in nurse/applicator tanks usually pulled behind farm pickup trucks.

gallons of water on the nurse tank, which is usually in a red plastic tank on top of the nurse tank. This water is used for emergency flushing eyes or body when it has contacted ammonia.

Anhydrous ammonia is classified by the DOT as a Division 2.2 nonflammable gas (Figure 3.87). Unfortunately, this classification leaves three other very important hazards of anhydrous ammonia unidentified by the DOT placard and label system. Not only will anhydrous ammonia burn under certain conditions, but it is classified as a caustic (corrosive) liquid and poison gas in other parts of the world. MSDSs from the manufacturers of anhydrous ammonia in the United States identify its hazards as flammable, toxic and corrosive. Anhydrous means "without water." Anhydrous does not apply to just ammonia, but other chemicals may also have the term "anhydrous" in their name, which means the same thing, "without water."

Anhydrous ammonia (NH_3) is a colorless liquefied compressed gas that is free of water. Because anhydrous is free of water, it has a high affinity for water. Thirteen hundred gallons of ammonia vapor will dissolve in just 1 gallon of water. Ammonia has a very sharp, intensely irritating odor. Anyone in the area of a release will not want to stay! Ammonia is corrosive to galvanized metals, copper and copper alloys. Ammonia gas is lighter than air, but also very close to the weight of air with a density of 0.6, so cold vapors or dense aerosol clouds may stay close to the ground or in low-lying areas. It has an autoignition temperature of 1,204°F and a flammable range of 16%–25%.

Figure 3.87 Anhydrous ammonia is classified by the DOT as a Division 2.2 nonflammable gas.

DOT does not classify ammonia as flammable gas because it doesn't meet the definition used for flammable gases. According to DOT, a flammable gas has a lower explosive limit below 13 or a flammable range of greater than 12 percentage points. Ammonia misses the definition on both counts. Ammonia has a lower explosive limit of 16, 3 points more than the DOT requirement for flammable gas, and the flammable range is 10 percentage points, not the 12 required by DOT's definition. It does, however, burn and has caused injury and two known firefighter fatalities, in the past when it ignites. Normally, ammonia needs to be inside of a building or confined space to ignite. It does not usually burn outside in the open. However, during a tanker crash in Sacramento, CA, ammonia trapped under an overpass did ignite.

Anhydrous ammonia is considered to be toxic with a NIOSH IDLH of 300 ppm in air. According to the Centers for Disease Control (CDC), inhalation of concentrated fumes at the rate of 5,000–10,000 ppm for short periods may be fatal. Additionally, exposure to 2,500–6,000 ppm for 30 min or greater is considered dangerous to life. Responders will require appropriate PPE including chemical protective clothing and SCBA to protect themselves or to perform rescue. Most reference sources recommend Level A gas/vapor suits for chemical protection. Chemical protective clothing is not usually readily available to most rural departments.

Common procedure at anhydrous ammonia fertilizer facilities is to vent hoses with ammonia liquid or vapor into a plastic or steel bleed-off water tank to absorb ammonia into the water. Typical plastic tanks have a 275-gallon liquid capacity with a cap on top, but they are not pressure tanks. By venting the ammonia into the water, the amount released into the air is minimal. This process while legal has its hazards and limitations. First of all, there is a physical limit to the amount of ammonia that can be absorbed into the water. A gallon of water will only absorb 1–2 lb of ammonia before the water becomes saturated. If the water in the tank is not changed once this saturation occurs, little additional liquid or vapor can be absorbed. This may result in an overpressurization of the tank, which may result in an explosive tank failure if the tank is not vented. Second, when ammonia is absorbed by water, a caustic chemical called ammonium hydroxide or aqua ammonia is formed.

Mild exposure to anhydrous can cause irritation to eye, nose and lung tissues. When NH_3, combines with moisture in the lungs, it causes severe irritation. Ammonium hydroxide is actually produced in the lungs. Prolonged breathing of high ammonia concentrations can severely damage lung tissue and cause suffocation. The human eye is a complex organ made up of nerves, veins and cells. The front of the human eye is covered by membranes, which resist exposure to dust and dirt. None of these can keep out anhydrous ammonia, because the entire eye is about 80%

of water. A shot of ammonia under pressure can cause extensive, almost immediate damage to the eye. Ammonia extracts the fluid and destroys eye cells and tissue in minutes. If you get a shot of anhydrous ammonia in your eye, the first few seconds are crucial. Immediately flush the eyes with copious amounts of water. If wearing contact lenses (contacts are NOT recommended when working with ammonia), remove them. Your eyes will fight to stay closed because of the extreme pain, but they must be held open so the water can flush out the ammonia.

Eyes doused with ammonia close involuntarily. They must be forced open so water can flush the entire eye surface and inner lining of the eyelids. Continue to flush the eyes for at least 15 min. Get professional medical help as soon as possible to prevent permanent damage. If water is not available, fruit juice or cool coffee can be used to flush the eyes. Remove contaminated clothing and thoroughly wash the skin. Clothing frozen to skin by liquid ammonia can be loosed with liberal application of water. Wet clothing and body thoroughly and then remove the clothing. Never put an exposed victim into an ambulance without first thoroughly decontaminating them and any clothing with copious amounts of water.

Leave burns exposed to the air and do not cover with clothing or dressings. Immediately after first-aid treatment with water, get the burn victim to a physician. Do not apply salves, ointments or oils—these cause ammonia to burn deeper. Let a physician determine the proper medical treatment. Remove the victim to an area free from fumes if an accident occurs. If the patient is overcome by ammonia fumes and stops breathing, get them to fresh air and give artificial respiration. The patient should be placed in a reclining position with head and shoulders elevated.

Basic life support should be administered if needed. Oxygen has been found useful in treating victims who have inhaled ammonia fumes. Administer 100% oxygen at atmospheric pressure. Any person who has been burned or overcome by ammonia should be placed under a physician's care as soon as possible. Begin irrigation with water immediately. The rescuer should use freshwater if possible. If the incident is a farm accident, there is a requirement for water tanks for irrigation of the eyes and body on the anhydrous ammonia tank. Open water in the vicinity of an anhydrous ammonia leak may have picked up enough NH_3 to be a caustic aqua ammonia solution. This could aggravate the damage if used in the eyes or for washing burns. The victim should be kept warm, especially to minimize shock. If the nose and throat are affected, irrigate them with water continuously for at least 15 min. Take care not to cause the victim to choke. If the patient can swallow, encourage drinking lots of some type of citrus drinks such as lemonade or fruit juice. The acidity will counteract some of the effects of the anhydrous ammonia.

Response to anhydrous ammonia emergencies can present many challenges to emergency responders. Ammonia is a colorless gas, so there may not be any visual indications of where the gas is. There are things to watch for. Ammonia gas will quickly turn vegetation brown. If it's a time of year where the vegetation is expected to be green, then watch for brown vegetation. You can also watch for animal or bird kill, which may have resulted from exposure to the ammonia gas in a release. Ammonia also has a strong odor; you can smell it before reaching a lethal dose. The odor threshold is 5–20 ppm, well below dangerous levels. However, as with all hazardous materials, responders should not be in a position to smell materials. Firefighter turnouts do not provide any protection from ammonia gas or liquid. SCBA will protect the respiratory system. Ammonia vapors will seek out locations on the bodies of responders where there is moisture. The eyes are a major concern as they can be damaged or blindness can occur from ammonia contact. Areas in the groin and armpits are also potential moisture spots. However, firefighters in full turnouts can sweat, and moisture can be present on any part of the body, depending on ambient temperatures. First responders in firefighter turnouts should avoid contact with ammonia vapors or liquid. Because ammonia has a great affinity for water, first responders can use hose streams to decontaminate victims exposed to ammonia vapors or liquid. They can also use fog streams to dissolve ammonia gas from the air to protect victims or those in harm's way. Remember, however, that water and ammonia form ammonium hydroxide, which is a corrosive liquid, and after victims receive emergency decontamination, efforts should be made to control the runoff.

Anhydrous ammonia can cause corrosion on some metals, particularly copper, brass or galvanized steel. Many parts of fire apparatus and firefighting equipment are made of brass, which can be damaged if in contact with anhydrous. Like anhydrous ammonia, LPG is stored in the same type of steel tanks. A high percentage of agricultural anhydrous ammonia accidents are the result of using improper procedures, lack of training in equipment operation or failure to follow prescribed practices. Emergency responders can avoid additional injuries and death with the proper planning, training and equipment to effectively handle agricultural ammonia emergencies (*Firehouse Magazine*).

Historic Anhydrous Ammonia Incidents
Crete NE February 19, 1969 Derailment and Anhydrous Ammonia Release

February 18, 1969 set up as a "perfect storm" for a train derailment in Crete, Nebraska, a small college town of approximately 4,500 people. Temperature was 4°F, wind was calm, and relative humidity was 90%, approximately 14 inches of snow on the ground a temperature inversion was in place and ground fog. At 6:30 a.m. CST Chicago, Burlington and Quincy (CB&Q) Train #64, consisting of three locomotive units and 95 cars, were entering town. Train #824 with one locomotive and 49 cars were standing on a siding North of the main line. It contained three tank cars of Anhydrous Ammonia. Three Crete residents died during the accident, three died later in the hospital. Three unidentified transients riding the train were killed by trauma during the derailment. Injury reports varied, however, the Crete News, the local paper, lists approximately 25, although the NTSB reported 53 in its final report. Fog and anhydrous ammonia vapors made it difficult to see what was going on. They did not know what the fog was, but some firefighters indicated there was a smell of ammonia in the air. (Anhydrous ammonia is heavier than air and tends to pool in low places and on the ground.) Turnout gear was limited as were SCBA (*Firehouse Magazine*).

> **Author's Note:** *It is interesting to note that the house number 1005, 13th Avenue is located adjacent to the cleaners where two victims died and just west of 13th and Unona where one person died. The person at 13th and Unona who died on the street corner lived at 813, 13th Street. The house number "1005" is the UN/DOT designation for anhydrous ammonia in bulk quantities, which was not used at the time of the derailment. Had this incident happened at any other time of the year, it would have likely been much worse. In warmer weather, people would have had less clothing on, windows might have been opened in the dwellings and more people may have gone outside to see what happened.*

Tecumseh, NE March 2014 Anhydrous Ammonia Release

During March of 2014 a release of anhydrous ammonia occurred at the Midwest Farmers Cooperative facility in Tecumseh, Nebraska. An anhydrous ammonia semi-transport driven by an employee of Midwest had been positioned at the ammonia bulk plant in preparation for off loading the liquid ammonia into the bulk plant. During that off-loading process an explosion reportedly occurred which resulted in the release of approximately 100 pounds into the atmosphere. One employee of Midwest was killed and three other people were injured including a deputy sheriff. This incident was investigated by the Nebraska State Fire Marshal's Office. Anhydrous ammonia is considered an extremely hazardous substance by the Environmental Protection Agency (EPA) and releases of 100 pounds or more must be reported to the National Response Center (NRC) within 24 hours. This liquid is very corrosive to skin and can cause serious chemical burns Initial reports of the incident in Tecumseh indicated an explosion had occurred resulting in the release of the ammonia vapors (Nebraska State Fire Marshal Investigation Report).

Confined Space Gases

Chemical Notebook

Hydrogen Sulfide, H_2S, HF/HR Hydrogen sulfide is a nasty actor in terms of hazardous materials incidents. It is a colorless, flammable, and extremely hazardous gas with a "rotten egg" smell. It occurs naturally in crude petroleum and natural gas, and can be produced by the breakdown of organic matter and human/animal wastes (e.g., sewage). It is heavier than air and can collect in low-lying and enclosed, poorly ventilated areas such as basements, manholes, sewer lines and underground telephone/electrical vaults.

Odor Threshold

- It can be smelled at low levels, but with continuous low-level exposure or at higher concentrations, you lose your ability to smell the gas even though it is still present.
- At high concentrations, your ability to smell the gas can be lost instantly.
- DO NOT depend on your sense of smell for indicating the continuing presence of this gas or for warning of hazardous concentrations.

Health effects vary with how long, and at what level, you are exposed. Asthmatics may be at greater risk.

- **Low concentrations**—irritation of eyes, nose, throat or respiratory system; effects can be delayed
- **Moderate concentrations**—more severe eye and respiratory effects, headache, dizziness, nausea, coughing, vomiting and difficulty breathing
- **High concentrations**—shock, convulsions, unable to breathe, coma and death; effects can be extremely rapid (within a few breaths)

Before entering areas with possible hydrogen sulfide,

- The air needs to be tested for the presence and concentration of hydrogen sulfide by a qualified person using test equipment. This individual also determines if fire/explosion precautions are necessary.
- If gas is present, the space should be ventilated.
- If the gas cannot be removed, use appropriate respiratory protection and any other necessary PPE, rescue and communication equipment. Atmospheres containing high concentrations (greater than 100 ppm) are considered immediately dangerous to life and health (IDLH), and an SCBA is required.

Exposure limits for H_2S vary widely as a function of jurisdiction and workplace activity. The most widely recognized standards for H_2S reference an 8 h TWA of 10 ppm, and a 15 min STEL of no more than 15 ppm. The American Conference of Governmental Industrial Hygienists (ACGIH) TLV for H_2S is much more conservative. It consists of an 8 h TWA limit of 1.0 ppm and a 15-min STEL of 5.0 ppm. When in doubt, be conservative! Concentrations above 100 ppm should be regarded as immediately dangerous to life and health, with the potential for causing irreversible physiological harm to the exposed individual. Many monitoring programs use instruments with the alarms set to sound immediately if the concentration reaches 10 ppm, in which case the workers immediately leave the affected area. This approach essentially eliminates the potential for ever-reaching STEL or TWA exposure limits.

Carbon Monoxide Carbon monoxide is a colorless, odorless and highly toxic gas that is produced as a by-product of incomplete combustion. It bonds to the hemoglobin molecules in red blood cells, preventing them from properly transporting oxygen. CO is potentially present whenever combustion occurs. It is particularly associated with internal combustion engine exhaust. Carbon monoxide can be generated by hot work that involves combustion, operating internal combustion engines within the confined space, or introduced into the space by improper use of ventilation equipment. Vehicle exhaust has been implicated in many accidents. Verify that blowers and ventilation equipment introduce only fresh air into the space, and that atmosphere evacuated from the space is vented safely.

Carbon monoxide is a chronically toxic gas. Prolonged or repeated exposure to relatively low concentrations of CO can eventually lead to injury, illness or death. Although high concentrations of carbon monoxide may be acutely toxic and lead to immediate respiratory arrest or death, it is the long-term physiological effects due to chronic exposure at lower concentrations that take the greatest toll of affected workers. Even when exposure levels are too low to produce immediate symptoms, small repeated doses can reduce the oxygen-carrying capacity of the blood over time to dangerously low levels. This partial impairment of the blood supply may lead to serious physiological consequences over time.

OSHA permissible exposure limits (PELs are published in 29 CFR 1910 Subpart G, "Occupational Health and Environmental Control," or Subpart Z, "Toxic and Hazardous Substances." If the toxic gas concentration exceeds the PEL, the atmosphere is hazardous.

The OSHA PEL for carbon monoxide is 50 ppm calculated as an 8 h TWA limit. The NIOSH recommended exposure limit (REL) consists of a two-part definition, an 8 h TWA limit of 35 ppm and a ceiling limit of 200 ppm. The ACGIH TLV for CO is 25 ppm calculated as an 8-h TWA.

Methane Methane (CH_4) is a colorless, odorless, tasteless, and flammable gas that is widely distributed in nature. Methane gas is produced whenever organic material is decomposed by bacterial action in the absence of oxygen. The atmosphere contains about 2.2 ppm by volume of methane. Like radon, methane gas can migrate significant distances under the ground surface and be forced into adjacent buildings by the pressure gradient between the soil and the building interior. Points of entry include floor/wall cracks, floor drains, sewer pipe entry points as well as utility access penetrations. Gas can accumulate inside wall cavities, crawl spaces, inside sumps and in poorly ventilated basement areas.

Methane Toxicity/Fire/Explosion Risk Methane gas is relatively nontoxic; it does not have an OSHA PEL standard. Its health effects are associated with being a simple asphyxiant displacing oxygen in the lungs. Miners previously placed canaries in deep mines to check methane gas levels. Reportedly, canaries keeled over at about 16% oxygen, indicating that it was time to leave.

Methane is extremely flammable and can explode at concentrations between 5% (lower explosive limit) and 15% (upper explosive limit). These concentrations are much lower than the concentrations at which asphyxiant risk is significant. Reportedly, the most violent methane explosions occur at concentrations of about 9%; coal mines are thus kept well ventilated (pumped with fresh air) to maintain methane levels at or below 1%.

Oxygen Deficiency Oxygen-deficient atmospheres are the leading cause of confined space fatalities in the shipyard. Although normal atmosphere contains between 20.8% and 21% oxygen, OSHA defines any atmosphere that contains less than 19.5% oxygen as oxygen deficient and any atmosphere that contains more than 22% as oxygen enriched.

Oxygen-deficient atmospheres may be created when oxygen is displaced by inert gases, such as carbon dioxide, nitrogen or argon, or the ship's inert gas system or firefighting system. Oxygen can also be consumed by rusting metal, ripening fruits, drying paint, coatings, combustion or bacterial activities.

Oxygen-enriched atmospheres may be produced by certain chemical reactions, but in a shipyard they are typically caused by leaking oxygen hoses and torches. They present a significant fire and explosion risk (Hazardous Materials Chemistry for Emergency Responders).

Compressed Gas Containers

Because of the difficulty in controlling leaks and releases of gases, it is important to know the vapor density of the gas (Figure 3.88). Temperature can have a significant effect on a liquefied gas within a pressure container.

VAPOR DENSITY

Density of a gas or vapor compared to air. Air = 1

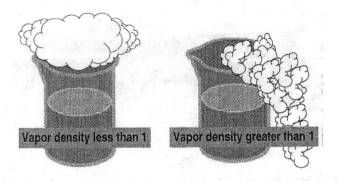

Figure 3.88 Vapor density of the gas.

As temperatures around the container increase, the temperature within the container increases. The boiling point of a liquefied compressed gas or cryogenic liquid determines the point at which a gas is produced from the liquid. Most boiling points of these materials are well below any ambient temperatures commonly encountered by responders. Therefore, with liquefied compressed gases, the liquid in the tank already exists above its boiling point, so the increased temperature increases the rate of boiling of the liquid.

This increase in boiling rate causes an increase in the pressure within the tank because more gas is being released from the liquid. Although most liquefied gas tanks have relief valves to relieve excess pressure, they are only designed to release pressure from changes in ambient temperature. If heat from a fire or direct flame impingement on a tank is the cause of the temperature increase, the relief valve may not be able to keep up with the increased pressure and the tank can fail violently, known as a BLEVE. Just because a relief valve on a tank or container is functioning, it does not mean that the incident is stabilized. When a relief valve functions on a pressure tank, it means there is a buildup of pressure within the tank. Under certain conditions, the pressure buildup can occur at a rate greater than what the relief valve can remove. At that point the tank can come apart.

Containers

Highway **MC 331** tanks are high-pressure containers used to haul liquefied gases including propane, LPG, chlorine anhydrous ammonia, and others (Figure 3.89). Propane and butane are two very common flammable

Figure 3.89 MC 331 tanks are high-pressure containers used to haul liquefied gases including propane, LPG, chlorine and anhydrous ammonia.

liquefied compressed gases and have boiling points of −42.5°C and −0.5°C, respectively. Both materials are above their boiling points under ambient temperature conditions in many parts of the country year-round. This makes the materials very dangerous when a leak or fire occurs, especially if there is flame impingement on the container. Because the materials are already above their boiling points, flame impingement, radiant heat transfer or increases in ambient temperature can cause the materials to boil faster. Faster boiling causes an increase in vapor pressure within the container.

Even though the containers are specially designed to withstand pressure and have relief valves to release excess pressure, there are limits to the pressure they can tolerate. If the pressure buildup in the container exceeds the ability of the tank to hold the pressure or the relief valve to relieve the pressure, the container will fail. MC 331 tankers are equipped with temperature, pressure and liquid level gauges. These can be helpful in determining what is going on in a tank that is not leaking. If a tank is leaking or on fire, it is too dangerous to worry about these gauges unless you have a drone. Gauges give you a visual indication that there is a leak. The temperature in the container can be monitored along with the tank pressure. Tanks can be cooled when there is no fire, and the temperature gauge can be used to monitor the cooling effectiveness.

Home Delivery Trucks (Bobtail)   Home delivery is often the final destination of the transportation of LPGs, utilizing bobtail trucks with a capacity of up to 3,500 gallons (Figure 3.90). These vehicles are equipped with the same safety features as their over-the-road counterparts. In addition to the emergency valve, shut off valves on the vehicle drivers often carry a

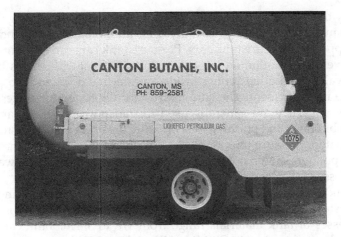

Figure 3.90 Home delivery is often the final destination of the transportation of LPGs, utilizing bobtail trucks with a capacity of up to 3.500 gallons.

remote shut off as well. When responding to bobtail truck accidents of fires, treat them as any other propane transport vehicle.

Railcar Amounts of hazardous materials in a tank car can range from a few hundred gallons to as much as 34,500 gallons (no tank car is considered empty unless it has been purged of the product and cleaned (Figure 3.91). Rail tank cars are assigned identification numbers, which

Figure 3.91 Amounts of hazardous materials in a tank car can range from a few hundred gallons to as much as 34,500 gallons (no tank car is considered empty unless it has been purged of the product and cleaned.

identify the type of tank in terms of pressure, nonpressure, cryogenic and miscellaneous. Liquefied gases anhydrous ammonia, chlorine and others are shipped in pressure tank cars. Pressure tank cars can be identified by the valve configuration on the top center of the car. These cars have all of the valves and other piping inside a protective dome to help prevent damage in an accident. They are hydrostatically tested for a pressure of 100–600 psi. They are generally used to transport flammable and non-flammable gases and poison gases. Pressure tank cars may be insulated or uninsulated. Pressure tank cars are top loaded through the dome assembly. Responders should not expect to find bottom connections or washouts on a pressure tank car.

Tank cars are constructed using a variety of materials including carbon steel, aluminum, stainless steel, nickel, chromium and iron. Single thicknesses of tank materials range from 1/8 to 3/4 in. Tank car design standards are found in 49 CFR Part 179 of the DOT Hazardous Materials Regulations. Modifications may occur to tanks that accommodate products transported because of product temperature, flammability or chemical reactivity. Tanks may be insulated externally to protect against the effects of ambient temperature. Insulation materials can include fiberglass, polyurethane and pearlite, and in some older tank cars, cork is used.

Tanks may also be provided with thermal protection designed to keep tank temperatures below 800°F during a 100 min pool fire or a 30 min torch impingement test (Figure 3.92). This protection is provided by a

Figure 3.92 Tanks may also be provided with thermal protection designed to keep tank temperatures below 800°F during a 100 min pool fire or a 30 min torch impingement test.

Figure 3.93 Tank cars are also protected from physical damage during an accident with the use of shelf-couplers and head shields.

layer of wool or ceramic fiber covered by a 1/8 in. steel jacket. Thermal protection can also be provided through a textured coating sprayed onto the tank's outer surface. Heat from a flame exposure is absorbed by the coating material and not transferred to the tank metal. Metal does not hold up well to direct flame contact and will fail in 15–20 min according to the information from the National Fire Protection Association. Tank cars are also protected from physical damage during an accident with the use of shelf-couplers and head shields (Figure 3.93). Shelf-couplers are designed to stay together so they do not puncture another car during a derailment or other accident. Head shields are provided to give an extra layer of metal to help prevent dents and punctures when a pressure tank is hit by another car or object during an accident.

With the exception of certain poison gases, pressure tanks have relief valves designed to relieve excess pressure caused by increases in ambient temperature. These relief valves are not designed to relieve the pressure created from radiant heat from a fire or other source or direct flame impingement. Many accidents have occurred over the years from BLEVEs when pressure containers have had excess pressure buildup. These accidents have resulted in many firefighter and civilian deaths and injuries when these rail tank cars experienced flame impingement and BLEVEd.

Fixed Facilities Anhydrous ammonia and LPG are common materials stored in fixed pressure tanks. Tanks range in size from the 5-gallon containers used with barbecue grills and 250-gallon tanks used for home heating to bulk storage tanks containing thousands of gallons of product at propane facilities. Propane tanks generally have rounded ends, which is a primary indictor of a pressure vessel. Pressure tanks are equipped

Figure 3.94 Pressure tanks are equipped with relief valves to vent excess pressure caused by increases in ambient temperature.

with relief valves to vent excess pressure caused by increases in ambient temperature (Figure 3.94). The normal relief valve height is between 6 and 10 in. They have extensions on the relief valves to extend vapors well above the tank in the event the vapors catch fire. This is done to keep flame impingement off of the top of the tank where the vapor space exists.

LPG valves are usually locked to prevent tampering. They have excess flow valves on the belly of the tank that is designed to shut off the flow if a pipe or valve is broken off. Fighting fire in a fixed storage tank is no different than any other propane container. Don't take BBQ tanks and other home heating tanks lightly. Just because the amount of LPG is smaller, it doesn't mean that the danger to responders is less. All of the same rules for fighting fires with LPG tanks apply regardless of the tank size (*Firehouse Magazine*).

Cryogenic Containers

Highway **MC 338** tanks are used for the transportation of cryogenic gases, sometimes referred to as refrigerated liquids (Figure 3.95). These materials are very cold with boiling points of –130 for carbon dioxide to –452 for liquid helium. Common cryogenics include oxygen, nitrogen, helium, argon, and others. Many of the materials carried in 338 tanks are considered inert gases. That is to say, they do not readily react chemically to other materials, are not flammable and are nonpoisonous. They do, however, have significant hazards when released as a liquid or a gas. Liquids are extremely cold and can cause frostbite and solidification of anything it contacts including body parts. They have large expansion

Figure 3.95 MC 338 tanks are used for the transportation of cryogenic gases, sometimes referred to as refrigerated liquids.

ratios, producing huge amounts of vapor from a small spill. In some cases, as little as 1 gallon of a cryogenic liquid can produce over 900 gallons of gas. Although these gases are inert in many cases, they can still displace the oxygen in the air and cause simple asphyxiation.

Great care should be taken when dealing with liquid or vapor leaks from 338 containers. MC 338 tanks are actually a tank within a tank, very much like a thermos bottle. Between the inner and outer tanks, there is a space for insulation to help prevent the material within the tank from heating up and vaporizing. The tanks are not refrigerated, and even with the insulation, a certain amount will vaporize, increasing the pressure in the tank and causing the excess pressure to be exhausted through the relief valve. This is a normal process and does not indicate a leak in the tank. Cryogenic tanks are constructed of steel on the outer tank and special alloys of steel on the inner tank to withstand the extreme cold temperatures and the internal pressure of 23.5–500 psi.

Railcars Cryogenic tank cars are usually constructed of nickel or stainless steel. (Figure 3.96). Cryogenic tanks may also be found within box cars. Because cryogenic liquids are very cold, insulation is placed between the two tanks and a vacuum pulled on the space to maintain the temperature. This process will allow the tank car a 30-day holding time.

Fixed Facility Vertical cryogenic storage tanks are often found next to manufacturing buildings, hospitals, bottled gas facilities and welding

Figure 3.96 Cryogenic tank cars are actually a tank within a tank.

supply houses. A heat exchanger next to the tank or tanks is confirmation that these are indeed cryogenic tanks. Heat exchangers are a series of silver-colored tubes with fins next to the tanks. The heat exchanger allows for the cold liquids to be turned back into gases for use. Ambient air around the tubes and fins of the heat exchanger warms the cold cryogenic liquids into their gas state. Cryogenic tanks have narrow circumferences and are very tall (Figure 3.97). These are high-pressure tanks used to store cryogenic liquids, which are very cold. Some tanks, particularly those found at cryogenic production facilities, may each hold as much as 400,000 gallons.

Tube Banks

Highway Tube trailers are used to transport high-pressure gases. Unlike the MC 331, there are no liquids in these containers and therefore no protection against flame impingement anywhere on the surface of the tank (Figure 3.98). Pressures in these tanks may be in excess of 3,000 psi. These trailers are actually a series of small pressure tanks placed on a flatbed trailer, then banded and cascaded together. Accidents have been reported where the tanks have come loose during an accident and come through the cab of the truck.

Railcars High-pressure tank cars (similar to the highway tube trailer) are approximately 40 ft long and contain a series of 25–30 steel cylinders or individual tanks, which are tested to 4,000 psi. High-pressure tanks are uninsulated and equipped with pressure relief valves, and are usually used to transport helium or hydrogen. Another type of pressure tank car

Figure 3.97 Cryogenic tanks have narrow circumferences and are very tall.

Figure 3.98 Tube trailers are used to transport high-pressure gases. Unlike the MC 331, there are no liquids in these containers and therefore no protection against flame impingement anywhere on the surface of the tank.

is actually only pressurized during unloading. This car is a pneumatically unloaded, covered hopper car. Pressure is applied during the unloading.

Fixed Facilities Tube banks are a type of ultra-high pressure tanks that are often found at compressed gas companies and distribution facilities (Figure 3.99). The pressures in this type of tanks can be in excess of 3,000 psi. Tube banks are actually a series of individual tanks stacked together

Figure 3.99 Tube banks are a type of ultra-high pressure tanks that are often found at compressed gas companies and distribution facilities.

with valves and piping into a single outlet/inlet; similar to a fire department cascade system. Unlike the liquefied gases in high-pressure tanks, the materials in tube bands are gases. There is no liquid level or space to absorb heat. Under fire conditions, these tanks can fail very quickly.

Bulk Containers

Highway Types of bulk containers warrant discussion because of dangers of the products inside. Dry bulk tanks carry dry materials which have a very fine particle size. These materials are rarely placarded and are not regulated by the DOT. However, if released in air in the presence of an ignition source, they could produce a dust explosion. The only dry bulk tanks I have ever seen placarded were those carrying ammonium nitrate, with an oxidizer placard, and those carrying ammonium nitrate fuel oil mixture, which is placarded as a blasting agent/Class 1.5 explosive.

Another type of truck used for hazardous materials is the blasting agent mixer (Figure 3.100). This truck looks very much like an agricultural feed truck. It has three tanks: one for dry ammonium nitrate, one for fuel oil and one for a mixture of the two. Once mixed, the blasting agent is off-loaded through a mechanical arm, much like grain from a feed truck. This vehicle usually carries three placards: oxidizer, flammable liquid and explosive 1.5, also known as blasting agent.

Railcar Hopper cars do not always contain hazardous materials, but the physical state of the materials in the container may present a hazard. Fine powders and dusts when suspended in air can become a dust explosion

Figure 3.100 Another type of truck used for hazardous materials is the blasting agent mixer.

hazard. An accident could cause the materials to be air borne, and if an ignition source is present, a dust explosion can occur. Another type of pressure tank car is actually only pressurized during unloading. This car is a pneumatically unloaded, covered hopper car. Pressure is applied during the unloading process, and the tank is tested to between 20 and 80 psi. This type of tank is used for dry caustic soda.

Fixed Facility Fixed facilities may have bulk containers to handle dry bulk materials which present dust explosion hazards. These materials include ammonium nitrate grain, saw and flour dusts. Numerous dust explosions have occurred in these types of facilities. Care should be taken by responders when responding to reported explosions in a dry bulk facility (*Firehouse Magazine*).

Hazard Class 3 Flammable Liquids (HF/LR)

Flammable liquids are the third most hazardous material in the U.S. Department of Transportation (DOT) hierarchy. According to the DOT, flammable liquids are involved in over 50% of all hazardous materials incidents. This shouldn't be surprising because flammable liquids are used as motor fuels for passenger vehicles, highway transport vehicles, railroad locomotives, marine vessels and aircraft. In addition, many flammable liquids are used to heat homes and businesses. Class 3 materials are liquids that are flammable or combustible. Flammable liquids, according to the DOT, "have a flash point of not more than 141 degrees F, or any material in a liquid phase with a flash point at or above 100 degrees F that is intentionally heated and offered for transportation or transported at or above its flash point in bulk packaging." Combustible liquids are "materials that do not meet the definition of any other hazard class specified in the DOT flammable liquid regulations, and have flash points above 141 degrees F and below 200 degrees F." The National Fire Protection Association (NFPA) also has a classification system for flammable and combustible liquids in fixed storage; the system does not, however, apply to transportation of hazardous materials.

Chemistry of Fire

Fire is a rapid oxidation of a material in the exothermic chemical process of combustion, releasing heat, light and various products of the combustion. Certain components need to be in place for a fire to occur in the first place. These can be illustrated using the Fire Triangle. The fire triangle shows that there must be three things for fire to occur: a fuel, oxygen and heat. As time passed, research revealed that the fire triangle did not tell the whole story of combustion. It is believed that there is also a chemical chain reaction that takes place until one of the three components of fire is removed. For example, if the fuel is consumed, the fire will go out. Removing oxygen or the heat will also cause the fire to go out. This new theory of combustion is called the Fire Tetrahedron and (Figure 3.101) remains in use today.

The science of fire extinguishment also utilizes the Fire Tetrahedron to remove components so the fire will go out. Removing the fuel from a fire is not a practical task unless the fuel is a liquid or gas and a valve can be shut off to stop the flow of fuel. Foam can blanket the surface of a flammable liquid and keep oxygen from reaching fuel and the fire will go out.

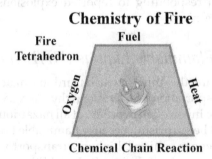

Chemistry of Fire

Figure 3.101 Certain components need to be in place for a fire to occur in the first place. These can be illustrated using the Fire Tetrahedron.

Chemical fire extinguishers work by interrupting the chemical chain reaction and the fire goes out. Finally, water will work in some instances to cool the heat to the point the fire goes out.

Flammability is not the only hazard associated with Class 3 materials. They can also be poisons or corrosives; thus, don't forget about the "hidden" hazards of all placarded hazardous materials. For all general purposes, there are no UN/DOT Divisions of flammable liquids. From an emergency responder's view, all materials with red placards burn under certain conditions and appropriate precautions should be taken. The most important precaution is to control ignition sources at the incident scene. To effectively deal with flammable liquids at an incident scene, emergency responders should have a basic understanding of the physical characteristic of flammable liquids.

Make sure you know the temperature scale being used when you are doing research because there is a big difference between the Fahrenheit and Centigrade scales. Fahrenheit scale is most familiar to emergency responders because it is the temperature scale most commonly used in the United States. Centigrade scale is predominantly used throughout the rest of the world and in the scientific community of the United States. It is also used in many of the reference books used by emergency responders to obtain the information on hazardous materials.

Alkyl Halide (Halogenated Hydrocarbons)

Alkyl halide compounds can be found as liquids or gases. As a family the alkyl halides are toxic to varying degrees and some are flammable. Alkyl halides are composed of one of the hydrocarbon radicals with varying numbers and combinations of the halogen family of elements attached: fluorine (F), chlorine (Cl), bromine (Br) and iodine (I). The toxicity of the compound is derived from the halogens and the flammability from the hydrocarbons, specifically the hydrogen. The more halogens present the more toxic the compound; the more hydrogens the more flammable they become. The simplest of the alkyl halides is formed from methane. One hydrogen atom is replaced by one chlorine atom forming the compound methyl chloride, also called chloromethane. Hydrocarbon is represented in the name as well as chlorine. Methyl chloride is a colorless compressed gas or liquid with an ether-like odor. Fire risk is significant with a flammable range of 10.7%–17% in air. Methyl chloride has narcotic effects, with a threshold limit value (TLV) of 50 ppm in air. Uses include a refrigerant, topical anesthetic, solvent and herbicide. Adding two more chlorines, it becomes trimethyl chloride, with the trade name of chloroform. Now the flammability is gone, but the toxicity has increased to 10 ppm in air.

Chloroform is used as an anesthetic, fumigant, solvent and insecticide. When four chlorines are attached to methane, the compound formed is carbon tetrachloride, one of the first halon compounds and a former fire extinguishing agent. It was found that when carbon tetrachloride comes in contact with a hot surface, phosgene gas is given off, so it is discontinued as a fire extinguishing agent. Other halogens can be added to methane in combination to form compounds. For example, when one chlorine and two fluorine atoms are added to methane, chloro trifluoro methane is formed. It is a nonflammable refrigerant with low toxicity. Methane is not the only hydrocarbon radical used to form alkyl halides. But whatever the radical, the family characteristics of the alkyl halides will prevail.

Ether (Oxide)
Ether is represented by a single oxygen –O–. Since oxygen must have two connections to complete the bonding requirements, there will be two radicals in ether compounds. They may be the same radical as in methyl ether, or they may be different as in methyl, ethyl ether. Ethers are usually encountered as liquids; however, there are some ether compounds that are gases. Ether compounds have a wide flammable range of 2%–48% in air, which means they burn rich; fire may occur inside a container. Ether compounds are extremely flammable. Additionally, they are anesthetic and form explosive peroxides inside the container after exposure to the oxygen in the air. Ether is very dangerous in storage longer than 6 months. Ether containers should be dated when they are received in a facility and when they are opened. They should be disposed of after 6–12 months when opened, depending on the compound. If responders encounter ether containers in response or fire inspection situations, extreme caution should be used. Containers over 6–12 months old should be assumed to contain peroxides and handled by the bomb squad!

Ethyl ether has two ethyl radicals, one on each side of the oxygen. Ether compounds where the radicals are the same are named using the radical and the word "ether." Common sense would be to call the compound diethyl ether because there are two ethyl radicals. However, in practice, the "di" is dropped and it is simply ethyl ether. (Ethers are researched in reference books and computer databases may be listed either way.) Hydrocarbon derivative ethers should not be confused with petroleum ethers, the type used to start cars in the winter. This type of ether does not form explosive peroxides. Ethers formed from the aromatic radicals generally do not form peroxides either. Ethers, in general, have the same types of hazards. Some common ether compounds are methyl ether, ethyl ether and isopropyl ether.

Amine
Amines are a hydrocarbon derivative family. As a family, they are toxic and flammable, and they may be gases or liquids. The name "amine"

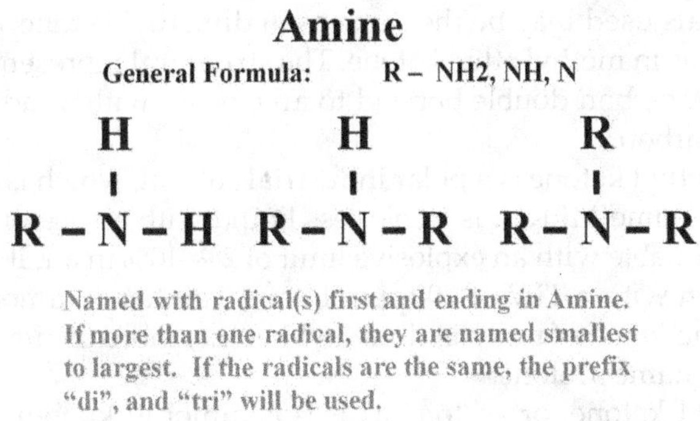

Amine

General Formula: R – NH2, NH, N

```
    H              H              R
    |              |              |
R – N – H     R – N – R      R – N – R
```

Named with radical(s) first and ending in Amine. If more than one radical, they are named smallest to largest. If the radicals are the same, the prefix "di", and "tri" will be used.

Figure 3.102 Amines have a very distinctive odor which is very unpleasant, similar to rotting flesh, or bowl odor.

will follow the radical name of the compound. Amines have a very distinctive odor which is very unpleasant, similar to rotting flesh, or bowl odor (Figure 3.102). Nitrogen is the primary component of amines and is attached to one, two or three hydrocarbon radicals depending on the compound. It comes from ammonia that is used in the chemical process for making amines. NH_3 is the molecular formula for ammonia. Once one or more of the hydrogen atoms are removed, it is no longer ammonia; however, when hydrocarbon radicals are attached, there still may be some ammonia characteristics. Methyl amine is a simple compound with just one hydrocarbon radical attached to nitrogen and hydrogen filling the other two nitrogen connections. You may remember that nitrogen should have three connections to satisfy the bonding requirements to become stable. Methyl amine is a colorless gas or liquid with strong ammonia-like odor. The flammable range is 5%–21% in air, and the TLV is 10 ppm in air. Methyl butyl amine contains two hydrocarbon radicals. It is a water white liquid which is a dangerous fire risk. Methyl diethyl amine has three hydrocarbon radicals attached. It is a combustible liquid with fairly low toxicity.

Ketone

The first of the carbonyl hydrocarbon derivatives is the ketone family. Ketone compounds have two hydrocarbon radicals and are named by identifying the radicals attached to the ketone and ending with the word "ketone." The smallest radical is named first, then the larger one. Ketone compounds are flammable liquids that have narcotic effects on the body if they are inhaled and can be toxic if inhaled long enough in a high concentration. Ketone is represented by "CO," which is also the carbonyl symbol. It has a hydrocarbon radical on each side of the carbon in the structural form. Radicals used may be the same as in dimethyl ketone, or they may be different as in methyl ethyl ketone. The structural representation of the ketone is the carbon double bonded to an oxygen with a radical on each side of the carbon.

Methyl ethyl ketone is a polar industrial solvent, which is also known by the trade name MEK. It is a colorless liquid with an acetone-like odor. MEK is flammable with an explosive limit of 2%–10% in air. It is a narcotic by inhalation with a TLV of 200 ppm in air. Another common industrial solvent of the ketone family is dimethyl ketone, more commonly known by the trade name acetone.

Dimethyl ketone, or acetone as it is commonly known, is a widely used industrial solvent. Acetone is a colorless and volatile liquid with a sweet odor. The explosive limit is 2.6%–12.8% in air. It is narcotic in high concentrations and has a TLV of 750 ppm in air. Acetone is toxic by ingestion and inhalation. All of the ketones are flammable polar solvents and require the use of polar solvent or alcohol-type foam when fighting fires. Dealing with all other ketone compounds, responders need to remember that as a family the ketones are flammable and have narcotic effects when in contact with the human body. When ketone is in the name and you don't know anything else about the compound, you should know what the potential hazards are.

Aldehyde

Aldehyde has a wide flammable range of 7%–73% in air. Its flash point is below 100°F. Formaldehyde, a common aldehyde, exists normally as a gas that is readily polymerized with a strong pungent odor; however, it is also found in aqueous solutions with a concentration of 37%–50%. It is sometimes found with 15% methyl alcohol mixed in. Aldehyde and organic acid use an alternate naming convention for one and two carbon compounds. Form is used instead of Meth for one carbon and Acet is used for two carbons. Only one radical is present in aldehyde compounds. So the radical is named first and the compound name ends in aldehyde. Thus, formaldehyde and acetaldehyde. Acetaldehyde is also sometimes known as ethanal. The "al" ending is the hint that this is a two carbon aldehyde. Aldehydes are polar compounds with a carbonyl structure. Alcohols and aldehydes are both polar so they mix well together. There is a rule in chemistry that like materials mix with like materials. Other aldehyde compounds include acetaldehyde, crotonaldehyde and acrylaldehyde. Aldehydes are used as solvents, disinfectants, preservatives and pesticides and in the manufacture of plastics. Formaldehyde is designated as a carcinogen. Polarity is also important when selecting the right type of firefighting foam. Polar compounds require polar solvent foam, also known as alcohol-type foam.

Alcohol

Alcohol is another derivative family. Alcohols have only one radical and are named by identifying the radical and ending with the word "alcohol." Alcohols can be found as liquids or solids and as a group; they are all flammable with a wide flammable range. Materials with wide flammable ranges can be dangerous because they could burn inside a container. Alcohols burn with a pale blue to invisible flame and are difficult to see when burning. Most of the alcohols are toxic while ethyl alcohol is used in alcoholic beverages and can have varying levels of toxicity depending on the individual and the amount consumed. Alcohols do not have the carbonyl "CO" in their formula, so they are not carbonyls. They are, however, polar because of the "OH" or hydroxyl formulation in the compound. The OH in a hydrocarbon derivative family indicates polarity just as the CO in the carbonyl compounds. Alcohols and organic acids have the OH in their compounds. Organic acids are also carbonyls and thus have a kind of "double dose" of polarity. Polar compounds may have different degrees of polarity. Organic acids are the most polar because of the "double dose"; alcohols are next because the OH creates more polarity than the CO does. All of the rest of the polar compounds are considered to have the same level of polarity.

Alcohol compounds have one hydrocarbon radical attached to the oxygen in the OH. If the methyl radical is attached to an OH, the compound is called methyl alcohol. Alcohols are also sometimes named with an alternate system using the radical name and ending with "ol." In the case of methyl alcohol, it is also called methanol. Methyl alcohol is a colorless, highly polar liquid and mixes well with water, other alcohols and ether. Methanol is highly flammable in the range of 6%–35% in air. It is toxic by ingestion and causes blindness and has a TLV of 200 ppm in air. Ethyl alcohol, also known as ethanol, is the type used in alcoholic beverages and as a motor fuel in combination with varying amounts of gasoline. The ethyl radical is added to the hydroxyl OH. It is sometimes referred to as grain alcohol and is a clear, colorless and volatile liquid. Ethyl alcohol is classified as a depressant drug and, when ingested in very large amounts, can cause death. The flammable range is 3.3%–19% in air with a TLV of 1,000 ppm in air.

Phenol is technically an alcohol; however, its hazards are more severe than most alcohol compounds. Phenol is toxic by ingestion, inhalation and skin absorption. The TLV is 5 ppm in air. It is listed as a strong irritant to tissue; in reality, it has a severe corrosive effect on skin, much like acids. Phenol does appear to be an alcohol when looking at the structure and formula.

Phenol is formed from the benzene ring with one hydrogen removed and the OH added to the compound. The radical of benzene used in phenol is called the phenyl radical. Other common alcohols include propyl, isopropyl, butyl, secondary butyl and tertiary butyl. The hazards are all similar, with a wide flammable range and toxicity.

Ester

Esters are hydrocarbon derivatives. They are flammable and toxic, and may undergo polymerization. Esters are members of the carbonyl family and are all polar. They are formed by combining, in a process vessel, an alcohol and an organic acid. Water is a by-product in the chemical reaction along with the resulting ester. The finished compound does not have ester in the name. Esters end in "ate." Esters are named by first identifying the radical and then ending in acetate or acrylate. How all of this occurs is beyond the scope of Applied Chemistry. So when you encounter a compound in the field that that is a liquid and the name ends in "ate," there is a good chance it is an ester.

Esters are found as liquids that may undergo polymerization in uncontrolled situations such as hazardous materials incidents. Polymerization is a chemical reaction that occurs within the compound where double bonds break and the molecules connect to themselves forming a long-chain polymer. The uncontrolled polymerization can be a violent reaction that can involve container failure. The compound subject to polymerization is shipped in a tank with an inhibitor which prevents the polymerization from occurring under normal shipping conditions. The molecules in their normal state are referred to as monomers. However, in an accident the inhibitor can become separated from the monomer, and then an uncontrolled polymerization can occur.

Ester compounds are also monomers and form polymers when the double bonds in the ester break to form a polymer. This process is carried out under controlled conditions in a chemical plant to form plastics. Esters have two radicals connected to a single carbon with two oxygens COO. One oxygen is double bonded to the carbon, and the other is single bonded to the carbon and one of the radicals. Methyl acetate is a colorless and volatile liquid with a sweet smelling odor. It has a flammable range of 3%–16% in air and a TLV of 200 ppm. Other esters include vinyl acetate, ethyl acrylate and ethyl acetate, all ending in "ate." (*Hazardous Materials Chemistry for Emergency Responders*)

Plastics and Polymerization

Plastics have been around for more than 100 years. The first plastic developed was cellulose nitrate, which was a replacement ivory in billiard balls. Since World War II, however, the plastics industry has been one of the fastest developing technologies. The forms, variations and applications of plastics have developed at a tremendous rate, producing a family of materials that are unusually complicated and diverse. Plastics are composed of organic materials that are part of a group of materials known as polymers. Polymers can be subdivided into two groups: naturally occurring and man-made. Common naturally occurring polymers include leather, wood, paper, silk, cotton and wool. Man-made polymers are created from organic materials found in nature.

The American Society for Testing Materials (ASTM) defines a plastic as "a material that contains as an essential ingredient one or more organic polymeric substances of large molecular weight, is solid in its finished state, and at some stage in its manufacture or processing into finished articles can be shaped by flow." The word "polymer" comes from the Greek *poly* ("many") and *mer* ("part"). Therefore, a polymer is a compound with many parts. In reality, a polymer is a long-chained molecule composed of many smaller parts called monomers ("mono" is Greek for "one"). Monomers, however, are able to hook together into long chains of hundreds or thousands of parts. In polymerization, monomers that have double bonds are broken down by heat or chemical reaction to single bonds. The single bonds attach to each other creating the self-reaction, which in turn creates the long-chained polymer. This reaction usually takes place in a reactor vessel in a chemical plant under controlled conditions. But it can also take place during transportation or in storage, creating a hazard for emergency responders.

Polymers

Man-made and naturally occurring polymers behave much the same way during reactions, especially when exposed to fire. In fact, some synthetic polymers contain the same elements as natural polymers, exhibit the same burning characteristics, and produce the same products of combustion, many of which can be toxic. Naturally occurring polymers are standard in their identities; wood, for example, is not mistaken for cotton or wool. Synthetic polymers, however, are produced to conform to specific properties, and it may be difficult to distinguish between them. For example, polypropylene and polyethylene are very similar, as are styrene and styrene-acrylonitrile. Many terms are interchangeable when referring to synthetic polymers. The terms "plastic," "polymer," "resin compound" and "high-polymer macro-molecular substance" are often used interchangeable.

Plastics are such a common part of our everyday lives that we don't give them much thought. They are found everywhere. Look around the room you are in—how many things are made of natural polymers and how many are synthetic? As I type this volume on my computer, I see that the housings for the computer as well as the keyboard, printer and monitor are made of plastic; the table the computer sits on is plastic; the chair I am sitting on is largely made of plastic and plastic fibers; the carpeting on the floor is made from plastic fibers; the telephone on the table is plastic; the cabinets next to the computer look like wood, but are actually plastic; the list could go on and on. The jobs of emergency responders are made easier and safer because of plastics. Nearly all components of firefighter turnouts are made of plastic and plastic fibers; chemical suits, gloves and boots worn by hazmat technicians are made of plastic; ropes we use for rescue are made of plastic fibers; many EMS supplies are either plastic or packaged in plastic; again the list could go on and on.

Polymer Family Tree

The polymer family tree has three distinct branches or divisions, and each branch may be identified by particular characteristics. Those branches are thermoset, thermoplastic and elastomer. Thermoset are plastics that are hardened into their final shape in the manufacturing process by heating and usually cannot be softened again by heating without losing their identity. If attempts are made to re-form them by heat, they will start to burn. Those that can be softened by heating cannot be remelted or returned to their original form before they harden once again. Thermoplastics are compounds formed by heat and pressure, then cooled into their final shape. They can be re-formed again if they are placed under heat or pressure, or both. In some cases, they can be heated and re-formed up to 20 times without losing their properties.

Elastomers are sometimes called "synthetic rubbers" because of their elastic or rubberlike features. Natural rubber is also classified as a polymer that is in the elastomer branch of the family tree. Synthetic rubbers are created to duplicate or surpass the most desirable properties of natural rubber. Elastomers are substances that at room temperature can stretch to at least twice, and in some cases many times, their original size and upon release return to their previous form with notable force. Elastomers cannot be heated and re-formed.

Manufacturing Plastics

The process of making plastics involves many different compounds and hazardous materials that are commonly shipped in transportation and stored in manufacturing facilities. One of the most common compounds used in the making of plastics is the monomer. Monomers can be found as solids, liquids or gases (*Firehouse Magazine*).

Chemical Notebook **Ethylene**, which is used to make polyethylene plastic, is a colorless gas with a sweet odor and taste. It is also a highly flammable gas with a wide flammable range of 3%–36% in air. It is not toxic but can displace oxygen in the air and create an asphyxiation hazard to response personnel. (Ethylene is also the gas that is produced naturally by ripening fruit and is used in orchards to hasten fruit ripening.)

Styrene is another monomer from the aromatic hydrocarbon family, along with benzene, toluene and xylene. Also called vinylbenzene, it is a colorless, oily, aromatic liquid. Styrene is a moderate fire risk with a narrow flammable range of 1.1%–6.1% in air. It is toxic by inhalation and ingestion and has a TLV of 50 ppm in air. Styrene monomer is used to make polystyrene, which is the rigid plastic from which soft drink cup covers are made.

Polyacrylamide is a solid monomer which is white in color. It is used to make the clear plastic from which compact discs are made.

Butadiene is another common but dangerous monomer. It is a gas that is highly flammable and has a flammable range of 2%–11% in air. It is used in making elastomers and neoprene. Nylon is a polymer having a high molecular weight. It is formed from adipic acid and hexamethylene diamine, which produces a material with high strength, elasticity, abrasion resistance and solvent resistance. Nylon is used to make tire cords and rope apparel, and has other industrial and automotive uses.

Kelvar is a form of nylon that is used in some firefighter gloves and other safety products because of its great strength and resistance to puncture. It also has a high thermal endurance and doesn't start melting until it tips 950°F. U.S. DOT regulations prohibit the transportation of most monomers without the material being inhibited (Hazardous Materials Chemistry for Emergency Responders).

The *Emergency Response Guidebook* identifies materials in transportation that may undergo polymerization creating a danger to response personnel. When researching a chemical by its four-digit identification number in the yellow section or alphabetically in the blue section of the book, you will be referred to a three-digit guide page in the orange section of the book. That number will have a "P" after it if the material in question has the potential to undergo polymerization. You may also notice that next to the common shipping name of the material in the ERG will appear the words "inhibited" or "stabilized."

Once monomers and other chemicals are processed into plastic materials, they generally do not present any danger to civilians or emergency responders. Many plastic items, in fact, are used to sustain and improve the quality of life: artificial hearts, replacement joints, medical tubing and packaging. The greatest concern for emergency responders, and particularly firefighters, is the burning of plastics. All plastics that contain carbon will burn. As with any materials that will burn, some are more combustible than others. Chemicals can be combined with plastics during manufacture to reduce combustibility. Other plastic materials are formulated to be self-extinguishing. Plastics that contain carbon and hydrogen–such as polyethylene, polypropylene, polybutylene and polystyrene—burn very well.

Burning polyethylene melts, smells like wax and produces dripping of the melted flaming material that could spread the fire. Polystyrene burns much slower, producing large soot particles, and smells like vinegar. Styrene is an aromatic hydrocarbon and, as a family, burns sooty with incomplete combustion. Plastics that are composed of carbon, hydrogen and oxygen burn slower than the others mentioned. Thermosetting plastics can produce burning smells like charred wood or formaldehyde. Plastics that contain nitrogen and sulfur will produce very toxic gases when they burn. Plastics containing nitrogen burn with no smell, and those with sulfur produce a choking sulfur dioxide smell. Plastics that contain the halogens (fluorine and chlorine) produce acrid, choking odors when they burn. With fluorine- and chlorine-based plastics, the flame must be continuous or the plastic will self-extinguish. Almost all plastics can be made flame or smoke retardant by adding other chemicals during manufacturing. Adding any of the halogens will retard burning. Halogens are used with carbon to produce halon fire extinguishing agents. Silicone-based plastics will not burn at all.

Combustion Products

When plastics burn, the rate of burn and the amount of smoke generated can vary widely depending on the type of plastic and flame-retardant chemicals present. Cellulose nitrate motion picture film was used for many years in Hollywood. Cellulose nitrate is highly flammable and ignites readily. Cellulose nitrate is formed by mixing sulfuric acid and nitric acid on cellulose materials such as cotton. Cellulose nitrate plastics exhibit dangerous burning characteristics unlike any other plastics. When exposed to heat, the physical makeup of cellulose nitrate is changed in such a way that it may become subject to spontaneous combustion. Combustion products produced, when cellulose nitrate burns, can be very toxic. Use of cellulose nitrate is not as widespread as it once was, but if encountered it should be handled very carefully. New safer plastics have been developed to replace cellulose nitrates in such products as photographic and motion picture film and house wares. Teflon, on the other hand, another popular plastic, is not combustible at all and experiences heat damage only at extremely high temperatures.

Some plastics may exhibit unusual burning characteristics compared to building materials made from natural polymers such as wood. Plastics as a group generally have higher ignition temperatures than wood and other cellulose-building products. Plastics have been reported to have very high flame spread characteristics, as high as 2 ft/s, or 10 times that of wood on the surface. Vinyl, when tested in a solid form in the laboratory, has been shown to burn slowly. However, when in the form of a thin coating on wall coverings, it spreads rapidly contributing to flame spread.

Nylon has a tendency to self-extinguish when a flame is removed. When nylon is in the form of carpet fiber under certain conditions, it burns with great enthusiasm. Polyurethane foam that has not been treated with a flame retardant is very flammable. It is used as an insulating material in construction and burns with a very smoky flame. Because of its burning characteristics, it has contributed to rapid flame spread in several fatal fires. Burning of plastics may produce large quantities of thick, black smoke. When chemicals are added to retard burning, they may actually expand the amount of smoke that is produced. Because of their ability to melt and run, plastics can spread fires in ways that could mislead fire investigators. When skylights and light fixture diffusers are ignited by high ceiling temperatures, they may soften and sag. They can fall into combustible materials below and start fires in several isolated locations. This could lead an investigator to suspect an incendiary fire when in fact it was not.

Hazards to Responders

Products of combustion produced from burning plastics and other materials are the most significant hazard to both occupants of a building and firefighters during a fire. Plastic materials that contain only carbon and hydrogen in their formulation will generate only carbon, carbon monoxide, carbon dioxide and water as they burn. Intermediate products of combustion, however, are also produced and can include acrolein, formaldehyde, acetaldehyde, propionaldehyde and butraldehyde. Members of the aldehyde hydrocarbon family are irritants and flammable, with wide flammable ranges. Acrolein and carbon monoxide are lethal poisons. In addition to being toxic, carbon monoxide is also extremely flammable. Examples of carbon- and hydrogen-based plastics include polyethylene, polypropylene and polystyrene. Combustible products produced from burning plastics containing only carbon and hydrogen are the same as natural polymers such as wood, paper and other Class A combustible materials. Plastics containing carbon, hydrogen and oxygen produce the same combustion products as those containing just carbon and hydrogen.

The major difference is that these plastics tend as a group to be less combustible than those with just carbon and hydrogen. Plastics in this group include acetal, acrylics, alyls, cellulosics, some epoxies, ethylene vinyl acetate, ionomers, phenolics, polycarbonate, polyesters and polyphenylene oxide. When nitrogen is added to the plastic compound, an additional product of combustion occurs while burning. Another product of combustion is hydrogen cyanide, an extremely toxic and flammable gas. Just like carbon monoxide, much of the hydrogen cyanide is consumed by the fire as it is produced. (It must be noted that hydrogen cyanide is also generated by the burning of the natural polymers including leather, wood, silk and some types of paper.) Plastics that contain fluorine and chlorine are generally less combustible than other plastics. As they decompose when exposed to the heat of a fire, they produce hydrogen fluoride and hydrogen chloride gases, respectively. Both are corrosive gases that can form acid when in contact with moisture, such as in the eyes, skin or lungs.

Plastics containing fluorine include fluoroplastics, of which Teflon (polytetrafluoroethylene) is the most common. Chlorine containing plastics include polyvinyl chloride (PVC), polyvinyl dichloride and polyvinylidene chloride. Plastic compounds can also contain sulfur, which will decompose to sulfur dioxide during the combustion process. It is a strong irritant and can be lethal when exposed to concentrations of 100 ppm for more than 10 min. Plastics that contain sulfur include polysulfone and polyethersulfone. Firefighters who wear self-contained breathing apparatus (SCBA) will be protected from carbon monoxide and other toxic fire gases. Many of these gases are invisible and remain present even after the smoke is "cleared." Firefighters should continue to wear SCBA during overhaul operations and should not remove them until they are outside the fire building or until the air is monitored within the building and shown to be safe. It is also important to remember that the same toxic materials inside the fire and smoke can become impregnated in firefighter turnouts. Each time the unwashed turnouts are worn, the firefighter is reexposed to the toxic substances in the turnouts. Firefighter turnouts should be washed or decontaminated to help prevent this continued exposure to toxic materials (*Firehouse Magazine*).

Organic Acid

Organic acid is a hydrocarbon derivative family. They can be flammable and toxic and all are corrosive. In chemistry, there are two general types of acids: organic and inorganic. Inorganic acids do not burn, but they can be oxidizers and support combustion. Some organic acids burn, whereas others do not. If a flammable placard is found on an acid container or the MC 312/DOT 412 acid tanker, it is an organic acid. Organic acids are polar; in fact, of all the hydrocarbon derivatives, they are the most polar. They have both the carbonyl structure **CO** like the ketones, aldehydes and esters and the **OH** structure like the alcohol. The formula for a generic organic acid is **COOH**. One radical is attached and the name comes from the radical with an "ic" added and with the ending word "acid." Acetic acid is formed when a methyl radical is attached to the generic COOH of the organic acid. This does not follow the naming conventions of other derivatives, however, the reasons why are not important in terms of Applied Chemistry. It can be found in varying concentrations from strong, as in glacial acetic acid which is the pure compound (99.8%), to weak solutions known as vinegar. Acetic acid is a clear, colorless liquid with a pungent

Applied Chemistry and Physics

odor. It is one of the organic acids that are flammable, and are toxic by inhalation and ingestion in higher concentrations, with a TLV of 10 ppm in air. Other types of common organic acids include formic acid, acrylic acid, butric acid and propionic acid.

Organic acids are used in plastics production, in water purification, as solvents and in decalcification. Hydrocarbon and hydrocarbon derivative families all have particular hazards associated with each other. If responders become familiar with the hazards of the family, they will have an idea of the hazards facing them during an incident, because all compounds in a particular family will have similar hazards (Hazardous Materials Chemistry for Emergency Responders).

Flammable Liquid Containers

MC/DOT 306–406 tanker is primarily an atmospheric pressure, non-insulated, flammable liquid container, which is hydrostatically tested to 3 psi (Figure 3.103). Its capacities vary from 2,000 to 9,000 gallons. It generally has an elliptical shape, although some manufacturers do make a round version, and is used to haul gasoline, diesel, ethanol, aviation fuel and other flammable liquids. Materials used to construct these tanks include aluminum, steel or stainless steel. Baffles installed within the container limit product movement during transportation. These tanks may have multiple compartments and may be carrying several different flammable materials.

Specification plates are usually located on the right frame rail of the trailer. Specification plates are small and difficult to read. They may be

Figure 3.103 MC/DOT 306-406 tanker is primarily an atmospheric pressure, non-insulated, flammable liquid container.

of more use during training and preplanning than during an incident. The plate contains information about the type of tank, manufacturer, construction material, date built, design and test pressures and number of compartments and capacities. Specification plates for other bulk tanks contain similar information and are located in the same general area.

Relief valves on 306/406 tanks are spring loaded and remain closed during transportation. Valve operation can be mechanical, pneumatic or hydraulic. All valves are equipped with an automatic heat-activated closure system. This system is usually a fusible link, but could operate by some other means which operates at temperatures up to 250°F. There is also a secondary closure system which is separated from the fill and discharge valves and mounted inside the tank. Manual controls for the secondary system are usually on the left front of the container. This type of tank is unloaded on the bottom. Bottom valves are designed to shear off in the event of a collision of such force that the valve would be damaged.

There may be up to 10 gallons of fuel in the valve and piping system under the tank. Fill openings on top of the tank are protected by manhole covers that are securely closed. Tanks are provided with rollover protection and a safety device that prevents the covers from releasing when excessive internal tank pressure exists. Vacuum and relief vents are located on top of the barrel of the tank or internally. The vacuum vent is set to open at 6 oz of vacuum, and the relief vents open with as little as 1 psi. Both vents are designed to prevent the release of product during a rollover. If mounted on the outside of the tank, valves must be protected from rollover damage.

Some 306/406 tanks have vapor recovery systems in place to prevent vapors from reaching the atmosphere. Bumpers extend 6 in. from any vehicle part, serve to protect the vehicle and provide a method of gauging the impact of an accident on the vehicle. Response personnel can use the information during damage assessment. If the bumper is significantly damaged, the tank's baffles are also likely to have been damaged and their integrity compromised.

MC/DOT 307/407 is a low-pressure tank that is used for the transport of flammable liquids, Class 6.1 poisons (poison liquids) and light corrosives (Figure 3.104). Working pressure in the tank is 25 psi and not greater than 40 psi. Tank capacities range from 2,000 to 8,000 gallons, and they are constructed of aluminum or stainless steel. This tank may be insulated or uninsulated. Insulated tanks have a horseshoe shape when viewed from the rear. Uninsulated tanks are generally round, and some have reinforcing rings similar to the 312/412, but the tank diameter is much larger. Insulated tanks that no longer meet low-pressure specifications are sometimes used to transport molten solids such as asphalt. These materials have temperatures in excess of 300°F and can be a thermal hazard.

Vehicles carrying molten materials are placarded with a HOT and, in addition, sometimes a miscellaneous hazardous materials designation.

Figure 3.104 MC/DOT 307/407 is a low-pressure tank that is used for the transport of flammable liquids, Class 6.1 poisons (poison liquids) and light corrosives.

Valve location and information are the same as for the 306/406. Pressure lines may be present on the 307/407 which contains inert gases that may be injected into the tank to absorb moisture or used to assist in off-loading the product. Pressure vents are installed in the 307/407 tanks to limit the internal pressure to 130% of the maximum allowable working pressure. Vents are pressure activated by a spring-loaded mechanism. Fusible and/or frangible (breakable) venting may be provided with fusible vents activating at 250°F. Frangible disks are designed to burst at not less than 130% or more than 150% of the maximum allowable working pressure.

> *Hazmatology Point:* DOT/MC 412 acid containers may also be found with flammable corrosive materials carried in highway transportation. Organic acids do burn. So, don't be confused by a red flammable liquid placard on a corrosive tanker. Remember what the container tells you as well as the placard.

Nonpressure/Low-Pressure Rail Cars

Nonpressure tank cars are hydrostatically tested from 35 to 100 psi (Figure 3.105). They are commonly used to transport flammable and combustible liquids, flammable solids, oxidizers, organic peroxides, poison liquids and corrosives. Amounts of hazardous materials in a tank car can range from a few hundred gallons to as much as 34,500 gallons (no tank car is considered empty unless it has been purged of the product and cleaned). They may be found in small containers in box trucks and in fixed facilities in many different locations. There are also many different

Figure 3.105 Nonpressure tank cars.

sizes and shapes of fixed facility storage tanks and pipelines that have various flammable liquids stored and transported in them. Nonpressure tank cars can have unprotected valves and piping on tops of the tank or, in some cases, on top of dome. Nonpressure tanks may also have bottom fittings and washouts. With this information in mind, a quick glance by response personnel can determine whether a tank car is a liquid car or not. Nonpressure tank cars are hydrostatically tested from 35 to 100 psi.

Fixed Facility Flammable Liquid Bulk Storage Tanks

Flammable liquids are also stored in bulk petroleum storage tanks. These include tanks with cone roofs, floating roofs, open floating roofs and retrofitted floating roofs. Many of these tanks are associated with tank storage facilities or "tank farms" where there are multiple tanks at one facility. Tank farms are often connected with pipelines as well as highway, rail and waterway transportation. Fixed storage tanks for flammable liquids have pressures ranging from atmospheric (0–5 psi) to low (5–100 psi). Bulk petroleum tanks are generally considered atmospheric pressure tanks.

Cone-Roof Tanks

Cone-roof tanks get their name from the inverted cone-shaped construction of their roofs (Figure 3.106). They are atmospheric and low-pressure tanks with cylindrical outer walls supporting the cone roof. American Petroleum Institute Specification 650 calls for the roof-to-shell seam for

Figure 3.106 Cone-roof tanks get their name from the inverted cone-shaped construction of their roofs.

this tank to be designed to fail as the result of a fire or explosion, reducing the possibility of a pressure buildup. Cone-roof tanks may be used to store gasoline, diesel fuel and corrosive liquids. Some of these chemicals, such as gasoline, are very volatile and easily produce vapor at normal atmospheric temperatures. Because a cone-roof tank is open inside, the surface of the liquid is exposed to air and the product will be lost to vaporization. Therefore, a cone-roof tank is primarily used for nonvolatile materials. Contents of cone-roof tanks change frequently: daily, even hourly under some circumstances. Cone-roof tanks are recognized by the cone-shaped roof and the lack of wind girders or external vent around the top of the sidewall.

Open Floating Roof Tanks

An open floating roof tank has a roof that literally floats on the liquid product in the tank. The outer walls of these tanks are vertical and cylindrical, with the floating roofs eliminating the vapor space in the tan (Figure 3.107). Drains in place are designed to remove accumulations of moisture on the surface of the tank. An articulating arm-type device on the top of the tank functions as a ladder for examination of the interior. As the product level and floating roof go down, so does the end of the arm, which rests on top of the roof. The location of this arm, when viewed from the outside, can be an indicator of the approximate amount of product in the tank. If the arm is completely visible, the tank is near full. If the arm is completely inside the tank, it is near empty. A wind girder is located near the top and on the outside of the tank. Primarily a retaining band, the wind girder provides necessary rigidity to the container wall when

Figure 3.107 An open floating roof tank has a roof that literally floats on the liquid product in the tank.

liquid levels are lowered. An external staircase is also located on the tank wall to provide access to the articulating arm (walkway) on top of the floating roof.

Internal Floating Roof Tank

Another version of the floating roof tank is the covered or internal floating roof (Figure 3.108). These tanks may resemble cone-roof tanks but are distinguishable by the vents around the tank near the roof to sidewall

Figure 3.108 Another version of the floating roof tank is the covered or internal floating roof.

seam. The cone roof provides protection from rain and snow, whereas the internal floating roof eliminates the vapor space above the liquid and helps prevent the loss of vapor into the air. This style of tank is sometimes used for polar solvent materials, which are miscible with water. If water or snow entered the tank, it would dilute the product. Materials stored in the internal floating roof tank are generally flammable and combustible liquids because of the vapor loss protection of the floating roof on top of the liquid. A floating roof tank may also be covered by a geodesic dome that prevents outside weather from affecting the floating roof in the tank. This form of internal floating roof tank is also used for flammable and combustible liquids.

Horizontal Tanks

Horizontal tanks are another kind of tank used for chemical storage in many parts of the country, particularly in rural areas (Figure 3.109). These tanks range in size from hundreds of gallons to thousands of gallons with low or atmospheric pressure. The ends of the tanks are flat, which usually indicates a limited amount of pressure. Many horizontal tanks are supported on stands constructed of metal. Unprotected metal can be dangerous during fire conditions, as the metal will be stressed by heat and fail very quickly. In the 1950s, six Kansas City, Kansas, firefighters were killed when steel supports of a horizontal tank at a gasoline service station failed, sending burning gasoline into their location on the street.

NFPA 30 Flammable and Combustible Liquids Code requires that the steel supports of horizontal tanks be protected from flame impingement by encasement in concrete or other nonflammable material. Horizontal

Figure 3.109 Horizontal tanks are another kind of tank used for chemical storage in many parts of the country, particularly in rural areas.

tanks may be used to store flammable and combustible liquids, corrosive liquids and many other types of hazardous materials. A direct result of the Kansas City fire was the requirement for gasoline tanks at service stations frequented by the public to have underground storage tanks. According to NFPA, there has never been a fire or explosion involving an underground fuel storage tank. During the 1970s and 1980s, it was discovered that underground storage tanks, which were usually constructed of steel, eventually corroded and leaked fuel into the ground.

Gasoline and other flammable fuels have a specific gravity less than water, so they float on the surface of water. Moisture that accumulates inside underground storage tanks settles to the bottom of the tank and causes the metal to corrode. Because of the leakage, the U.S. Environmental Protection Agency (EPA) requires that all underground tanks be replaced and safeguards put in place to prevent and detect leaks. An alternative to the placement of tanks underground was allowed in the form of an aboveground vault. The vault is constructed of concrete, and the tank(s) are placed within the vault, partially underground with the vault top above the ground. There is access to the vault from the aboveground. Vaults for flammable and combustible liquids must have safety precautions such as monitors installed and must be able to hold the contents of the largest tank if they should leak (*Firehouse Magazine*).

Emergency Response to Ethanol Spills and Fires

Emergency responders have been dealing with spills and fires involving gasoline and diesel fuel for over 100 years. I think it is safe to say that most responders understand the hazards of gasoline and diesel fuel as motor fuels and in some cases find dealing with them pretty routine. On the other hand, ethanol and its blends are fairly new in terms of use as a motor fuel and likely to present a challenge to emergency responders. As with any hazardous material, response personnel need to be able to recognize when they are potentially dealing with an incident involving ethanol and its blends or just plain gasoline. Containers used to transport ethanol and its blends are called bulk containers. Bulk containers are required to carry placards with the United Nations (UN) four-digit identification number in the center that identifies the product. This number can be looked up in the DOT *Emergency Response Guidebook* to obtain emergency information.

Following placards, one of the best sources of information is the shipping paper. MSDSs sheets can also be very helpful if available. There are similarities and differences of gasoline, diesel fuel and ethanol and its blends. Although all are flammable liquids according to the DOT, they do have some important differences in terms of chemical and physical characteristics. Firefighters need to become as familiar with these as they are with gasoline and diesel fuels. Ethanol burns with a pale blue flame and may not be visible in daylight. Ethanol and blends that are less than 15%

gasoline will burn with little or no smoke. Ethanol blends with 10% or less ethanol will burn with thick, black carbon smoke like gasoline.

During the manufacturing process and to a lesser degree in storage at the manufacturing facility, ethanol is pure190-proof grain alcohol. As a pure alcohol, ethanol is placarded by the DOT as a flammable bulk liquid and assigned the four-digit identification number of 1170. This material is also referred to as E100. When it is shipped from the manufacturing facility, ethanol is denatured with 2%–5% natural gasoline, also known as E-98 and E95, respectively. A blend of 95% ethanol and 5% gasoline has been assigned a 4-digit identification number of 1987 for denatured alcohol or alcohol n.o.s (not otherwise specified). Mixtures of E95 through E99 are also assigned a 4-digit identification number. Additionally, E95 may utilize the 4-digit identification number 3475. Ethanol is ultimately blended with petroleum gasoline to form a motor fuel in various concentrations depending on whether it is used as an additive/oxygenator or blended motor fuel. Ethanol and gasoline mixtures are assigned the 4-digit identification number 3475 including E11–E99. E1–E10 blends are assigned the 4-digit identification number 1203, which is also used for gasoline. Pure ethanol, (E100), E10, E85, E95 and gasoline are all assigned an NFPA 704 designation of flammability—3, health—1 and reactivity—0.

Once ethanol is blended with gasoline, the resulting blend has physical and chemical characteristics somewhere in between pure ethanol and gasoline. E10 is the most common fuel blend of ethanol and gasoline, and is widely available across the country at automotive service stations. E10 (Gasohol) is 90% gasoline and 10% ethanol. Any motor vehicle can operate using E10 without any special modifications. E85 is the next most common blend of gasoline and ethanol and is used in flex fuel vehicles that can burn gasoline or E85. E85 is 15% gasoline and 85% ethanol.

Gasoline has a flash point of −40°F, and pure ethanol has a flash point of 55°F. The flash point of E85 is −20°F to −4°F. The lower flash point of gasoline lowers the higher flash point of pure ethanol. At lower temperatures (<32), E85 vapors are more flammable than gasoline. However, at higher temperatures, E85 vapor is less flammable than gasoline because of the higher autoignition temperature of E85. Because of a lower vapor pressure and a lower heat of combustion, E85 is generally less of a fire risk than gasoline. Ethanol does have a much wider flammable range than gasoline, which means that it will burn in a greater number of concentrations with air than gasoline. It is possible for materials with wide flammable ranges to burn inside containers under the right conditions.

The main point I want to make here is that we should not get into a line of thinking that says ethanol or gasoline or blends are any more flammable than the other. The fact is that all are considered Class 3 flammable liquids by the DOT and have a flammability of 3 on the NFPA 704 system. If there is a spill, control ignition sources and prevent fire from occurring. If spilled fuels are already on fire, understand that different types of foam may be required to extinguish fires involving gasoline, ethanol and their blends.

Small fires involving ethanol and its blends can be extinguished with a Class B-type fire extinguisher (dry chemical). Generally, large fires involving flammable liquids are best contained and extinguished using firefighting foam. There are two basic firefighting foams: one for hydrocarbon fires and the other for alcohol or polar solvent-type fires. Fires involving ethanol/gasoline mixtures with greater than 10% alcohol (E85 for example) should be treated differently than traditional gasoline fires. DOT recommends that emergency responders refer to Orange Guide 127 of the *Emergency Response Guidebook* when responding to incidents involving fuel mixtures known to contain or potentially contain more than 10% alcohol. Orange Guide 127 specifies the use of alcohol-resistant foam. Ethanol mixtures above 10% are polar/water-miscible flammable liquids and degrade the effectiveness of non-alcohol-resistant firefighting foams.

Denatured alcohol fires E-95 can only be extinguished by using alcohol-resistant aqueous film-forming foam (ARAFFF) or alcohol-resistant film-forming fluoroprotein (ARFFFP) foam (Figure 3.110). Conventional aqueous film-forming foam (AFFF) and film-forming fluoroprotein foam (FFFP) will not work on alcohol fires because foams contain water. Water is polar and alcohols are also polar. Alcohol mixes with water because they are both polar, and the conventional AFFF and other hydrocarbon foams will break down and not be effective against the alcohol fire. Gasohol (E10) fires can be extinguished by using conventional foams because E10 contains 90% gasoline.

Figure 3.110 Denatured alcohol fires E-95 can only be extinguished by using ARAFFF or ARFFFP.

ARAFFF can also be used to extinguish E10 fires, but increased application rates may be required. AR-type foams need to be applied to ethanol fires using Type II gentle application techniques. Direct application to the surface of the fuel will likely be ineffective unless fuel depth is very shallow. Because ethanol is a polar solvent and mixes readily with water, some thought might be given to dilution of the ethanol or blended mixture. On the surface this seems logical; however, ethanol mixtures with up to 80% water content still burn quite well. Dilution would not be a successful way to extinguish an ethanol fire.

Most responder encounters with ethanol and its blends will occur at the manufacturing facility or in transportation involving an MC/DOT 306/406 atmospheric highway tanker truck or a train of multiple railcars. Ethanol has become the largest volume hazardous material shipped in the United States. It is also shipped by barge and sea-going tanker. Very little ethanol or its blends are currently transported by pipeline. E98, E95 or denatured ethanol is the most common form transported. The primary transportation mode for ethanol is the railcar. Trains with ethanol railcars may contain large numbers of tank cars with potentially hundreds of thousand of gallons of the flammable liquid. Each ethanol tank car carries approximately 30,000 gallons. They are liquid tank cars at atmospheric pressure. Each tank car has a pressure relief valve set to go off at 75 psi overpressure inside the tank. It is, however, possible that pressure may build up inside a tank faster than the relief valve can discharge it. This may result in the tank rupturing violently. Ethanol blends with gasoline are generally not stored at fixed sites. Denatured ethanol is stored in bulk closed floating roof tanks at storage facilities. The blending process takes place at the loading rack where ethanol and gasoline are blended as they are loaded into tanker trucks. These vehicles carry approximately 9,200 gallons of fuel. Barges may carry from 420,000 gallons to 4.2 million gallons of fuel. Most incidents involving ethanol and its blends with gasoline occur during transportation or transfer of product.

I have read where people have referred to explosions involving ethanol tank cars as BLEVEs. A BLEVE is a boiling liquid expanding vapor explosion. Tanks that contain boiling liquids subject to a BLEVE are pressure tanks that contain materials such as propane, butane and liquefied petroleum gases (LPGs). Ethanol tanks are liquid atmospheric pressure tanks. Under normal conditions, the tanks are at atmospheric pressure. The types of materials that are subject to a BLEVE are gases which have been liquefied in order to ship larger quantities. Liquefied gases remain a liquid in the tank above their boiling point because of continued pressure in the tank. They continue to be liquids as long as the pressure container is intact. If a breach occurs in the container, all of the liquid in the tank car immediately turns back into a gas at once. This occurs explosively, often rocketing parts of the disintegrating tank car over 1,000 ft. Thus, the term

BLEVE, boiling liquid in the tank, vapor expands as the tank is breached, occurring explosively. The main factor to bear in mind is that the power of a BLEVE lies in the release of **LIQUID** product that quickly (if not instantaneously) boils or evaporates into the vapor phase and combines with the surrounding atmosphere to form a combustible and explosive mixture. Tanks containing ethanol and its mixtures with gasoline exposed to fire can rupture because of excess pressure built up by flame impingement, but they do not BLEVE. Since 2000 there have been reports of at least 25 incidents involving ethanol and its blends at fixed facilities and in transportation (*Firehouse Magazine*).

CASE STUDY

CHERRY VALLEY, IL, 2009

On June 17, 2009, 18 cars of a Canadian National Railway train containing denatured ethanol derailed, and 14 of them caught fire in Cherry Valley, IL, near Rockford. Of the train's 114 cars, 74 contained denatured ethanol. A civilian setting in their vehicle at the railroad crossing was fatally burned by the fire which engulfed her vehicle. Six others were also injured and taken to area hospitals. Approximately 600 homes in the area of the derailment were evacuated. Firefighters from 26 local departments responded to the fire that was allowed to burn itself out over several days (NTSB Investigation Report).

Crude Oil and Its Response Challenges
With the advent of new technologies, production of crude oil from shale formations in North Dakota, Colorado, Pennsylvania and Texas has significantly increased the transportation of crude oil by rail and highway over the past several years. Correspondingly, there have been an alarming number of train derailments in the United States and Canada involving crude oil (Figure 3.111). One of the reasons why the magnitude of these incidents has been so great is the number of cars of crude oil that are on these trains. Trains can have over 100 cars carrying 30,000 gallons each of crude oil to refineries. Derailments have also shown that typical tank cars used for transportation of crude oil have flaws that increase the risk to emergency responders and the public.

Most crude oil in the United States is shipped through pipelines to the refineries. Currently, limited pipeline infrastructure is available to ship crude oil from some of these new oil fields to the refineries, so it has to be shipped by a combination of truck and rail. This increase in the production of crude oil in North Dakota and also at other locations in the United

Figure 3.111 Numerous train derailments have occurred in the United States and Canada involving crude oil.

States and Canada comes from developments in technology involving horizontal drilling and hydraulic fracturing. The boom in oil production in North Dakota is primarily from the discovery of the Parshall Oil Field in Mountrail County in 2006.

The Bakken oil boom has propelled North Dakota into the top ranks of oil-producing states. As recently as 2007, North Dakota ranked eighth among the states in oil production. In 2008, the state overtook Wyoming and New Mexico; in 2009 it outproduced Louisiana and Oklahoma. North Dakota surpassed California in oil production in December 2011, then overtook Alaska in March 2012, to become the number 2 oil-producing state in the country, exceeded only by Texas.

Petroleum or crude oil is a mixture of hydrocarbons (compounds composed of carbon and hydrogen) and other chemicals. Crude oil hydrocarbons should not be confused with refined hydrocarbons with many commercial uses in our everyday life such as propane, butane and bitumen, which all come from the same barrel of crude oil. The color and viscosity of petroleum vary markedly from one place to another. Most petroleum is dark brown or blackish in color, but it also occurs in green, red or yellow. The elemental composition of crude oil includes the following: carbon (83%–87%), hydrogen (10%–14%), nitrogen (0.1%–2%), oxygen (0.05%–1.5%), sulfur (0.05%–6%) and metals <0.1%.

Hydrocarbons found in crude oil range from having only 1 carbon (methane or natural gas) to having 50 (pentacontane). Four main types of hydrocarbons are found in crude oil: paraffins (15%–60%), naphthenes (30%–60%), aromatics (3%–30%) and asphaltics (the balance). Paraffins, also known as alkanes, are single-bonded hydrocarbons considered to be saturated. They contain from 20 to 40 carbons in the compounds.

Naphthenes are cyclic single-bonded hydrocarbons that end in "-ane," such as cyclohexane. Aromatics are an organic molecule containing the benzene ring. Asphaltics is a term associated with asphalts, a class of solid to semisolid hydrocarbons derived from crude oil. Other chemicals found in crude oil besides the hydrocarbons include sulfur, nitrogen, oxygen and metals. The most common metals are iron, nickel, copper and vanadium.

Crude oil is a Class 3 flammable liquid under the DOT Hazard Classification System. Because it is transported in bulk, it has a UN four-digit identification number of 1267, which appears in the center of a red flammable liquid placard or in an orange rectangle near a flammable liquid placard. NFPA 704 fixed facility marking system designation for crude oil is fire—3, health—2 and reactivity—0. Crude oil flammable characteristics can vary greatly depending on where the crude oil is found in nature. Because crude oil is a mixture of hydrocarbons rather than a pure compound, and mixture components can vary, it is difficult to determine the exact physical properties of any one crude oil mixture.

Boiling point, flash point, flammable range, vapor pressure, vapor density and specific gravity are all important physical characteristics to know when dealing with emergencies involving flammable liquids. But because crude oil is a mixture of hydrocarbons, information on physical characteristics of a given mixture may not be available. In general, the flash point of a crude oil is below 10°C (50°F) and the boiling point is below 35°C (95°F). Vapors from crude oil spill will likely be heavier than air and pool in low-lying areas. Response resources such as the *Emergency Response Guidebook* (ERG), MSDS Sheets if available, CHEMTREC and the National Response Center should be used initially to obtain the information about crude oil incidents.

Crude oil from the Bakken Region of North Dakota is considered by some to be more volatile than "normal" crude oil from other areas. However, when responding to an accident involving crude oil, you need to understand that it is a flammable liquid. That means if the crude oil is not already on fire, it will burn, and you need to take precautions like you would with any flammable liquid that has been released from its container. If a flammable liquid is already on fire, none of the physical characteristics really matter at that point. Focus needs to turn to tactics to deal with the fire conditions. One of the "hidden" characteristics of crude oil is that it is considered a heavy hydrocarbon. Heavy hydrocarbons are usually difficult to get to burn, but once burning have a high heat output and are very difficult to extinguish. Water does not work well by itself for extinguishing flammable liquid fires. The specific gravity of flammable liquids and specifically crude oil is less than one and is not miscible with water, so it will float on the surface of water. This may allow burning crude oil to spread. Water should be used to cool tanks, protect exposures and extinguish exposure fires if these things can be accomplished safely.

Unmanned monitors should be used for water and foam applications. According to the NFPA a minimum of 500 gallons of water flow per minute is required for tank cooling operations. Effective extinguishment of crude oil fires requires the use of hydrocarbon foam concentrates and appliances for application of foam. Fire extinguishment using water requires a particular volume of water to extinguish a given volume of fire. The same is true for foam firefighting operations; it requires a certain amount of foam to extinguish a given volume of flammable liquid fire.

Once extinguishment is accomplished, a blanket of foam needs to be maintained on spilled liquids as well to prevent reignition. The bigger the fire, the more foam required. Many fire departments do not immediately have access to large quantities of foam. This will limit the ability of extinguishment until enough foam concentrate can be obtained and brought to the incident scene. It is a waste of time and foam concentrate to apply foam to a flammable liquid on fire or to lay down a foam blanket if there is not enough concentrate available to accomplish the task. Initial tactical objectives may be limited to defensive actions including protection of personnel and the public through evacuations and letting the fires burn. Offensive tactics will often place emergency response personnel at greater risk.

Trains carrying crude oil can be over a mile long with hundreds of tank cars. If a derailment occurs, thousands of gallons of product can be released from derailed tank cars. Several possible scenarios exist when a train derailment occurs which involves crude oil tank cars.

- Pool fire, where a leak occurs, and liquid pools and ignites
- Flame impingement on tanks
- Tank rupture with possible fragmentation
- Fire balls
- Flash fires, where fuel vapor ignites at a point beyond the release

Most emergency response agencies will not have immediate resources or trained personnel to safely respond to or have the capability to intervene with such an event. This type of incident will require specialized equipment and trained personnel to mitigate, and these resources may be hours away. Railroads should be contacted immediately.

Propane suspended in oil may have caused railcar explosions
As federal regulators continue investigating why tank cars on three trains carrying North Dakota crude oil have exploded in the past 8 months, energy experts say that part of the problem might be that some producers are deliberately leaving too much liquid propane in their product, making

the oil riskier to transport by rail. Sweet, light crude from the Bakken Shale formation straddling North Dakota and Montana has long been known to be especially rich in volatile natural gas liquids such as propane. Much of the oil is being shipped in railcars designed in the 1960s and identified in 1991 by the National Transportation Safety Board as having a dangerous penchant to rupture during derailments or other accidents.

While there's no way to completely eliminate natural gas liquids from crude, well operators are supposed to use separators at the wellhead to strip out methane, ethane, propane and butane before shipping the oil. A simple adjustment of the pressure setting on the separator allows operators to calibrate how much of these volatile gases are removed. The worry, according to a half-dozen industry experts who spoke with *Inside Climate News*, is that some producers are adjusting the pressure settings to leave in substantial amounts of natural gas liquids.

"There is a strong suspicion that a number of producers are cheating. They generally want to simply fill up the barrel and sell it—and there are some who are not overly worried about quality," said Alan J. Troner, president of Houston-based Asia Pacific Energy Consulting, which provides research and analysis for oil and gas companies. "I suspect that some are cheating and this is a suspicion that at least some refiners share." Harry Giles, a now-retired, 30-year veteran of the Department of Energy whose duties included managing the crude oil quality program for the Strategic Petroleum Reserve, said that there's "a distinct possibility" that propane has been intentionally left in Bakken oil.

"I think there is such a large focus on what's happening in the Bakken...that no one really cares to talk about these issues," Giles said. Producers might be tempted to leave in some of the natural gas liquids because there aren't enough gas-processing facilities or pipelines in the Bakken to handle all the methane, ethane, propane and butane that is suspended in the crude when it comes out of the ground. Without sufficient infrastructure, operators are left with few options. They can flare or vent the volatile gases into the North Dakota sky, although they risk being penalized for violating emission limits. Or they can leave some of the gases, especially propane, suspended as liquid in the crude oil they send to refineries, where gas-processing facilities already exist.

Some drillers might also be purposefully selling their crude "fluffed up" with propane and small amounts of butane to boost the volume of oil in the railcar and maximize their profits, according to the experts, some of whom spoke on the condition they not be identified because of pending lawsuits triggered by recent accidents. The Bakken, a vast crude reservoir lying about 2 miles beneath the Earth's surface, has been tapped since 1953. It was only in recent years that new fracking technologies allowed the volume of crude taken from the ground to explode, jumping from a negligible amount in 2007 to one million barrels a day currently.

Energy companies have been scrambling to install the infrastructure they need to support the boom. But they face awkward economics. Constructing gas plants and pipelines is expensive and involves a lengthy permitting process. By the time the facilities are in place, production at many Bakken wells might be in decline. Lynn Helms, head of the North Dakota Department of Mineral Resources and the state's chief oil well regulator, said in a statement emailed by his spokesperson that "at this time we are investigating what, if any, issues there may be surrounding separation of Bakken streams."

At the federal level, the movement of crude oil by rail is regulated by the Federal Railroad Administration and the Pipeline and Hazardous Materials Safety Administration (PHMSA), both housed within the DOT. PHMSA officials did not respond to questions about whether the agency is investigating Bakken oil companies for deliberately leaving too much propane in their crude. The American Petroleum Institute, which has been assisting PHMSA in its effort to determine what new rules or testing methods are needed, declined to comment.

PHMSA began testing Bakken crude to see what was making it so volatile after an oil train from North Dakota derailed and exploded in Canada in July, killing 47 people and generating up to $2 billion in liabilities. In response to questions from *Inside Climate News* about what is making the Bakken crude explosive, PHMSA spokesman Gordon "Joe" Delcambre said in a February 14 email that the agency is "still awaiting the final report of the test results on the crude oil samples submitted to the lab. Keep checking back periodically." As of last week, PHMSA had provided no update (*Firehouse Magazine*).

CASE STUDIES

ALICEVILLE, AL, NOVEMBER 8, 2013, CRUDE OIL DERAILMENT AND FIRE

On November 8, a train carrying Bakken oil derailed and burned near Aliceville, AL, en route to a refinery in Mobile. On December 30, another train hauling Bakken oil train collided with a derailed grain train outside Casselton, ND, setting off a series of explosions that sent large, mushroom-shaped fireballs into the sky. In response to the accidents, PHMSA issued a safety alert on January 2 saying, "recent derailments and resulting fires indicate that the type of crude oil being transported from the Bakken region may be more flammable than traditional heavy crude oil." On February 4, the agency proposed fining three Bakken producers for shipping oil that was more hazardous than shipping documents indicated.

On February 25, the DOT ordered Bakken operators to begin testing their rail shipments "with sufficient frequency and quality" to ensure the shipping papers properly reflect the oil's flammability. At a congressional hearing 2 days later, PHMSA administrator Cynthia Quarterman said that oil companies would have wide discretion in determining what constituted "sufficient frequency and quality" when it came to testing. "We specifically left those terms to be determined by the shippers based on their operations," Quarterman said. "We did not want to say each and every instance before a shipment occurs that testing needed to occur. It may be that a shipper, if they are a producer, are producing from one play and that play is consistent and over time the test results would be the same."

Giles, the retired Department of Energy official, expressed doubts about the testing program. "The number of railcars they're loading each day, the number of tank trucks that are going into each of these, the wide range of oil quality across the Bakken area are creating challenges for a sampling and analysis program," he said. "It's not like some of the established fields in Oklahoma or Louisiana or Texas where the quality is fairly constant across the producing area. Here in the Bakken it ranges from a light crude to a fairly heavy crude. The amount of light constituents that are being produced are creating real challenges. I have reservations about what is being done and how it's being done." However, Giles believes that the fiery explosions have already led producers to be more careful with their shipments (DOT Reader).

"There are numerous operators there and I am confident that they are going to do whatever is needed and prudent to minimize any increase in the volatility of the crude oil," he said. All the industry experts interviewed by *Inside Climate News* say that a key step in preventing future rail explosions is to properly strip out propane and other natural gas liquids from the crude.

Oil comes out of the ground mixed with natural gas, natural gas liquids and water. The first step is to put this mixture through a series of separators that reduces the pressure of the fluid, separating the ingredients into distinct streams. Raising the final separator's pressure setting leaves more propane dissolved within the crude. As an oil train shakes, rattles and rolls toward the refinery, which in the case of Bakken oil can be thousands of miles away, the propane begins separating from the liquid and turning into gas. A typical tank car carries about 30,000 gallons of flammable liquid at the start of its journey. Some oil trains pull more than 100 cars for a total of more than three million gallons of propane-rich crude.

If one of those cars ruptures, the propane gas inside will likely make contact with outside air. If the gas is ignited perhaps by a spark thrown off when the car rips open or maybe a spark thrown up from steel wheels scraping over steel tracks—the car can explode. Then the burning car can act like a blowtorch on the tanker next to it, even if that car is upright and intact. Eventually, the metal shell of the second tanker would fail from the heat and explode like the first one. Engineers have a clunky technical term for such a disaster. They call it a BLEVE (rhymes with levee). At that point, railcars can explode in a domino fashion. Experts say that the volatility of the Bakken oil lies in its chemistry.

"It's typical of this type of oil. So it's not surprising. There's no mystery to it... especially if it were in a tanker not meant to carry that type of fluid," said Ramanan Krishnamoorti, a professor of petroleum engineering at the University of Houston.

Krishnamoorti was referring to the much-criticized DOT-111, a black, torpedo-shaped railcar designed in the 1960s that has become the workhorse of the crude rail industry during the nation's drilling boom. The U.S. National Transportation Safety Board (NTSB) ratcheted its long-standing push for sturdier railcars after a 2009 accident in which a DOT-111 ethanol train derailed and exploded at a railroad crossing in Cherry Valley, IL, killing one motorist and injuring others. The NTSB's investigation of that accident prompted the Association of American Railroads (AAR), an industry trade group, to petition PHMSA to expedite tougher standards for future railcars. Expressing doubts that PHMSA would act in a timely way, the AAR adopted its own voluntary standards for newly built railcars in July 2011. The issue of the railcar standards remains a point of contention between the government agencies, PHMSA and NTSB. In May 2012, Quarterman told NTSB Chair Deborah Hersman that implementing the NTSB's recommendations would have financial consequences for industry.

"Requiring all new and existing DOT-111 rail tank cars to comply with enhanced design standards will no doubt be a very costly endeavor," she said in a letter to Hersman, adding, "We invite and encourage NTSB to comment as we proceed through the regulatory process."

When Quarterman wrote that letter, it had been 21 years since the NTSB had first warned of the dangers of using the DOT-111 to carry hazardous materials. PHMSA's defenders note that the NTSB doesn't have to worry about pushback from industry stakeholders or the rough and tumble of regulatory rulemaking.

"NTSB has a very broad mandate: Investigate causes and potential remedies," said former PHMSA administrator Brigham McCown. "But unlike all the other executive branch agencies, they're not constrained by cost-benefit analysis or any of the other regulatory restrictions that are typically placed on other executive branch agencies."

Eric Weiss, an NTSB spokesperson, agreed with McCown's interpretation of the NTSB's mission. "Our focus is on safety, on representing the American people when it comes to safety," Weiss said. "The Cherry Valley accident happened in 2009. We issued our strong recommendations in March 2012. And it's now 2014." PHMSA's consideration of upgrade railcar requirements will continue at least through early next year, according to its timetable. It will take even longer to get the new or retrofitted railcars onto the tracks, since manufacturers are stuck in a holding pattern until new standards are determined. In the meantime, PHMSA's Delcambre said that the agency would have inspectors in North Dakota "performing unannounced inspections and taking crude oil samples at crude oil loading and handling facilities."

On Thursday, the U.S. Senate Committee on Commerce, Science and Transportation conducted a hearing on rail safety, including the transportation of Bakken crude oil. The hearing came just days after two trains carrying oil from North Dakota derailed in New York State. None of the oil spilled, and nobody was hurt. But the accidents prompted New York Governor Andrew Cuomo to urge federal officials to do more to tighten safety regulations. "I am not convinced that federal regulations and oversight sufficiently protect New York's communities and natural resources from safety hazards in transporting this material," Cuomo said in a letter to U.S. Transportation Secretary Anthony Foxx and Homeland Security Secretary Jeh Johnson (Inside Climate News).

LAC-MEGANTIC, QUEBEC, CANADA, 2013

By far, the worst train derailment involving crude oil occurred in Lac-Megantic, Quebec, Canada on July 6, 2013. This derailment occurred in downtown Lac-Megantic and resulted in 47 civilian fatalities and destroyed over 30 buildings (Figure 3.112). Victims ranged in age from 4 to 93. Five victims were never found and presumed to have been vaporized by the fires. Fire departments from Canada and the state of Maine responded to the massive fires in downtown Lac-Megantic. The train had been parked on the main line 7 miles outside of town. During the night, a fire occurred in one of the locomotives on the

Figure 3.112 This derailment occurred in downtown Lac-Megantic and resulted in 47 civilian fatalities and destroyed over 30 buildings.

parked train. The fire department responded and extinguished the fire, and railroad employees declared the train safe. Brakes failed to hold the train in place following the engine fire, and sometime after midnight the train started to roll downhill toward Lac-Megantic. It is reported by witnesses that the train rolled into Lac-Megantic at a high rate of speed and derailed at 1:14 a.m. Sixty-three of the train's 72 cars derailed near a grade crossing at Frontenac Street, the main street in downtown Lac-Megantic. Derailed tank cars immediately caught fire, and four to six explosions were reported by witnesses (Wikipedia.org).

Crude oil is typically carried on trains utilizing "Legacy" DOT-111 and in some cases the newer CPC-1232 tank cars. The NTSB has flagged the DOT-111 tank car as unfit to haul crude oil and ethanol because the car is prone to puncture in the event of a derailment. CPC-1232 tank cars have half head shields installed to protect from punctures. However, the cars have been involved in derailments with resulting fires and thermally failed in at least four derailments resulting in explosions. NTSB has further determined that both the DOT-111 and CPC-1232, which are nonthermally protected, are vulnerable to thermal failures. NTSB wants these cars to be equipped with jackets, thermal protection and appropriately sized pressure relief devices, which allow for the release of pressure under fire conditions. NTSB believes that the 10-year deadline for retrofitting or removing from service CPC-1232 tank cars is excessively long. Crude oil production in North America is expected to increase by more than five million barrels per day by 2024.

According to DOT's PHMSA, "The size, scope and resources needed to successfully manage a crude oil rail transportation incident will overwhelm the capability of most emergency response agencies." They make the following transportation and planning recommendations for emergency response personnel for dealing with crude oil incidents. Responses to unit train derailments of crude oil require specialized outside resources that may not arrive at the scene for hours; therefore, it is critical that responders coordinate their activities with the involved railroad and initiate requests for specialized resources as soon as possible.

- These derailments likely require mutual aid and a more robust on-scene incident management system than responders may normally use. Therefore, preincident planning, preparedness and coordination of response strategies should be considered and made part of response plans, conduct drills and exercises that include the shippers and rail carriers of this commodity.
- Evaluate the risks of personnel intervening directly in the incident. Consider the limitations of people involved and the ability to have adequate resources available on-site (e.g., sufficient firefighting foam concentrate, water supplies, appliances, equipment, trained personnel and technical expertise) and the ability to sustain operations for extended periods of time (hours or days).
- For nonfire spill scenarios, have the concentrations of any flammable or toxic vapors present been determined using air monitoring instruments? What are the flammability and toxicity readings? Has the need for continuous air monitoring been properly evaluated and discussed with technical specialists? Can sources of ignition be removed and/or be eliminated? Are adequate foam supplies and equipment available for vapor suppression?
- Based on the results of the hazard assessment and risk evaluation process, are there adequate resources available to respond to the scene within a reasonable time frame so that intervention efforts will be successful?

If your agency is not fully prepared and capable in terms of resources, equipment and properly trained personnel to intervene, defensive or nonintervention strategies will likely be the preferred strategic options (NTSB).

Hazard Class 4 Flammable Solids (LF/LR)

Hazard Class 4 is divided into three divisions: 4.1 Flammable Solids, 4.2 Spontaneously Combustible and 4.3 Water Reactive. Solid materials may take on many physical forms including fine powder, filings, chips and various sized solid chunks. Some solid materials such as white phosphorus may be pyrophoric, that is, they spontaneously combust on exposure to air. Others such as calcium carbide must be wet before they become a hazard by releasing acetylene gas. Some of the materials release poisonous gases such as phosgene, nitrous oxide or chlorine. There are solid materials that when wet release oxygen such as sodium peroxide. This can present added dangers to firefighters because the primary extinguishing agent they use is water, which can release oxygen from the water-reactive materials that will accelerate the combustion process and make the fire more difficult to extinguish.

Other solid materials spontaneously combust when exposed to water, and the reaction is exothermic or heat producing; if combustible materials are present, combustion will occur. Some materials such as phosphorus are shipped under water, whereas others such as picric acid are shipped with 10%–50% water in the container. Materials such as picric acid are considered wetted explosives from Class 1, but present a limited hazard as long as the water is present. It is important that emergency responders have a thorough understanding of each of the divisions, the types of materials that make them up and the hazards posed in a release.

Division 4.1 Flammable Solids—LF/LR

The U.S. Department of Transportation (DOT) defines 4.1 materials as (1) wetted explosives that when dry are explosives of Hazard Class 1; (2) self-reactive materials that are liable to undergo, at normal or elevated temperatures, a strongly exothermal decomposition caused by excessively high transport temperatures or by contamination; and (3) readily combustible solids that may cause a fire through friction such as matches show a burning rate faster than 0.087 in./s or any metal powders that can be ignited and react over the whole length of a sample in 10 min or less.

Chemical Notebook

Ammonium picrate, yellow crystals with not less than 10% water by mass, is highly explosive when dry and flammable when wet.

Barium azide is a crystalline solid, with not less than 50% water by mass, and explodes when shocked or heated. It decomposes and gives off nitrogen at 120°C and is soluble in water.

Camphor, synthetic, is composed of colorless or white crystals, granules or easily broken masses. It has a penetrating aromatic odor and sublimes slowly at room temperature. It has a flash point of 150°F and an autoignition temperature of 871°F. It evolves flammable and explosive vapors when heated. Camphor is a material that under goes sublimation. Sublimation is the process of changing physical states. In the case of camphor, it goes from a solid to a vapor without becoming a liquid first. These materials are also known as flash point solids. The flash points are generally all above 100°F.

Cobalt naphthenates are brown, amorphous powders or bluish-red solids, and are combustible.

Fusee are flares used on the highway or rail as warning devices and are considered flammable solids, which are ignited by friction.

Magnesium is found as a silvery soft metal, pellets, turnings or ribbons. It is flammable and a dangerous fire hazard. It reacts violently with water when burning. In the form of powder, pellets or turnings, it can explode when in contact with water.

Matches are shipped as flammable solids and ignite by friction.

Picric acid is shipped with not less than 10% water and is composed of yellow crystals. It is a severe explosion risk when shocked or heated, reacts with metals or metallic salts and toxic by skin absorption. It has caused disposal problems in school and other chemistry laboratories where the moisture has escaped from the container as the materials age. When the picric acid dries out, it becomes a high explosive closely related to TNT. Picric acid has been found in various amounts in school laboratories across the country. It is very dangerous in a dry condition and should be handled by the bomb squad.

Sulfur is composed of pale yellow crystals. It is a dangerous fire and explosion risk in finely divided form. It is an elevated temperature material, which displays the HOT marking on the container. Molten sulfur is above 300°F and can cause severe burns, and the vapors are toxic.

Titanium is a silvery solid or dark gray amorphous powder. It is a dangerous fire and explosion risk, and burns in a nitrogen atmosphere. Water and carbon dioxide are ineffective extinguishing agents for fires involving titanium.

Urea nitrate is a colorless crystal shipped with not less than 20% water by mass. It is a dangerous fire and explosion risk.

Zirconium is found as a grayish, crystalline scale or gray amorphous powder. It is flammable or explosive in the form of a powder or dust, and it is a suspected carcinogen. The powder should be kept wet in storage.

Diethyl zinc is an organo-metal compound and a dangerous fire hazard. It spontaneously ignites in air and reacts violently with water, releasing flammable vapors and heat. It is a colorless pyrophoric liquid with a specific gravity of 1.2, which is heavier than water, so it will sink to the bottom. It decomposes explosively at 248°F. It has a boiling point of 243°F, a flash point of –20°F and a melting point of –18°F. The four-digit identification number is 1366. The National Fire Protection Association (NFPA) 704 designation is health—3, flammability—4 and reactivity—3. The white space at the bottom of the diamond has a "W" with a slash through it, indicating water reactivity. The primary uses of diethyl zinc are in the polymerization of olefins, high-energy aircraft and missile fuel and the production of ethyl mercuric chloride.

Pentaborane is a nonmetallic, colorless liquid with a pungent odor. It decomposes at 300°F if it has not already ignited and ignites spontaneously in air if impure. It is a dangerous fire and explosion risk, with a flammable range of 0.46%–98% in air. The boiling point is 145°F, the flash point is 86°F and the ignition temperature is 95°F, which is extremely low. Any object that is 95°F or above can be an ignition source. Ignition sources can be ordinary objects on a hot day in the summer, such as the pavement, metal on vehicles and even the air. The 4-digit identification number is 1380. The NFPA 704 designation for pentaborane is health—4, flammability—4 and reactivity—2. It is primarily used as fuel for air-breathing engines and as a propellant.

Aluminum alkyls are colorless liquids or solids. They are pyrophoric and may ignite spontaneously in air. They are pyrophoric materials in a flammable solvent. The vapors are heavier than air, water reactive and corrosive. Decomposition begins at 350°F. The 4-digit identification number is 3051. The NFPA 704 designation is health—3, flammability—4 and reactivity—3. The white space at the bottom of the diamond has a "W" with a slash through it, indicating water reactivity. They are used as catalysts in polymerization reactions.

Aluminum phosphide (AlP) is a binary salt. These salts have the specific hazard of giving off poisonous and pyrophoric phosphine gas when in contact with moist air, water or steam. They will also ignite spontaneously in contact with air. This compound is composed of gray or dark yellow crystals and is a dangerous fire risk. Aluminum phosphide decomposes when in contact with water and has a specific gravity of 2.85, which is heavier than water. The 4-digit identification number is 1397. The NFPA 704 designation is health—4, flammability—4 and reactivity—2. The white section at the bottom of the diamond has a "W" with a slash through it, indicating water reactivity. Aluminum phosphide is used in insecticides, fumigants and semiconductor technology.

Potassium sulfide (K_2S) is a binary salt. It is a red or yellow–red crystalline mass or fused solid. It is deliquescent in air, which means it absorbs water from the air, and it is also soluble in water. Potassium sulfide is a dangerous fire risk and may ignite spontaneously. It is explosive in the form of dust and powder. It decomposes at 1,562°F and melts at 1,674°F. The specific gravity is 1.74, which is heavier than air. The 4-digit identification number is 1382. The NFPA 704 designation is health—3, flammability—1 and reactivity—0. Potassium sulfide is used primarily in analytical chemistry and medicine.

Sodium hydride (NaH) is a binary salt that has a specific hazard of releasing hydrogen when in contact with water. It is an odorless powder that is violently water reactive. The 4-digit identification number is 1427. The NFPA 704 designation is health—3, flammability—3 and reactivity—2. The white space at the bottom of the diamond has a "W" with a slash through it, indicating water reactivity.

There are other materials that may become flammable solids because of their physical state. These materials are considered flammable dusts, which are very finely, divided particles of some other ordinary Class A combustible materials. Some of these materials may include such ordinary materials as sawdust, grain dusts, flour and coal dust. These materials may not be considered hazardous materials by the DOT. They may sometimes be shipped in dry bulk transportation containers. The combustible dusts become a problem when they are suspended in air in an indoor environment, such as a grain elevator or flourmill, and an ignition source is present and an explosion occurs. These materials may also be suspended in air in a transportation accident and in the presence of an ignition source could create an explosion. Even though combustible dusts are not considered in any of the DOT hazard classes, they present a significant fire and explosion hazard under certain conditions and responders should be aware of this hazard.

Division 4.2 Spontaneously Combustible—LF/LR

DOT defines them as a pyrophoric material. These materials may be found as solids or liquids. Spontaneous combustion can occur when a hazardous material is exposed to air, such as phosphorus (Figure 3.114), or slow spontaneous combustion of animal vegetable oils or agricultural products such as hay. Spontaneously combustible materials are a concern not only to hazmat teams but to firefighters and fire investigators. These materials can start fires without any external ignition sources, and if not carefully investigated be deemed a suspicious or arson fire.

Liquids
Liquids are included in a solid category because they are spontaneously combustible. There is no other hazard class that they would fit into. Liquids can ignite without an external ignition source within 5 min after coming in contact with air.

Solids
There are other 4.2 materials that may be self-heating. That is, when in contact with air and without an energy supply (ignition source), they

are liable to self-heat, which can result in a fire involving the materials or other combustible materials nearby. Bulk transport and storage may result in spontaneous combustion of agricultural products, such as hay, straw and cotton. Organic materials that are prone to spontaneous ignition at ordinary ambient temperatures must have access to atmospheric oxygen or carry oxidizer in some reactive form and be raised to its autoignition temperature by the exothermic reaction. In some cases of spontaneous combustion, moisture is essential for the resulting reaction. Hay, for example, will not spontaneously combust if the moisture level is below 33%. Some materials subject to spontaneous ignition are shown below.

Chemical Notebook
Spontaneous combustion of stored coal.

According the to the United States Department of Energy, "Spontaneous combustion has long been recognized as a fire hazard in stored coal. Spontaneous combustion fires usually begin as "hot spots" deep within the reserve of coal. The hot spots appear when coal absorbs oxygen from the air. Heat generated by the oxidation then initiated the fire."

Animal or vegetable oils such as linseed oil, cooking oil and cottonseed oil undergo spontaneous combustion when in contact with rags or other combustible materials. The oxidation reaction that occurs with animal or vegetable oils is different than the reaction with hydrocarbon petroleum-based materials. The oxygen from the air trapped in the mass reacts with the double bonds present in the animal or vegetable oils. The braking of the double bonds creates heat, which ignites the materials when they reach their respective ignition temperatures.

Petroleum products that contain saturated alkanes do not contain these double bonds and therefore cannot undergo this type of spontaneous heating and cause a fire. Fires started by this spontaneous heating process can be difficult to extinguish because they usually involve deep-seated fires. In order for enough heat to be sustained to cause combustion, there must be insulation. This insulation can be the material itself or in the form of some other combustible material such as rags or other fabrics (Hazardous Materials Chemistry for Emergency Responders).

CASE STUDIES

PHILADELPHIA, PA ONE MERIDIAN PLAZA, LINSEED OIL SOAKED RAGS

The high-rise fire at One Meridian Plaza in Philadelphia that killed three firefighters was started by rags soaked with linseed oil during construction operations and improperly disposed of (Figure 3.113).

Figure 3.113 The high-rise fire at One Meridian Plaza in Philadelphia that killed three firefighters was started by rags soaked with linseed oil during construction operations and improperly disposed of.

The rags underwent slow heating and eventually spontaneously ignited and started the fire. (Hazardous Materials Chemistry for Emergency Responders).

GETTYSBURG, PA, MARCH 23, 1979
PHOSPHORUS TRUCK FIRE

An incident occurred in Gettysburg, PA, involving phosphorus being shipped under water in 55-gallon drums. One drum developed a leak and the water drained off. This allowed the phosphorus to be exposed to air, which caused it to spontaneously ignite. The fire spread to the other containers and eventually consumed the entire truck. The ensuing fire was fought with large volumes of water and in the final stages covered with wet sand. Clean-up created problems

Figure 3.114 An incident occurred in Gettysburg, PA, involving phosphorus being shipped under water in 55-gallon drums.

because as the phosphorus and sand mixture was shoveled into overpack drums, the phosphorus was again exposed to air and reignited small fires (Figure 3.114) (*Firehouse Magazine*).

BROWNSON, NE, 1978

A train derailment in Brownson, NE, resulted in a tank car of phosphorus overturning and the phosphorus igniting upon contact with air. Phosphorus is shipped under water so there was water inside the tank car. CHEMTREC was called and responders were told correctly that the phosphorous would not explode. However, the water inside the tank car was turned to steam from the heat of the phosphorus fire. The pressure from the steam caused a boiler-type of explosion that had nothing chemically to do with the phosphorus! This is just another example of the actors that emergency responders must be aware of when dealing with hazardous materials. Not only do the hazardous materials have to be considered, but also the container and any "inert" materials that may be involved (*Firehouse Magazine*).

Phenomenon of Spontaneous Combustion

Spontaneous combustion is a sometimes mysterious and often misunderstood or misdiagnosed combustion of flammable materials. I am sure that many fires have been blamed on spontaneous combustion that shouldn't have been and many fires that were caused by spontaneous combustion were not properly attributed. Certainly, fire investigators and hazardous materials personnel—if not all firefighters—should know about the concept of spontaneous combustion, the types of materials that may undergo such a reaction and the conditions necessary for it to happen.

Spontaneous combustion, according to the *Handbook of Fire Prevention Engineering*, "is a runaway temperature rise in a body of combustible material that results from heat being generated by some process taking place within the body."

Spontaneous combustion may be rapid or slow. It can result from reactions of a susceptible material with air or water or from a chemical reaction. Materials involved can be chemicals, elements or hydrocarbon compounds or a mixture. Slow spontaneous combustion can occur in two general ways: biological processes of some microorganisms and slow oxidation. Biological processes occur within organic materials such as hay and grass clippings. The activity of biological organisms within the material generates heat that is confined by the materials themselves until the ignition temperature is reached and ignition occurs.

Slow oxidation is a chemical reaction. Chemical reactions may produce heat; reactions that produce heat are considered exothermic. If the heat is insulated from dissipating to the outside of the material, it will continue to build up. As the heat builds, the material is heated from within. The process continues until the ignition temperature of the material is reached and ignition occurs.

Hydrocarbon compounds usually undergo slow spontaneous combustion. Generally, hydrocarbon compounds are considered saturated or unsaturated. Within saturated compounds, all of the chemical bonds among the elements in the compound are single. Single bonds are "full," for there isn't room or a method for another element to attach. Double or triple bonds are unsaturated. In other words, the bond can break, and when it does, another element can attach. Spontaneous ignition occurs when this double or triple bond breaks and creates heat and the heat is confined. The material itself produces heat sufficient to reach its own ignition temperature. Some materials subject to spontaneous combustion are not considered hazardous in transit by the DOT and may not be placarded or labeled. Materials may have other hazards considered more severe. DOT has assigned a hazard class to materials that are shipped in transportation and subject to spontaneous combustion by chemical nature.

Flammable solids can ignite without an external ignition source within 5 min after coming into contact with air. There are other 4.2 materials that may be self-heating, that is, in contact with air and without an energy supply (ignition source); they are liable to self-heat, which can result in a fire involving the material or other combustibles nearby. This type of spontaneous combustion is considered rapid.

Just some of the materials subject to spontaneous heating are listed below:

- Alfalfa meal
- Castor oil
- Coal
- Cottonseed oil
- Fertilizers
- Fish meal
- Fish oil
- Lanolin
- Lard oil
- Linseed oil
- Soybean oil
- Used burlap

Some types of combustible liquids, such as animal and vegetable oils, have a hidden hazard: they may burn spontaneously when improperly handled. They have high boiling and flash points, narrow flammable ranges and low ignition temperatures, and are nonpolar. Animal or vegetable oils, such as linseed oil, cooking oil, cottonseed oil, corn oil, soybean oil, lard and margarine, can undergo spontaneous combustion when in contact with rags, cardboard, paper or other combustibles. These unsaturated compounds can be dangerous when combustible materials containing residue are not properly disposed of or they come in contact with other combustible materials. The key to this spontaneous combustion is that the materials in contact with the animal vegetable oils must be confined for the heat to build up.

There is a double bond in the chemical makeup of animal and vegetable oils that react with oxygen in the air. The oxygen from the air trapped in the mass reacts with the double bonds present in the animal and vegetable oils. The breaking of the double bonds creates heat. If the heat is allowed to build up in a pile of rags, for example, spontaneous combustion will occur over a period of hours. Fires started by this spontaneous heating process can be difficult to extinguish because they usually involve deep-seated fires. In order for enough heat to be sustained to cause combustion, there must be insulation. This insulation can be the material itself or may be in the form of some other combustible material.

Spontaneous heating cannot occur in the case of petroleum oils or other hydrocarbon materials that are saturated. Ordinary petroleum products, such as motor oil, grease, diesel fuel and gasoline, are not pure compounds, but they are mixtures. They do not have double bonds in their chemical makeup. For that reason, the oxidation reaction that occurs with animal and vegetable oils and the oxygen in the air does not happen with saturated hydrocarbons. This fact may come as a surprise to some people because there have been numerous fires blamed on oily rags with those products on them. The fact is that saturated flammable liquids do not spontaneously ignite and cannot start to burn without some other ignition source (*Firehouse Magazine*).

CASE STUDIES

A series of fires have occurred in laundries around the country. One in every six commercial, industrial or institutional laundries reports a fire each year, which results in over 3,000 fires. The primary cause is thought to be spontaneous combustion. Chemicals, including animal and vegetable oils, may be left behind in fabrics after laundering. The heat from drying may cause the initiation of the chemical reaction that causes spontaneous ignition.

A fire in a nursing home laundry in Litchfield, IL, caused $1.5 million in damage. The cause was determined to be spontaneous ignition of residual chemicals in the laundered fabric reacting to heat from the dryer. In Findlay, OH, a fire destroyed a commercial laundry and caused over $5 million in damage. Traces of linseed oil were found in a pile of clean, warm garments in a cart.

Fires in restaurants have also occurred involving residual animal or vegetable oils in cleaning rags. The oils are never completely removed by laundering. When placed in the dryer, the rags are heated. When they are put away on a storage shelf, this heat can become trapped, along with the oil remaining on rags, when confined. The spontaneous combustion process begins very slowly, and the heat of the reaction increases until combustion occurs.

PREVENTION TIPS

- Cooking oil-contaminated materials should be treated the same as all other oily rags, if disposing of these uses a metal drum with a self-closing lid.
- If such materials must be washed, ensure items that were contaminated with combustible substances such as animal/vegetable oils and grease from cooking operations should be washed in very hot water with adequate detergent and rinsed thoroughly to completely remove the contamination.
- Ensure that the cooldown cycle of the tumble dryer is adequate to reduce the temperature of the items and that the items are cooled before folding.
- Ensure that lint filters in the tumble dryers are cleaned before use and that lint is not allowed to accumulate around the appliance.
- Articles should be removed from the dryer as soon as the drying/cooling cycle is completed.

VERDIGRIS, OK

A fire occurred in an aircraft hangar at a small airport. The owner's living quarters were on the second level of the hangar. Workers had been polishing wooden parts of an airplane in the afternoon. The rags used to apply linseed oil were placed in a plastic container in a storage room in the hangar, just below the living quarters. At around 2 A.M., the rags with the linseed oil spontaneously ignited and the fire traveled up the wall into the living quarters. Fortunately, the owner had smoke alarms; the family was awakened and the fire department was called promptly. The fire was quickly extinguished with a minimum of damage. The V-pattern on the wall led right back to the box where the linseed oil-soaked rags had been placed. There was little doubt what had happened; the confinement of the pile allowed the heat to build up as the double bonds were broken in the linseed oil releasing heat, which combined with oxygen in the air and spontaneous combustion occurred.

LINCOLN, NE

Another fire occurred at an auto shop in Lincoln, NE, when a van inside caught fire. Investigation indicated that the cause of the fire was spontaneous combustion from failure to properly dispose of linseed oil-soaked rags.

Some common materials that by chemical nature are subject to spontaneous combustion are presented here for informational purposes. During an incident involving these or any other hazardous materials, the compounds should be looked up in reference resources to determine hazards and proper tactics.

Chemical Notebook

Aluminum alkyls are colorless liquids or solids. They are pyrophoric and may ignite spontaneously in air. Aluminum alkyls are pyrophoric materials in a flammable solvent. The vapors are heavier than air, water reactive and corrosive. Decomposition begins at 350°F. The 4-digit identification number is 3051. The NFPA 704 designation is health—3, flammability—4 and reactivity—3. The white space at the bottom of the diamond has a "W" with a slash through it, indicating water reactivity. They are used as catalysts in polymerization reactions.

Aluminum phosphide (AlP) is a binary salt. These salts have the specific hazard of giving off poisonous and pyrophoric phosphine gas when in contact with moist air, water or steam. They will also ignite spontaneously in contact with air. This compound is composed of gray or dark yellow crystals and is a dangerous fire risk. Aluminum phosphide decomposes when in contact with water and has a specific gravity of 2.85, which is heavier than water. The 4-digit identification number is 1397. The NFPA 704 designation is health—4, flammability—4 and reactivity—2. The white section at the bottom of the diamond has a "W" with a slash through it, indicating water reactivity. Aluminum phosphide is used in insecticides, fumigants and semiconductor technology.

Diethyl zinc is an organo-metal compound and a dangerous fire hazard. It spontaneously ignites in air and reacts violently with water, releasing flammable vapors and heat. It is a colorless pyrophoric liquid with a specific gravity of 1.2, which is heavier than water, so it will sink to the bottom. It decomposes explosively at 248°F. It has a boiling point of 243°F, a flash point of –20°F and a melting point of –18°F. The 4-digit identification number is 1366. The NFPA 704 designation is health—3, flammability—4 and reactivity—3. The white space at the bottom of the diamond has a "W" with a slash through it, indicating water reactivity. The primary uses of diethyl zinc are in the polymerization of olefins, high-energy aircraft and missile fuel, and the production of ethyl mercuric chloride.

Pentaborane is a nonmetallic, colorless liquid with a pungent odor. It decomposes at 300°F if it has not already ignited and will ignite spontaneously in air if impure. It is a dangerous fire and explosion risk, with a flammable range of 0.46%–98% in air. The boiling point is 145°F, the flash point is 86°F and the ignition temperature is 95°F, which is extremely low. Any object that is 95°F or above can be an ignition source. Ignition sources can be ordinary objects on a hot day in the summer, such as the pavement, metal on vehicles and even the air. The 4-digit identification number is 1380. The NFPA 704 designation for pentaborane is health—4, flammability—4 and reactivity—2. It is primarily used as fuel for air-breathing engines and as a propellant.

Phosphorus (P), also known as yellow phosphorus, is a white, nonmetallic element that is found in the form of crystals or a waxlike transparent solid. It ignites spontaneously in air at 86°F, which is also its ignition temperature. White phosphorus should be stored and shipped under water and away from heat. It is a dangerous fire risk, with a boiling point of 536°F and a melting point of 111°F. The 4-digit identification number is 2447. The NFPA 704 designation is health—4, flammability—4 and reactivity 2. The primary uses are in rodenticides, smoke screens and analytical chemistry.

Potassium sulfide (K_2S) is a binary salt. It is a red or yellow–red crystalline mass or fused solid. It is deliquescent in air, which means that it absorbs water from the air, and it is also soluble in water. Potassium sulfide is a dangerous fire risk and may ignite spontaneously. It is explosive in the form of dust and powder. It decomposes at 1,562°F and melts at

1,674°F. The specific gravity is 1.74, which is heavier than air. The 4-digit identification number is 1382. The NFPA 704 designation is health—3, flammability—1 and reactivity—0. Potassium sulfide is used primarily in analytical chemistry and medicine.

Sodium hydride (NaH) is a binary salt that has a specific hazard of releasing hydrogen when in contact with water. It is an odorless powder that is violently water reactive. The 4-digit identification number is 1427. The NFPA 704 designation is health—3, flammability—3 and reactivity—2. The white space at the bottom of the diamond has a "W" with a slash through it, indicating water reactivity.

Division 4.3 Dangerous When Wet—LF/LR

The DOT definition is "a material that, by contact with water, is liable to become flammable or to give of flammable or toxic gas at a rate greater than 1 liter per kilogram of the material, per hour." Examples of water-reactive materials are zinc powder, trichlorosilane, sodium phosphide, sodium aluminum hydride, sodium metal, rubidium metal and the metallic elements such as potassium, sodium and lithium. These materials come from family 1 on the periodic table known as the alkali metals. They are in the first column on the table, and as with other families on the chart, they have similar chemical characteristics. They are silvery soft metals that are reactive with air and violently reactive with water. Contact with water causes spattering, the release of free hydrogen gas and the production of heat. The heat can be so great that it ignites the hydrogen gas. Other metallic elements such as magnesium are from family 2 on the periodic table. These materials are known as the alkaline earth metals. Magnesium, unlike the alkali metals, must be burning before it reacts with water or must be in a finely divided form such as filings and powder. Water in contact with burning magnesium produces a violent explosion. Water in contact with magnesium fillings or powder can produce a spontaneous explosion.

CASE STUDY

A fire in a warehouse in Chicago involved barrels of oil-soaked magnesium shavings and filings. Fighting the fire with water produced violent explosions resulting in whiplash injuries to firefighters. The facility was in a residential neighborhood so firefighters had to try to control the fire with water even though the material is water reactive when burning.

Chemical Notebook
Calcium carbide is a grayish-black hard solid and reacts with water to produce acetylene gas and a corrosive solid. Acetylene gas is manufactured by reacting calcium carbide with water. Because acetylene is so unstable, it is not shipped in bulk quantities. Calcium carbide is shipped to the acetylene-generating plant where it is reacted with water in a controlled reaction. After the reaction process, the acetylene gas is then placed into specially designed containers for shipment and use.

Phosphorus pentasulfide is a yellow to greenish-yellow crystalline mass with an odor similar to hydrogen sulfide. It is a dangerous fire risk and ignites by friction or when in contact with water. When it contacts water, it liberates poisonous and flammable hydrogen sulfide gas. Class 4.3 materials are water reactive. When large amounts of these materials are involved in fire, water is the only extinguishing agent available in quantities large enough to extinguish the fires. Just understand that when water is used, there may be violent reactions and explosions, and make preparations for the safety of personnel.

Small fires of water-reactive materials, especially metallic-based materials, can be extinguished with dry powder-extinguishing agent. For other flammable solid materials, water is also the agent of choice in most cases. It is important, however, to make sure that there is a positive identification of the product as with all hazardous materials. Once the product is identified, the proper extinguishing agent for any individual material can be identified through reference materials such as the DOT *Emergency Response Guidebook*, the CAMEO computer database, CHEMTREC or some other reference source (Hazardous Materials Chemistry for Emergency Responders).

Hazard Class 5 Oxidizers (LF/HR)

Class 5 is separated into two divisions: 5.1 and 5.2. Division 5.1 materials are solids and liquids that "by yielding oxygen, can cause or enhance the combustion of other materials." Division 5.2 materials are "organic peroxides." These are organic compounds that contain carbon in their formula. These materials contain oxygen in the bivalent –O–O– structure and may be considered a derivative of hydrogen peroxide. In these compounds, one or more of the hydrogen atoms have been replaced by organic radicals. Oxidizers can have other hazards like most hazardous materials. Hazardous materials in other hazard classes may also be oxidizers even though they are not the most significant hazard of the material.

Hypergolic Combustion

When oxidizers come into contact with materials called chemical activators, they immediately react with each other, releasing heat, oxygen and flammable vapors. This is quickly followed by a very rapid and intense fire. These materials possess the ability to spontaneously react with oxygen from the air or a chemical oxidizer at room temperature without any outside heat being applied. Because they have a tendency to react with oxygen at room temperature, we say these materials have slow oxidation potential. They can be metals (Na, K, Mg, etc.), nonmetals (P, C, etc.) or non-salts (linseed oil, turpentine, etc.). This oxidation reaction also liberates heat. Therefore, when these materials contact a chemical oxidizer, they immediately begin to break down the oxidizer, causing the release of oxygen and a considerable amount of heat. Because these chemical activators will also burn, this reaction almost always produces a very accelerated combustion. You may recall that this reaction is called a "hypergolic reaction." It is the type of combustion that occurs in rocket engines. Thus, these materials both activate the oxidizer and act as the fuel.

Hypergolic Propellant

Hypergolic propellant combination used in a rocket engine is one whose components spontaneously ignite when they come into contact with each other. The two propellant components usually consist of a fuel and an oxidizer. The main advantages of hypergolic propellants are that they can be stored as liquids at room temperature and that engines which are powered by them are easy to ignite reliably and repeatedly. Although commonly used, hypergolic propellants are difficult to handle due to their extreme toxicity and/or corrosiveness.

In contemporary usage, the terms "hypergolic" or "hypergolic propellant" usually mean the most common such propellant combination, dinitrogen tetroxide plus hydrazine and/or its relatives monomethylhydrazine and unsymmetrical dimethylhydrazine.

Oxidizers

Oxidizers can be classed into families with specific hazards associated with each family. There are elements from the periodic table that are oxidizers such as oxygen, chlorine, fluorine and bromine. Oxysalts are combinations of metals and nonmetal oxy radicals. Oxysalts do not react with water, but they are soluble in water. In the process of mixing with water, they may liberate oxygen or chlorine. Calcium hypochlorite, an oxysalt, is a common swimming pool chlorinator. When it mixes with water, it releases chlorine into the water. If the container becomes wet in storage, it can cause an exothermic reaction. If combustible materials are present, a fire may occur. The chlorine in the compound will then accelerate the combustion process. Nitrates are salts that can be very dangerous oxidizers. They may explode if contaminated, heated or shocked.

Inorganic peroxides are salts that contain a metal and two oxygen molecules. These oxidizers are water reactive and when in contact with water produce free oxygen and heat. The heat produced is sufficient to ignite nearby combustibles. Some acids at higher concentrations can be strong oxidizers and cause combustion in contact with organic materials. Nitric acid is a very dangerous oxidizing acid, it is corrosive and the vapors are poisonous.

Oxidizers, like many of the other hazard classes, can have more than one hazard. They can be corrosive, poisonous and explosive under certain conditions. Many are water reactive, and some may react violently with other chemicals, particularly organic materials. From an emergency response standpoint, oxidizers should be treated with the same respect as Class 1 explosive materials. They can be just as dangerous under the right conditions. Oxidizers themselves do not burn; however, they do support combustion. Enough heat can be produced in a reaction that combustible materials can be ignited.

Oxygen is one of the elements of the fire triangle and more recently the fire tetrahedron. The combustion process requires heat, oxygen and fuel, and also involves a chemical chain reaction. When we think about combustion, we usually think about atmospheric oxygen allowing combustion to occur. Without oxygen combustion cannot occur. Some materials that contain oxygen in their compound can support combustion even in the absence of atmospheric oxygen. When extra oxygen is added over and above atmospheric oxygen, the process of combustion is accelerated.

Some oxidizers are water reactive. The reaction with water can be exothermic, that is, it gives off heat. The heat and oxygen present represent two sides of the fire triangle or tetrahedron. All that is needed is fuel. If combustible materials are present, combustion can occur spontaneously. Some oxidizers mix with water, and when the mixture dries into a combustible material, the material will burn with great intensity if it catches on fire. Firefighter turnouts can become impregnated with oxidizer materials and place firefighters in danger when exposed to heat or fire.

There are some oxidizers that can cause spontaneous combustion. This can be slow in the case of carbon-based materials such as charcoal coming into contact with water. If the heat produced is accumulated in the material, combustion may occur. This can also occur with animal or vegetable oils when they are present in combustible materials such as rags. The corrosive action of strong acids will generate heat. When the corrosive material is also an oxidizer, the oxidizer will contribute to the acceleration of the combustion process. An explosion is nothing more than a very rapid combustion. A chemical explosion requires a chemical oxidizer to be present. Without the chemical oxidizer, the combustion is not accelerated fast enough to allow an explosion to occur.

Chemical Notebook

Ammonium nitrate is a common agricultural and home fertilizer. It is also a strong oxidizer, which is a colorless crystal that is soluble in water. It decomposes into nitrous oxide at 210°C. When mixed with fuel oil or contaminated with organic materials, it can become a dangerous explosive. It is also used as an oxidizer in solid rocket fuels; as an ingredient in herbicides, insecticides and pyrotechnics; and in the manufacture of nitrous oxide.

Ammonium perchlorate is a white crystalline material that is soluble in water. It is a strong oxidizing agent. It is shock sensitive and may explode when exposed to heat or by spontaneous chemical reaction. Oxidizers can be very dangerous explosive materials even though they are not classified as explosives (Hazardous Materials Chemistry for Emergency Responders).

CASE STUDIES

HENDERSON, NV, PEPCON

In Henderson, Nevada, a fire occurred in the PEPCON chemical plant. The fire heated ammonium perchlorate to the point that an explosion occurred. Ammonium perchlorate is an oxidizer used in the manufacture of solid rocket fuel. The material, which is an oxidizer, detonated creating a shock wave that blew the windshields out of responding fire apparatus. Several firefighters were injured by flying glass and debris (NASA).

KANSAS CITY, MO, 1988, AMMONIUM NITRATE

Firefighters were responding to a construction site where explosives were being used and stored. There was a fire involving a storage building. The firefighters may have been unaware that the explosives were stored there and fought the fire. The resulting explosion destroyed one fire engine and heavily damaged another in addition to the deaths of six firefighter. As a result of that tragic explosion in Kansas City, OSHA has issued a new regulation involving the use of U.S. Department of Transportation (DOT) placards and labels in fixed storage. All hazardous materials that require DOT placards and labels in transportation must continue to be placarded and labeled in fixed storage. The placards and labels must remain on the materials until they are all used up, and the containers have been purged or properly discarded. Ammonium nitrate fertilizer can be made resistant to flame and detonation by a proprietary process involving the addition of 5%–10% ammonium phosphate (*Firehouse Magazine*).

Chemical Notebook

Nitric acid is an oxidizer above 40% concentration. Above 40%, it will be placarded with an oxidizer placard; however, it is still a very dangerous corrosive.

Ammonium chlorate is a colorless or white crystal that is soluble in water. It is a strong oxidizer. When contaminated with combustible materials, it can ignite. It is shock sensitive and can detonate when exposed to heat or vibration. It is used to make explosives.

Chlorine, fluorine and bromine in their elemental form are all strong oxidizers even though they are placarded and labeled as poisons.

Chlorine is a dense, greenish-yellow gas. It is not combustible, but it will support combustion just like oxygen. Chlorine does not occur freely in nature; it is found in compounds within minerals, such as halite (rock salt), sylvite and carnallite, and as a chloride ion in seawater.

Fluorine is the most powerful oxidizing agent known. It is a pale yellow gas or liquid. It reacts violently with a wide range of organic and inorganic compounds. It is a dangerous fire and explosion risk when in contact with these materials. Like chlorine it is also classified as a 2.3 poison gas by the DOT.

Bromine is a dark reddish-brown liquid with irritating fumes. It attacks most metals and reacts vigorously with aluminum and explosively with potassium. It is a strong oxidizing agent and may ignite combustibles on contact.

Calcium chlorite is a white crystalline material. It is a strong oxidizer and a fire risk when in contact with organic materials.

Chromic acid is composed of dark purplish-red crystals that are soluble in water. It is a powerful oxidizing agent and may explode when in contact with organic materials. It is also a poison and corrosive to the skin.

Metal nitrates as a group have a wide range of hazards. Common to many of them, however, is the fact that they are oxidizers that are heat and shock sensitive.

Aluminum nitrate is a white crystal material soluble in cold water. It is a powerful oxidizing agent. It should not be stored near combustible materials.

Sodium nitrate is a slightly yellowish or white crystal, pellet, stick or powder. It oxidizes when exposed to air and is soluble in water. This material explodes at 1,000°F, certainly much lower than temperatures encountered in a fire. It is used as an antidote for cyanide poisoning, and in the curing of fish and meat.

Potassium nitrate (saltpeter) is a colorless or white crystalline powder or crystals. It is a dangerous fire and explosion risk when heated or shocked. It is a strong oxidizing agent. It is used in the manufacture of pyrotechnics, explosives and matches. It is often used in the illegal manufacture of homemade pyrotechnics and explosives.

Sodium peroxide is an inorganic peroxide. It is a yellowish-white powder that turns yellow when heated. It absorbs water and carbon dioxide from the air. It is soluble in cold water. It is a strong oxidizing agent. It is a dangerous fire and explosion risk when in contact with water, alcohol, acids, powdered metals and organic materials. It is used as a bleach and an oxygen-generating material for diving bells and submarines.

Division 5.2 Organic Peroxide LF/HR

Organic Peroxide is a hydrocarbon derivative family. Peroxide compounds have the same oxygen-to-oxygen double bond that is present in the nitro compounds –O–O–. This is also the type of peroxide that forms in an aging container of ether. The hazard of organic peroxides is that they will explode and undergo polymerization. Some may also be flammable and oxidizers. Peroxides may be found as liquids or solids. They are used in industry as oxidizing agents, bleaching agents and initiators of polymerization. Peroxides are shipped with an inhibitor chemical to prevent polymerization from occurring during transportation and storage. This inhibitor may become separated during an accident. Once the polymerization reaction starts, nothing will stop it, even if you have a tanker truck full of inhibitor to apply to the reaction!! (Which most of us do not have any way).

Organic peroxides have a characteristic known as Self Accelerating Decomposition Temperature (SADT). These temperatures may be near 0°F or above 50°F. Organic peroxides must be stored and shipped under refrigeration to prevent them from undergoing this rather violent process. Once the decomposition begins, it is irreversible. Refrigeration may be accomplished by the use of liquid nitrogen, which may also present responders with a frostbite or asphyxiation hazard. Just like the ethers, the organic peroxides have two radicals in the compound that may be the same or different. For example, if two methyl radicals were attached to the oxygens in the peroxide, the compound could be called dimethyl peroxide. Peroxides may also be found in combination with other hydrocarbon derivative functional groups, such as methyl ethyl ketone peroxide (Hazardous Materials Chemistry for Emergency Responders).

CASE STUDY

During Hurricane Harvey near Houston, power was lost at a plant where organic peroxide compounds are stored and transported. Backup generators failed as well. It didn't take long for the organic peroxide compounds to reach their SADT. Storage facilities and trucks loaded with the organic peroxides soon started to burn. Emergency responders could not reach the site because of the flood waters, so it was allowed to burn out. There were no explosions.

Organic peroxides are assigned to seven generic types of organic peroxides (see Table 3.1). They are classified by the extent to which they will detonate or deflagrate. Widely used in the plastics industry as polymerization reaction initiators, all organic peroxides are combustible. Decomposed can occur by heat, shock or friction. Methyl ethyl ketone peroxide can detonate. Organic peroxides are liquids or solids and are usually dissolved in a flammable or combustible solvent. Organic peroxides can be dangerously explosive materials. Some organic peroxides are so unstable that they are forbidden in transportation (Chemical Safety Board).

Table 3.1 Generic types of organic peroxides

Organic peroxides that can detonate or deflagrate rapidly as packaged for transport. Transportation of:

Type A	Organic peroxide is forbidden.
Type B	Organic peroxide as packaged for transport neither detonates nor deflagrates rapidly, but can undergo a thermal explosion.
Type C	Organic peroxide as packaged for transport neither detonates nor deflagrates rapidly and cannot undergo a thermal explosion.
Type D	Organic peroxide which: i. Detonates only partially, but does not deflagrate rapidly and is not affected by heat when confined; ii. Does not detonate, but deflagrates slowly, and shows no violent effect if heated when confined; or iii. Does not detonate or deflagrate, and shows a medium effect when heated under confinement.
Type E	Organic peroxide neither detonates nor deflagrates and shows low, or no, effect when heated under confinement.
Type F	Organic peroxide that will not detonate in a cavitated state does not deflagrate; shows only a low, or no, effect if heated when confined, and has low, or no, explosive power.
Type G	Organic peroxide that will not detonate in a cavitated state will not deflagrate, shows no effect when heated under confinement, has no explosive power and is thermally stable.

Chemical Notebook

Benzoyl peroxide, in its pure form, ignites very easily and burns with great intensity. It is very similar to black powder in its burning characteristics. Benzoyl peroxide decomposes very rapidly when heated, and if the material is confined, detonation will occur.

Ether peroxide. Most ether when stored for more than 6 months will form explosive ether peroxides in their containers. These peroxides are very sensitive to shock and heat and can detonate.

Hydrogen peroxide is an organic peroxide that is a dangerous fire and explosion risk. It is a strong oxidizing agent.

Perchloric acid, more than 50% but less than 72% by volume, is placarded as an organic peroxide. It is a colorless, fuming liquid. It is a strong oxidizing agent. It will ignite vigorously when in contact with organic materials or detonate by shock or heat.

Styrene undergoes polymerization at ordinary temperatures. As the temperature is increased, so is the rate of polymerization. The polymerization reaction is exothermic, and as the heat is generated, the reaction will become violent. When shipped, inhibitors are added to styrene to prevent the polymerization.

Vinyl chloride is a toxic and flammable gas that can undergo polymerization when exposed to elevated temperatures. Under fire conditions, the container can rupture violently. When shipped, inhibitors are added to retard the polymerization.

Dangers presented by all oxidizers are similar, but they are dangerous when in contact with organic materials, accelerate combustion and can be a serious fire and explosion hazard. Even though some are water reactive, water is the extinguishing agent of choice for fires involving oxidizers. Extinguishing agents that work by excluding atmospheric oxygen will not work with oxidizers. Oxidizers have their own oxygen supply and do not need atmospheric oxygen to support combustion. Emergency responders should treat oxidizers with respect. They should be handled in the same way that incidents involving explosives are handled because they can be just as dangerous.

Containers for Oxidizers

Oxidizers can be solids, liquids or gases. They can be found as cryogenic in typical cryogenic containers such as the MC 438 highway tanker, railcars, fixed vertical cryogenic tanks and dewars, cylinders and kegs. Oxidizers shipped in MC 412 tankers are also corrosive, but being an oxidizer is the most significant hazard (Figure 3.115). They may also be found in open and closed bulk highway and rail containers (Hazardous Materials Chemistry for Emergency Responders).

Figure 3.115 Oxidizers shipped in MC 412 tankers are also corrosive.

Hazard Class 6 Poisons and Infectious Substances HF/HR

Division 6.1 Poisons

Hazard Class 6 materials are solids and liquids, as well as some liquids that produce vapors and are inhalation hazards. Class 6 is divided into two Divisions: 6.1 and 6.2.

The U.S. Department of Transportation (DOT) defines a Division 6.1 poison as a material, other than a gas, known to be so toxic to humans as to afford a hazard to health during transportation or, in the absence of adequate data, presumed to be toxic to humans because it falls within any one of the following categories when tested on laboratory animals:

Oral toxicity. A liquid with a lethal dose (LD) of 50 of not more than 500 mg/kg or a solid with an LD_{50} of not more than 200 mg/kg of body weight of the animal. (LD_{50} is the single dose that will cause the death of 50% of a group of test animals exposed to it by any route other than inhalation.)

Dermal toxicity. A material with an LD_{50} for acute dermal toxicity of not more than 1,000 mg/kg.

Inhalation toxicity. A dust or mist with a lethal concentration (LC) of 50 for acute toxicity on inhalation of not more than 10 mg/kg. (LC_{50} is the concentration of a material in air that, on the basis of laboratory tests through inhalation, is expected to kill 50% of a group of test animals when administered in a specific period.) Cadmium and cadmium compound dusts are toxic by inhalation and listed as 6.1 poisons by the DOT.

Allyl alcohol, with the 4-digit identification number of 1098, is toxic by absorption, inhalation and ingestion. It also has multiple hazards. Allyl alcohol is placarded as a 6.1 poison primary hazard, but it is also a flammable liquid. Materials in Division 6.2 are infectious substances, meaning that they are viable microorganisms or toxins which may cause diseases in humans or animals. This section also includes regulated medical wastes. Poisons are among the most dangerous materials for emergency responders. Effects of poisons do not always present themselves right away; in fact, the toxic effect may not appear for days, months or even years. Because the effects may not present themselves right away, responders may be led to believe that there is no danger. One of the main reasons why decontamination is done for hazardous material incidents is to prevent the spread of toxic materials away from the "hot zone."

Toxicology is the study of toxic or poisonous substances and their effects on the body. It relates to the physiological effect, source, symptoms and remedial measures for the materials. Poisons can be divided into seven general categories:

Asphyxiants displace the oxygen in the air. In the case of Class 6 poisons, asphyxiants interfere with the blood's ability to convert or carry oxygen in the bloodstream; these are known as chemical asphyxiants. Death results from a lack of oxygen. Examples of chemical asphyxiants include hydrogen cyanide, benzene, toluene and aniline.

Carcinogens are materials that cause cancer. This information has been obtained by studying populations exposed to materials for long periods, usually in the workplace. The information can also be obtained through tests on laboratory animals. This does not mean that the material will actually cause cancer in humans, but it may be a good indication. Examples of known carcinogens include benzene, asbestos, arsenic, arsenic compounds, vinyl chloride and mustard gas.

Corrosives are acids or bases that are toxic to skin and other organs. They damage tissue in much the same way as a thermal burn, but they are much more damaging. The type of damage is the same whether exposed to acids or bases. Examples are nitric acid, sulfuric acid, phosphoric acid, sodium hydroxide and potassium hydroxide.

Irritants are materials that cause irritation to the respiratory system or other body organ or surface. Effects may range from minor discoloration of skin to rashes. Examples are tear gas and some disinfectants.

Mutagens cause mutations or changes in genic material. Changes may affect the current generation that was exposed or future generations. Examples include arsenic, chromium, dioxin and mercury.

Sensitizers, on first exposure, cause little or no harm in humans or test animals, but on repeated exposure they may cause a marked response not necessarily limited to the contact site. This is similar to the process of allergies—a person who moves into an area that has high pollen counts and other airborne allergens may not experience any effects at first, but the longer the exposure, the more symptoms present themselves.

Teratogens cause one-time birth defects when the pregnant woman is exposed or the father is exposed and passes the effect on to the mother. These types of materials do not permanently damage the reproductive system. Normal children can be produced as long as repeated exposures do not occur. Examples include thalidomide and O-benzoic sulfimide (the artificial sweetener saccharin).

Several factors influence toxic effects. One is the concentration of the toxic material. Concentrations are expressed in terms of parts per million or billion, percentage or milligram per kilogram. The higher the concentration, the more serious the effects. Types of exposure include the following:

- **Acute**—A one-time exposure. Depending on the concentration and duration of exposure, there may or may not be toxic effects.
- **Subacute**—It involves multiple exposure with a period of time between exposure.
- **Chronic**—Multiple exposures. All of these usually concern workplace exposures. An emergency responder can have an acute exposure to a toxic material at the scene of a hazmat incident. A single exposure may not produce a problem, whereas multiple exposures may cause damage. It is important to monitor exposures to personnel to determine any illness several days after the incident

Target organs. It takes some toxic materials several hours to several days to reach the susceptible target organs and cause damage. The response of the body to a hazardous material is based largely on concentration, length of exposure, type of exposure, route of exposure, susceptible target organ and other health-related variables such as age, sex, physical condition and size. This is referred to as dose/response. The type of response depends on which part of the body is affected by a particular toxic material. This is referred to as the "target organ." The target organ should not be confused with routes of exposure. A pesticide may enter the body through inhalation. But it may have an effect on the central nervous system. Target organs include the skin, brain, heart, lungs, liver and kidneys.

Routes of Exposure

There are four routes in which toxic materials can enter the body and cause damage:

- **Inhalation:** It involves toxic material entering the body through the respiratory system and that's where damage usually occurs.

- **Absorption:** Toxic material enters the body through the skin, eyes or some other tissue. Damage may occur to the point of contact or the material may travel to the susceptible target organ and cause harm there as well.
- **Ingestion:** It occurs when the material enters through the mouth and travels to the susceptible target organ; absorption can also occur after ingestion.
- **Injection:** It involves a sharp object that has been contaminated with a toxic material breaking the skin or the material entering through an open wound in the skin.

Exposure rates involve measurement of workplace exposures of hazardous materials based on tests conducted on animals. The results are then projected to humans based on weight ratios. The emergency responder's workplace is the incident scene, and the concentrations of toxic materials may be much higher than the values indicated through the following terms:

- **Threshold limit value**, a time-weighted average (TLV-TWA), is the maximum amount of acceptable exposure for 8h a day (40h a week) without ill effects.
- **TLV-STEL** (short-term exposure limit) is the maximum concentration averaged over a 15-min period to which healthy adults can be safely exposed for up to 15 min. Exposure should not occur more than four times per day with at least 1 h between exposures.
- **TLV-ceiling** is the maximum concentration to which a healthy adult can be exposed without injury.
- **PEL** (permissible exposure limit) is the maximum concentration averaged over 8h to which 95% of healthy adults can be repeatedly exposed for 8 h/day (40 h/week).
- **IDLH** (immediately dangerous to life and health) is the maximum amount of a toxic material that a healthy adult can be exposed to for up to 15 min and escape without irreversible health effects.

Not all people are affected by toxic materials in the same way. Variables include age, sex, weight, general health, body chemistry and physical condition. These can all affect the way individuals respond to a toxic material. Older people and younger children may be affected by a toxic material when other adult persons are not. Some materials affect females, but not males; some will affect both. Females who are pregnant may be affected by a toxic material that will cause damage to the fetus, and a male may not be affected. Toxicity is based on dose and size or weight of the test animal. This is then projected to humans based on the weight ratio of the dose, the weight of the animal and the weight and health of the human.

Cyanides and Isocyanates

Cyanides and isocyanates are derivative families that are extremely toxic. Industry does not usually call any of its compounds cyanide as that would cause some concern if this material was shipped through or stored in a community. So, they use the alternate naming system that uses the ending "nitrile." Few people would have any idea what nitrile was. The general formula of cyanide is –CN and the general formula of isocyanate is NCO. Examples of cyanides are vinyl cyanide (acrylonitrile) and methyl cyanide (acetonitrile). When naming cyanides, you name the radical hooked to the functional group and then the word "cyanide." When using the nitrile ending for cyanide, you also name the radical first followed by the nitrile ending. Just as with the aldehydes, esters and organic acids, you count the carbons, including the one in the cyanide functional group when naming the compounds. Vinyl is added to the cyanide radical and you get the compound vinyl cyanide, or the prefix for a three-carbon double bond which is acryl and the nitrile name for cyanide and you get acrylonitrile. It is flammable, toxic and corrosive.

Chemical Notebook

Acrylonitrile (vinyl cyanide), CH_2CHCN, is a colorless, volatile liquid, although commercial samples can be yellow due to impurities. It is flammable, reactive and toxic at low doses and will undergo explosive polymerization. It is used in the manufacturer of acrylic fibers, resins (acrylonitrile–butadiene–styrene, styrene–acrylonitrile and others) and nitrile rubbers (butadiene–acrylonitrile). It is toxic and flammable, and undergoes explosive polymerization. Combustion products are hydrogen cyanide and oxides of nitrogen. If you will notice, methyl and vinyl cyanide are the chemical names, but industry has changed the names to acetonitrile and acrylonitrile, respectively. I believe they do this to keep from having the name cyanide associated with their products.

Acetonitrile (methyl cyanide), CH_3CN, is a by-product of the manufacturer of vinyl cyanide and is metabolized rather slowly in the liver to hydrogen cyanide (2–12 h). It is flammable and mildly toxic in low doses.

Hydrogen cyanide (HCN) (prussic acid) is colorless, extremely toxic and extremely flammable liquid that boils slightly above room temperature. HCN is extremely flammable and extremely toxic liquid. It was used in World War I as a chemical agent on the battlefield.

Sulfur Compounds

"Thio" in the name of a compound is an indication that sulfur is present. Sulfides are sometimes referred to as thioethers or just thiols. The suffix "mercaptans," which means "mercury-seizing," is still sometimes used in place of "thiol." Primarily sulfur compounds are used as chemical intermediates or additives. Oxides of sulfur release hydrogen suflide (H_2S) when burning. They are highly toxic and severe skin irritants. H_2S toxicity is much the same as the cyanide compounds s (FEMA/NFA).

Chemical Notebook

Diphenyl trisulfice ($C_6H_5S_2$) is a colorless crystalline material and one of the more common organic sulfides. It is flammable and used in industrial processes.

Vinyl sulfide (C_4H_6S) is a colorless liquid with a faint odor and the product from hydrogen sulfide and acetylene. It is flammable and toxic.

Methyl mercaptan (CH_3SH) is a colorless gas with an unpleasant odor described as rotten cabbage. It is generally shipped as a liquefied compressed gas. Products of combustion are sulfur dioxide and flammable vapors. It irritates the respiratory system, airway, skin and eyes. It can be absorbed through the skin. It is used as an odorant in natural gas in certain regions of the United States. It is also used in the plastics industry, in pesticides and as a jet fuel additive. It occurs naturally in the blood, brain and other tissues of humans and animals. Methyl mercaptan is released from decaying organic matter in marshes. It is also present in animal feces. It occurs naturally in certain foods, such as some nuts and cheese.

Alkyl Halide

The first hydrocarbon derivative is the alkyl halide family. Alkyl halides can be found as liquids or gases. As a family, the alkyl halides are toxic to varying degrees. Alkyl halides are composed of one of the hydrocarbon radicals with varying numbers and combinations of the halogen family of elements attached: fluorine (F), chlorine (Cl), bromine (Br) and iodine (I). The toxicity of the compound is derived from the halogens and the flammability from the hydrocarbons, specifically the hydrogen(s).

The more halogens present, the more toxic the compound; the more hydrogens, the more flammable it becomes. The simplest of the alkyl halides is formed from methane. A single hydrogen is replaced by one chlorine forming the compound methyl chloride, also called chloro methane. As you can see, the hydrocarbon is represented in the name as well as the chlorine.

Methyl chloride is a colorless compressed gas or liquid with an etherlike odor. It is a significant fire risk with a flammable range of 10.7%–17% in air. It is narcotic, with a TLV of 50 ppm in air. It is used as a refrigerant, topical anesthetic, solvent and herbicide. By adding two more chlorines, trimethyl chloride, with the trade name of chloroform, is formed. Now the flammability is gone, but the toxicity has increased to 10 ppm in air.

Chloroform is used as an anesthetic, fumigant, solvent and insecticide. When four chlorines are attached to methane, the compound formed is carbon tetrachloride, one of the first halon compounds and a former fire extinguishing agent. It was found that when carbon tetrachloride comes in contact with a hot surface, phosgene gas is given off, so it was discontinued as a fire extinguishing agent. Other halogens can be added to methane in combination to form compounds. For example, when one chlorine and two fluorine atoms are added to methane, chloro trifluoro methane is formed. It is a nonflammable refrigerant with low toxicity. Methane is not the only hydrocarbon radical used to form alkyl halides. But whatever the radical, the family characteristics of the alkyl halides will prevail.

Aldehyde

The primary characteristic of the carbonyl compounds is polarity. Polarity of a compound also has an effect on how easily the human body can excrete toxic material that enters the body. Because the body is composed of primarily water, which is polar, polar compounds are more easily excreted through the body waste process than are nonpolar compounds. Formaldehyde exists normally as a gas that is readily polymerized with a strong pungent odor; however, it is also found as an aqueous solution from 37% to 50% concentrations. It is sometimes found with 15% methyl alcohol mixed in. Formaldehyde is toxic by inhalation with a TLV of 1 ppm in air. It is a strong irritant and a known carcinogen. Other aldehyde compounds include acetaldehyde and crotonaldehyde.

Ester

Ester is a hydrocarbon derivative family. They are flammable and toxic, and may undergo polymerization. Esters are members of the carbonyl family and are polar. They are used in the manufacturer of plastics. They are made from combining an organic acid and an alcohol with water as a by-product (Figure 3.116). Naming is based upon the acid and alcohol that

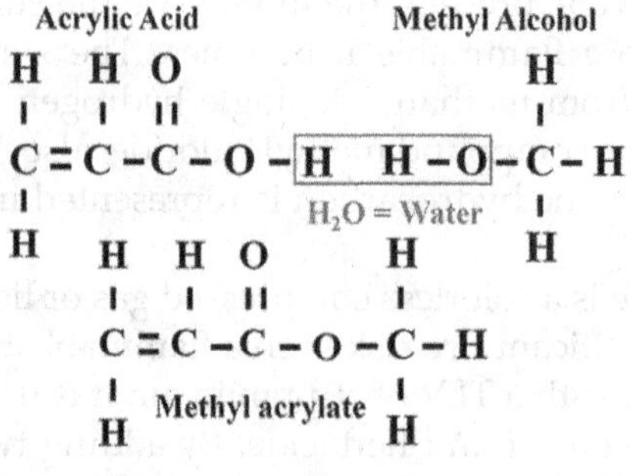

```
            Acrylic Acid              Methyl Alcohol
         H   H   O                          H
         |   |   ||                         |
         C = C - C - O - H       H - O  -  C - H
         |                                  |
         H               H₂O = Water        H
             H   H   O                  H
             |   |   ||                 |
             C = C - C - O -  C - H
             |                      |
             H    Methyl acrylate   H
```

Figure 3.116 Esters are made from combining an organic acid and an alcohol with water as a by-product.

is used, so you will not find the name ester in the compounds. Saturated esters are not generally classified as hazardous materials. Their uses include food additives, and saturated esters have a sweet odor and are used as candy flavorings. Unsaturated esters are extremely dangerous. They are flammable and very toxic and undergo explosive polymerization. They can also self-polymerize in storage. They are used widely in the production of plastics.

Because ester is not in the name of the compounds, be on the lookout for names such as acrylate and acetate. You might see vinyl acrylate or methyl acetate for example. When you see acrylate and acetate, you are dealing with esters and the hazards mentioned above apply to all of them.

Chemical Notebook
    Vinyl acetate ($CH_3CO_2CH=CH_2$) is a colorless liquid used as the precursor to polyvinyl acetate, an important industrial polymer. It is flammable, mildly toxic and reactive.
    Methyl acrylate ($C_4H_5O_2$) is a colorless liquid with a characteristic acrid odor. It is mainly used to produce acrylate fiber, which is used to weave synthetic carpets. It is also a reagent of various pharmaceutical intermediates. It is highly flammable.

Organic Acid
Organic acid is a hydrocarbon derivative family. It is toxic and corrosive. Corrosively is a form of toxicity to the tissues that the acid contacts. However, the organic acids have other toxic effects.

Chemical Notebook
    Formic acid is corrosive to skin and tissue. It has a TLV of 5 ppm in air and an IDLH of 30 ppm.
    Acetic acid is toxic by ingestion and inhalation. It is a strong irritant to skin and tissues. The TLV is 10 ppm in air, and the IDLH is 1,000 ppm.
    Propionic acid is a strong irritant, with a TLV of 10 ppm in air.
    Butyric acid is a strong irritant to skin and tissues.
    The degree of toxicity varies with the different organic acid compounds. Review reference sources and MSDS sheets to determine the exact hazards of specific acids.

Military Chemical Agents

Nerve Agents
Chemical nerve agents that could be used as weapons of terrorists are also compounds that have several derivative families along with some other elements such as sulfur and phosphorus (Figure 3.117). They belong to families called chlorinated sulfur compounds, fluorinated organ phosphorous compounds, suffocated organ phosphorous compounds and

Figure 3.117 Chemical nerve agents that could be used as weapons of terrorists are also compounds that have several derivative families along with some other elements such as sulfur and phosphorus. (Courtesy Center for Domestic Preparedness (CDP)).

organ phosphorous compounds. Looking at the elements that make up these compounds leaves little to wonder why they are so toxic. Organ phosphorous compounds themselves are pesticides and affect the central nervous system. Adding fluorine, chlorine and sulfur to the compound only increases the toxicity.

Chemical Notebook

VX, which is the most toxic of the nerve agents, is just such a compound. In addition to the hydrocarbons in the compound, oxygen, phosphorus, nitrogen and sulfur are present. It would be classified under the DOT system as a Division 6.1 poison and have an NFPA 704 designation of health—4, flammability—1, reactivity—1 and special—0. The four-digit UN 4-digit identification number for VX would be 2810, and Orange Guide 153 would be used from the *Emergency Response Guidebook*. VX has a high boiling point of 568°F and a high flash point of 315°F. The explosive limits have not been established.

Division 6.2 Infectious Substances LF/LR

An infectious substance is defined as a material known or reasonably expected to contain a pathogen (Figure 3.118). A pathogen is a

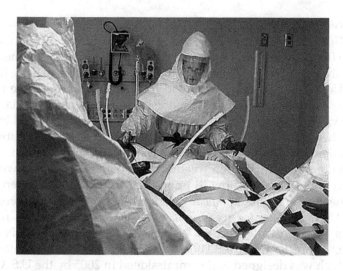

Figure 3.118 Special protective equipment is required to protect medical personnel from exposure to infectious substances. (Courtesy University of Nebraska Omaha Medical Center)

microorganism (including bacteria, viruses, rickettsiae, parasites and fungi) or an agent, such as a proteinaceous infectious particle (prion), that can cause disease in humans or animals. An infectious substance must be assigned the identification number 2814, 2900, 3373 or 3291 as appropriate and must be assigned to one of the following categories:

Category A: An infectious substance in a form capable of causing permanent disability or life-threatening or fatal disease in otherwise healthy humans or animals when exposure to it occurs. An exposure occurs when an infectious substance is released outside of its protective packaging, resulting in physical contact with humans or animals. A Category A infectious substance must be assigned to the 4-digit identification number is 2814 or 2900, as appropriate. Assignment to 2814 or 2900 must be based on the known medical history or symptoms of the source patient or animal, endemic local conditions or professional judgment concerning the individual circumstances of the source human or animal.

Category B: An infectious substance that is not in a form generally capable of causing permanent disability or life-threatening or fatal disease in otherwise healthy humans or animals when exposure to it occurs. This includes Category B infectious substances transported for diagnostic or investigational purposes. A Category B infectious substance must be described as "Biological Substance, Category B" and assigned to the identification number 3373. This does not include regulated medical waste, which must be assigned to the identification number 3291 (Hazardous Materials Chemistry for Emergency Responders).

Biological Notebook

Ebola Outbreaks During 2014 the African outbreak of Ebola virus disease (EVD) was brought to the United States for the first time. A U.S. citizen living in Dallas, Texas, was the first person diagnosed with EVD in the United States. The person had traveled to Liberia where the disease was contracted. Following unsuccessful treatment in a Dallas hospital, the victim died on October 8, 2014, of complications caused by EVD. Several health-care workers who had been involved in the treatment of EVD patients also contracted EVD but were successfully treated. Health-care workers returning from treating EVD patients in Africa were also exposed and several became ill upon their return to the United States. Other health-care workers became sick while in Africa and were flown to the United States for treatment. One of the facilities that received patients was the University of Nebraska Medical Center (UNMC) in Omaha, Nebraska (Figure 3.119). UNMC has a 10-bed Biocontainment Unit which was designed and commissioned in 2005 by the U.S. Centers for Disease Control (CDC) to provide the first-line treatment for people affected by bioterrorism or extremely infectious naturally occurring diseases such as EVD.

The facility in Omaha is the largest of its kind in the United States. Highly contagious and deadly infectious conditions that can be handled in the unit include severe acute respiratory syndrome (SARS), smallpox, tularemia, plague, EVD and other hemorrhagic fevers, monkey pox, vancomycin-resistant *Staphylococcus aureus* (VRSA) and multidrug-resistant tuberculosis. Safety measures are built into the Biocontainment Unit including air-handling equipment, high-level filtration and ultraviolet

Figure 3.119 One of the facilities that received Ebola patients was the UNMC in Omaha, Nebraska. (Courtesy University of Nebraska Omaha Medical Center)

light that prevent microorganisms from spreading beyond patient rooms. The entire unit is isolated from the rest of the hospital with its own ventilation system and a secured access. A dunk tank is provided for laboratory specimens, and a pass-through autoclave is also in place to ensure that hazardous infections are contained. Also in the unit is a special sterilizer for laundry so that contaminated bed clothing is not removed from the unit. Patients are flown into Omaha's Eppley Airfield and transported to the UNMC Biocontainment Unit, in an individual isolation unit also called a BIOPOD, by Omaha Fire Department EMS. Biocontainment Unit staff receive specialized training and participate in drills throughout the year.

Outbreaks of EVD (previously known as Ebola hemorrhagic fever or Ebola HF) have occurred over 20 times since the first documented outbreak in 1976 (Figure 3.120). Ebola is named after the Ebola River in the African country of Zaire (now Democratic Republic of Congo), which is where the first case of EVD occurred. Ebola is one of three viruses in the Filoviridae family of viruses. The other two viruses are Cueva virus and Marburg virus. Four subspecies of Ebola virus are known to exist that cause disease in humans: Zaire, Sudan, Taï Forest, and Bundibugyo.

Fifth and sixth sub-viruses, Reston and Bumbali are believed to infect and cause disease in primates. Reston can also infect humans, but it has not been documented to have caused disease in humans at this time. Much investigation has been conducted to determine the host that carries Ebola virus; however, to date, no natural host has been proven. Researchers, however, believe that the virus is carried by animals, with fruit bats being the most likely host. The World Health Organization

Figure 3.120 Outbreaks of EVD (previously known as Ebola HF) have occurred over 20 times since the first documented outbreak in 1976.

(WHO) says that infection of humans has been documented through the handling of infected chimps, gorillas, fruit bats, monkeys, forest antelope and porcupines that have been found ill or dead in the rainforest. Pig farms in Africa are suspected to draw fruit bats and amplify outbreaks of EVD.

Once human infection has occurred, the Ebola virus can be transmitted from person to person through close contact with a person infected via contact with bodily fluids. Ebola virus spreads by direct contact (through broken skin or mucous membranes) with the blood, secretions, organs or other bodily fluids of infected people. It is also transmitted by direct contact with surfaces and materials (such as bedding and clothing,) and objects such as needles and syringes contaminated with these fluids. Ebola virus is not spread through the air or by water or in general by food. Ebola virus may be spread by handling wild animals hunted for food and contact with infected bats. There is no evidence that mosquitoes or other insects can transmit Ebola virus. Only mammals, for example, humans, bats, monkeys and apes, have shown the ability to become infected with and spread Ebola virus.

Health-care providers caring for EVD patients and their families and friends in close contact with EVD are at the highest risk of getting sick because they may come into contact with the infected blood or body fluids of sick patients. Health-care workers in Africa have frequently been infected while treating patients with suspected or confirmed EVD. Infections occurred through close contact with patients when infection control precautions were not strictly practiced. In Africa, burial ceremonies where mourners have direct contact with the body of the deceased person may also be a source of the transmission of Ebola virus.

People infected with Ebola virus remain infectious as long as their blood and body fluids, including, but not limited, to urine, saliva, sweat, feces, vomit, semen and breast milk contain the virus. This is usually thought to be 21 days and those who have been exposed to an infected person or surfaces are watched for that amount of time, after which if they haven't developed symptoms, they are believed to be free of the virus. Men who have recovered from EVD can still transmit the virus through their semen for up to 7 weeks after their recovery.

The incubation period for the EVD is 2–21 days from the time of infection to when symptoms appear. Humans are not infectious until they develop symptoms. Initial symptoms are the sudden onset of fever (>101°F/38.3°C), fatigue, muscle pain, headache and sore throat. This is followed by vomiting, diarrhea, rash, symptoms of impaired kidney and liver function and, in some cases, both internal and external bleeding (e.g., oozing from the gums, blood in the stools). Laboratory findings may include low white blood cell and platelet counts and elevated liver enzymes. It may be difficult to distinguish EVD from other infectious diseases such as malaria, typhoid fever and meningitis. Confirmation that symptoms are caused by EVD is made using the following investigations:

- Antibody-capture enzyme-linked immunosorbent assay (ELISA)
- Antigen-capture detection tests
- Reverse transcriptase polymerase chain reaction (RT-PCR) assay
- Electron microscopy
- Virus isolation by culture

Samples from patients are an extreme biohazard risk; laboratory testing on noninactivated samples should be conducted under maximum biological containment conditions.

Supportive care rehydration with oral or intravenous fluids and treatment of specific symptoms improve survival. There is as yet no proven treatment available for EVD. However, a range of potential treatments including blood products, immune therapies and drug therapies are currently being evaluated. No licensed vaccines are available yet, but two potential vaccines are undergoing human safety testing (*Firehouse Magazine*).

Anthrax One of the first diseases identified in the field of microbiology in 1876 was anthrax and the first disease for which an effective live bacterial vaccine was developed in 1881 by Louis Pasteur. Inhalation anthrax was discovered during the late 19th century. Natural outbreaks of inhalation anthrax occurred among woolsorters in England, which became one of the first occupational respiratory infectious diseases. Processing of contaminated goat hair and alpaca wool resulted in the generation of infectious aerosols. The largest human exposure of inhalation anthrax occurred in Sverdlovsk, Russia, in 1979. Anthrax spores were released accidentally from a military research facility upwind from the outbreak. Cases were also reported in animals located more than 50 miles from the site. There were 66 documented human deaths and 11 injuries as a result of the release. It is believed by many that the death count may have been much higher.

Bacillus anthracis, also called "woolsorters" disease, is a Gram-positive, spore-forming bacillus that can survive for over 100 years in the spore form. Bacteria are classified as Gram-positive or Gram-negative based upon their response to the Gram staining procedure. The primary difference between Gram-negative and Gram-positive bacteria occurs in the cell wall. Gram-positive cell walls are usually much thicker and more difficult to penetrate than Gram-negative cell walls. *Bacillus* is a genus of bacteria that is found everywhere in nature (soil, water and airborne dust). Spores are not formed in living tissue. When a host dies and the disease is exposed to oxygen during the decay of the corpse, the spores are formed. Development of spores is a survival system, which allows the bacteria to survive in nature until a suitable host is once again contacted. About 2,000–5,000 cases of naturally occurring anthrax are reported every year throughout the world. It most commonly occurs in South and Central America, Southern and Eastern Europe, Asia, Africa, the Caribbean and the Middle East. About five cases are reported in the United States each year and occur mostly in five states: Texas, Louisiana, Mississippi, Oklahoma, and South Dakota.

Considered to be an "occupational disease," people most likely to contract anthrax naturally work around animals, such as veterinarians, ranch and farm workers, and people who work with animal carcasses, hair and wool. Anthrax spores are very persistent and remain viable for years in soil, dried or processed hides, skins and wool. They can survive in milk for 10 years, on dried filter paper for 41 years, on dried silk threads for 71 years, and in pond water for 2 years. During World War II, the British conducted tests with anthrax on Gruinard Island off the coast of Scotland. The island remained contaminated with anthrax spores and basically uninhabitable until 1986 when tons of topsoil was removed. The island decontaminated by soaking the remaining soil with seawater spiked with large amounts of formaldehyde.

Cutaneous Anthrax It is not known exactly how many spores are necessary to cause cutaneous anthrax in humans. The only data available are obtained from animal tests. Ken Alibek, a defector from the former Soviet Union's biological weapons program, believes that 10–15 spores could cause disease. Once the anthrax spores enter through a break in the skin, they usually incubate for a period of 1–5 days before symptoms occur, although it is possible to take up to 60 days. The first indication of the disease is the appearance of a small papule (small, solid, raised spot on the skin) (Figure 3.121).

Within 1–2 days, a vesicle (liquid-filled sack) forms on the skin, containing a bloody or thick and viscous-like serum with many organisms and a small amount of leukocytes. Leukocytes are the parts of the immune system, which eat the bacteria, parasites, viruses, germs, fungus and other assorted bad characters. Vesicles may be 1–2 cm in diameter and, when ruptured, leave a dead tissue ulcer. Lesions are painless and have varying degrees of fluid around them. Fluid development can be massive in some cases and involve the entire face or limb. Victims will have a high fever, overall sick feeling and headache. Headaches become more severe with increased fluid presence. The base of the ulcer develops a characteristic black hard plaque, indicating extensive tissue death. After a period of 2–3 weeks, the plaque separates and may leave a scar. With treatment, the mortality rate from cutaneous anthrax is around 1%. If untreated, the rate increases to ~20%.

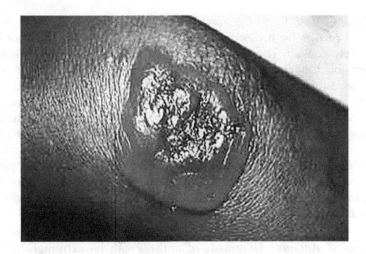

Figure 3.121 The first indication of anthrax is the appearance of a small papule (small, solid, raised spot on the skin).

Inhalation Anthrax Inhalation anthrax results from exposure to a dose of 8,000–10,000 aerosolized spores of a size ranging from 1 to 5 μm. The size is important to determine if spores will reach into the alveoli of the lungs to do their damage. The incubation period usually ranges from 1 to 6 days, although it is possible to take up to 60 days. Diagnosis is difficult because of nonspecific symptoms, which include an overall feeling of sickness, fatigue, muscle aches and fever. A nonproductive cough may be present along with mild chest discomfort. Symptoms persist for 2–3 days, often followed by a short period of improvement. The improvement is sometimes referred to as the "anthrax eclipse."

Following the period of feeling better, there is an onset of respiratory distress with breathing discomfort, shortness of breath, noisy, crowing respiratory sounds usually upon inspiration, bluish color to skin, increased chest pain and sweating. There may also be fluid retention in the chest and neck. Chest X-ray usually reveals a characteristic widening of the space between the lungs and often an accumulation of fluid in the plural space. Although not common, pneumonia may develop in some patients. Research suggests that people with existing pulmonary disease may have increased susceptibility to inhalation anthrax. Meningitis also occurs in as many as 50% of the cases, which may cause seizures in some patients. Onset of respiratory distress is followed by the rapid onset of shock and death with 24–36 h. Inhalation anthrax is almost always fatal unless treatment is started prior to the onset of symptoms. The current treatment involves the antibiotics Cipro, doxycycline and penicillin.

Oropharyngeal and Gastrointestinal Anthrax This form of anthrax is very rare and results from the ingestion of infected meat that has not been adequately cooked. The incubation period is usually 2–5 days, although it is possible to take up to 60 days. Oropharyngeal anthrax begins with a severe sore throat or a local oral or tonsillar ulcer, associated with fever, toxicity and swelling of the neck from fluid buildup. Respiratory distress and difficulty in swallowing may also be present. Gastrointestinal anthrax begins with nonspecific symptoms of nausea, vomiting and fever, followed, in many cases, by severe abdominal pain. First signs may be an acute abdomen accompanied by vomiting blood, massive lymph and blood plasma loss from the liver, and diarrhea. In the absence of treatment, mortality in both forms may be as high as 60%, particularly with the gastrointestinal form.

Diagnosis of Anthrax Diagnosis of anthrax can be extremely difficult because it is such a rare occurrence. Little data exist on human exposure after the onset of clinical signs. Symptoms in the early stages resemble colds and the flu. One major difference between anthrax and the flu is that there is no nasal involvement with anthrax. There should not be runny or stuffy nose symptoms associated with the colds and the flu. Because of the recent anthrax outbreak in the United States potentially related to terrorism, every case that presents flu-like symptoms should be investigated to rule out anthrax.

Treatment Meningitis caused by anthrax is clinically indistinguishable from meningitis due to other bacteria. One key test result is the presence of blood in the cerebrospinal fluid. This can occur in as many as 50% of the cases. Until recently, the drug of choice for treatment of all forms of anthrax has been penicillin. The Food and Drug Administration has approved Cipro for treatment, and since it has become the drug of choice, Cipro has many more side effects than the other antibiotics; some of them are very dangerous. Doxycycline has also been used effectively with anthrax exposure. Recent reports indicate that the anthrax strain(s) found in the U.S. outbreak are sensitive to all three antibiotics.

Vaccine Several vaccines are available for anthrax. However, vaccination is not typically available for the general public. It is only recommended for people who are at an increased risk of coming into contact with or have already been exposed to *B. anthracis*.

Centers for Disease Control (CDC) recommends anthrax vaccination for:

People who are at an increased risk of coming into contact with the bacteria that cause anthrax because of their job.

People who have been exposed to the bacteria that causes anthrax. (Hazardous Materials Chemistry for Emergency Responders).

Anthrax Scares
The Bomb Threat of the 21st Century Terrorism has been a hot topic throughout the emergency response community since the 1990s. Acts of terrorism occurring within the United States are still a very real possibility. Bombings have been the weapons of choice in the past; however, biological weapons present a credible threat for the future. In particular, anthrax is a likely choice of the terrorist for a number of reasons. First of all, it is highly lethal with a mortality rate of 80%–90% from inhalation exposure. Deadly concentrations are much less than for chemical nerve agents. Anthrax is relatively easy to produce and occurs naturally in many parts of the world, including the United States. It is also easy to deliver in its most deadly form, aerosol. Anthrax weapons can be produced at a fraction of the cost of other weapons of mass destruction, with equal effectiveness. One kilogram (2.2 lb) of anthrax can be produced for less than $50.00, and if properly aerosolized, under appropriate weather conditions, it could kill hundreds of thousands of people in a metropolitan area.

Following 9/11 terrorism in the United States, it has taken on a new twist, the "anthrax hoax" (Figure 3.122). A rash of threats boasting the release of anthrax or the exposure of individuals has occurred in at least 17 states and continues to spread. The Federal Bureau of Investigation (FBI) reports that it investigates new anthrax hoaxes on a daily basis.

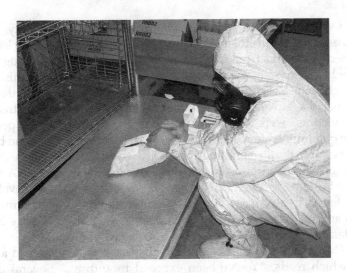

Figure 3.122 Following 9/11 terrorism in the United States, it has taken on a new twist, the "anthrax hoax".

Targets have included abortion and family planning clinics, federal and state buildings, churches, courthouses, public schools, dance clubs and office buildings. Threat mechanisms have included mailed letters with a white powder inside boasting "You've just been exposed to anthrax," threat letters, notes written on walls in buildings, telephoned threats and materials left in buildings.

These incidents are much like bomb threats that have occurred over the years. The potential for anthrax is very real, but all of the events that have occurred so far have turned out to be a hoax. Each time an anthrax threat occurs, it must be taken seriously. Response to these types of incidents causes much the same response from fire, police and EMS as a real anthrax release would cause. Response resources are tied up for hours and building activities are disrupted, as is traffic around the buildings and neighborhoods. Thousands of dollars are lost from business disruption and the cost of emergency responses to the incident.

Media coverage plays into the attention the responsible person(s) are trying to obtain and encourages copycats to create similar hoaxes in other locations. With the type of response the emergency community is giving to these hoax events, a terrorist wouldn't need to have anthrax or for that matter, any other biological material to create an act of terrorism, just the threat serves the purpose (*Firehouse Magazine*).

CASE STUDIES

On October 30, 1998, abortion clinics in Indiana, Kentucky, and Tennessee all received letters claiming to contain anthrax. The letters carried postmarks from Cincinnati. In Indiana, 31 people, thought to have been exposed, including a mail carrier, were decontaminated with chlorine bleach. They were then transported to hospitals, decontaminated again and given antibiotics. Hospital workers who treated the "victims" were also given antibiotics. In Los Angles, 91 people were quarantined for over 8h after a threat-reported anthrax had been released into the air-handling system of a federal building. Once again antibiotics were administered to those thought to be exposed. A state office building in Wichita, Kansas, was evacuated after an envelope was found in a stairwell with a white powder and note claiming it was anthrax. Between 20 and 25 people were believed to have been contaminated. Fifteen were decontaminated as a precaution.

On December 14, 1998, a school district secretary opened a letter which reads, "You've been exposed to anthrax." Several days later, a U.S. Bankruptcy Court in Woodland Hills, California, was targeted, and over 90 people were administered antibiotics. Three days later, telephoned threats caused two courthouses in Van Nuys, California, to be evacuated, resulting in 1,500 people being quarantined for several hours. It has been estimated that the incidents in California alone have cost the tax payers $500,000 per occurrence. The list of incidents and locations goes on and on, but the circumstances and outcomes are all very similar and they have all been hoaxes. As a result of the number of anthrax scares, the Weapons of Mass Destruction (Counter Terrorism For Emergency Responders).

(WMD) Unit of the FBI developed guidelines for emergency responders.

These guidelines are designed to reduce the impact of a hoax event on the response system, potential victims and the community.

- "Persons exposed to anthrax are not contagious and quarantine is thus not appropriate."
- "All first responders should follow local protocols for hazardous materials incidents involving biological hazards. Upon receipt of a threat, a thorough hazard risk assessment should be conducted. Upon notification, the FBI will coordinate a risk assessment in conjunction with the health department and other authorities on biological agents to ensure timely dissemination of appropriate technical advice."
- "Any contaminated evidence gathered at the scene should be triple-bagged. Individuals should be advised to await laboratory test results, which will be available within 48 hours. These individuals do **NOT** need to be placed on chemoprophylaxis (antibiotics) while awaiting laboratory test results to determine whether an infectious agent was present."
- "The individual needs to be instructed that if they become ill before laboratory results are available, they should immediately contact their local health department and proceed immediately to a predetermined emergency department, where they should inform the attending staff of their potential exposure."
- "Responders can be protected from anthrax spores by donning splash protection, gloves, and a full-face respirator with High Efficiency Particulate Air Filters (HEPA) (Level C) or self-contained breathing apparatus (SCBA) (Level-B). Victims who may be in the immediate area and are potentially contaminated should be decontaminated with soap and water: no bleach solutions are required.

A 1:10 dilution of household bleach (i.e., Clorox −5.25% hypochlorite) should only be used if there is confirmation of the agent and an inability to remove the materials through soap and water decontamination. Additionally, the use of bleach decontamination is **Only** recommended after soap and water decontamination, and should be rinsed off after 10 to 15 minutes. Technical assistance can be immediately provided by contacting the National Response Center (NRC) at (800) 424-8802."

- "If the envelope or package remains sealed (not opened), then first responders should not take any action other than notifying the FBI and packaging the evidence. Quarantine, evacuation, decontamination, and chemoprophylaxis efforts are **NOT** indicated if the envelope or package remains sealed. Also, anthrax will likely be visible as a powder or powder residue. The absence of visible powder is a strong indicator that anthrax is not present.
- "The use or threatened use of a weapon of mass destruction (including anthrax) is a violation of federal law. See Title 18. United States Code, Section 175 and Section 2332a. It should be reported to the FBI immediately."
- "This information is provided by the WMD Operations Unit of the Federal Bureau of Investigation and the National Domestic Preparedness Office (NDPO), in coordination with the Centers for Disease Control, the Department of Health and Human Services/ Office of Emergency Preparedness, and the U.S. Army Medical Research Institute of Infectious Diseases (USAMRIID). The NDPO was established to coordinate the Federal Government's efforts to prepare the nation's response community for threats involving weapons of mass destruction. Contact your local FBI office if confronted by a WMD threat."

When responding to a suspected terrorist threat or incident involving anthrax or any other weapon of mass destruction, the first concern is containment. Remember the responsibilities of the first responder to any type of hazardous material: Recognition/ Identification, Isolation, Notification, and Protection. Limit the exposed area and the people allowed into the area. If the air-handling system is involved, have the system shut down. The entire building does not have to be evacuated unless it is suspected that the agent has been aerosolized and has gotten into the air-handling unit, spreading it throughout the building. In the case of a letter or package without aerosolization, contain the incident to the room involved. Evacuation of the building will not be necessary. Anthrax spores in a package or envelope do not present any particular hazard as long as no one touches, inhales or ingests the material. Have the person who opened the package or envelope close it with the material inside. Have people potentially exposed keep their hands away from their face, and do not touch eyes, nose or mouth. Hands should be washed with soap and water while waiting for decontamination facilities to be set up. Appropriate actions taken by first responders can reduce the impact these types of incidents will have on the facility, response personnel and the community. Downplaying the incidents will help to reduce the number of copycat occurrences and the development of complacency among responders and the public (FBI WMD Unit).

Pesticides

A pesticide is a chemical or mixture of chemicals used to destroy, prevent or control any living thing considered to be a pest, including insects (insecticides), fungi (fungicides), rodents (rodenticides) or plants (herbicides) (Figure 3.123). The definition of a pesticide from the Federal Insecticide, Fungicide, and Rodenticide Act (FIFRA) is "a chemical or mixture of chemicals or substances used to repel or combat an animal or plant pest. This includes insects and other invertebrate organisms; all vertebrate pests, e.g., rodents, fish, pest birds, snakes, gophers; all plant pests growing where not wanted, e.g., weeds; and all microorganisms which may or may not produce disease in humans. Household germicides, plant growth regulators, and plant root destroyers are also included."

Farming and commercial landscaping activities involve the use of restricted pesticides to control weeds, insects and fungus in agricultural crops and lawns and ornamental plants (Figure 3.124). Restricted use

Figure 3.123 A pesticide is a chemical or mixture of chemicals used to destroy, prevent or control any living thing considered to be a pest, including insects (insecticides), fungi (fungicides), rodents (rodenticides) or plants (herbicides).

Figure 3.124 Farming and commercial landscaping activities involve the use of restricted pesticides to control weeds, insects and fungus in agricultural crops and lawns and ornamental plants.

pesticides are those regulated by EPA and only allowed for commercial or agricultural use by trained and licensed applicators. They are usually more toxic, may be more environmentally sensitive, are found in greater quantities in storage and are packaged in larger containers. When a pesticide is classified as restricted, the label will state "Restricted Use Pesticide" in a box at the top of the front panel.

A statement may also be included describing the reason for the restricted use classification. Usually, another statement will describe the category of certified applicator that can purchase and use the product. Restricted use pesticides may only be used by applicators certified by the state or the EPA. Home yard work often involves the use of general pesticides to control weeds, insects and fungi in lawns and ornamental trees and bushes. Chemicals approved by the EPA for homeowner use do not require any special training or licensing by the homeowner to be used. Container size is usually small and the chemicals can be safely used by following label directions.

The EPA estimates that there are 45,000 accidental pesticide poisonings in the United States each year, where more than one billion pounds are manufactured annually. Pesticides can be found in manufacturing facilities, commercial warehouses, agricultural chemical warehouses, farm supply stores, nurseries, farms, supermarkets, discount stores, hardware stores and other retail outlets (Figure 3.125). They may also be found in many homes, garages and storage sheds across the country.

More than 1,000 basic chemicals, mixed with other materials, produce about 35,000 pesticide products. However, when an emergency occurs

Figure 3.125 In rural areas, pesticides are sprayed on crops via the use of specially designed aircraft.

involving different groups of pesticides, chemicals may become mixed that are not normally found together. This mixing of chemicals may provide toxicology and cleanup problems for emergency responders. Great care should be taken during firefighting operations including overhaul when pesticides and other chemicals are involved. If the fire is allowed to burn off, care should be taken to avoid smoke or fumes that evolve. Runoff can become contaminated with toxic materials and damage firefighting protective clothing, equipment, apparatus and the environment. If a decision is made to fight a pesticide fire, runoff should be kept to a minimum.

If possible, route runoff water to a holding area. Hazardous materials teams are called upon to respond to numerous incidents each year involving pesticides. Care should be taken so as not to overreact to a pesticide spill. A pint bottle of a pesticide broken on the display floor of a lawn and garden center does not necessarily require a full-blown hazmat response and a multi-hour operation to effectively mitigate. Keep in mind that those pesticides designed for consumer use involve the opening of a container and mixing with water for application by the end user, many times without a need for extensive protective clothing. On the other hand, larger quantities of restricted use pesticides, or unrestricted pesticides would need to be handled as any other serious chemical spill. The important thing is to evaluate the incident. Do a risk–benefit analysis to determine the level of response necessary to mitigate the incident safely.

Pesticides like many other groups of chemicals can be segregated into families based upon their chemical make-up and characteristics, including toxicity. Common pesticide families are organophosphates, carbamates, chlorophenols and organocholorines. Organophosphates are derivatives of phosphoric acid and are acutely toxic, but are not enduring. They are generally known as all insecticides, which contain phosphorus. Organophosphates break down rapidly in the environment and do not accumulate in the tissues. They are generally much more toxic to vertebrates than other classes of insecticides and are therefore associated with more human poisonings than any other pesticide. They are also closely related to some of the most potent nerve agents including sarin and VX.

Organophosphates function by overstimulating, and then inhibiting neural transmission, primarily in the nervous, respiratory and circulatory systems. Signs and symptoms of exposure include pinpoint pupils, blurred vision, tearing, salivation and sweating. Pulse rate will decrease and breathing will become labored. Intestines and bladder may evacuate their contents. Muscles will become weak and uncomfortable. Additional symptoms include headache, dizziness, muscle twitching, tremor or nausea.

Symptoms of most pesticide poisonings are similar. Examples of organophosphate pesticides include malathion, methyl parathion, thimete, counter, lorisban and dursban. The chemical formulas of the organophosphates contain carbon, hydrogen, phosphorus and at least one sulfur atom, and some may contain at least one nitrogen atom. Antidotes are available for organophosphate pesticide poisonings. Many hospitals in rural areas where organophosphates are used by farmers and others will have extra stocks of atropine, which is used to counter the effects of organophosphate pesticides. Carbamate pesticides are derivatives of carbamic acid and are among the most widely used pesticides in the world. Most are herbicides and fungicides, such as 2,4-D, paraquat and dicamba. Carbamate pesticides function in the body by inhibiting nerve impulses. The formula of carbamates contains carbon, hydrogen, nitrogen and sulfur. Examples of carbmates include furadan, temik and sevin.

Organochlorine pesticides are known as chlorinated hydrocarbons, chlorinated organics, chlorinated insecticides and chlorinated synthetics. The formula contains carbon, hydrogen and chlorine. They are neurotoxins, which function by overstimulating the central nervous system, particularly the brain. Examples of organochlorine pesticides include aldrin, endrin, hesadrin, thiodane and chlordane. The best-known organochlorine is Dichlorodiphenyltrichloroethane (DDT), which has been banned for use in the United States because of its tissue accumulation and environmental persistence. Organophosphates do not break down in the environment, which can affect the food chain. Chlorophenols contain carbon, hydrogen, oxygen and chlorine. They affect the central nervous system, kidneys and liver when in contact with the body.

Pesticide labels contain valuable information for the emergency responder and medical personnel treating a patient exposed to pesticides

(Figure 3.126). EPA regulates the information required for pesticide labels. Pesticide users are required by law to comply with all the instructions and directions for use in pesticide labeling. Information on pesticide labeling usually is grouped under headings to make it easier to find the information when needed. These headings may include Identifying Information, Restricted-Use Designation, Front-Panel Precautionary Statements, Hazards to Humans and Domestic Animals, Environmental Hazards, Physical or Chemical Hazards and Directions for Use. Information on pesticide labels which emergency responders may need includes product name, "signal word," a statement of practical treatment, EPA registration number and establishment number, a note to physician and a statement of chemical hazards. EPA registration numbers indicate that the pesticide label has been approved by the EPA. The establishment number identifies the facility where the product was made. Other information on the label useful to responders includes active and inert ingredients. "Inert" does not necessarily mean that the ingredients do not pose a danger; it means only that the inert ingredients do not have any action on the pest for which the pesticide was designed. Many times the inert ingredients are flammable or combustible liquids. They can also range in toxicity from extremely toxic to relatively nontoxic.

Formulation information may also be provided on the label or may be abbreviated, such as WP for wettable powder, D for dust or EC for emulsifiable concentrate. The label also contains information about treatment for exposure. This information should be taken to the hospital when someone

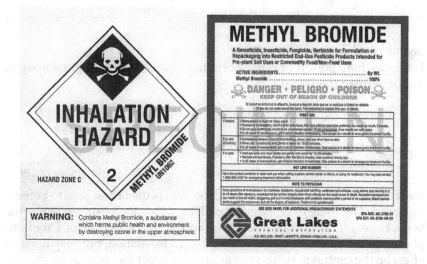

Figure 3.126 Pesticide labels contain valuable information for the emergency responder and medical personnel treating a patient exposed to pesticides.

has been contaminated with a pesticide. Do not, however, take the pesticide container to the hospital! Take the label or write the information down, take a picture of the label or use a pesticide label book. (Label books are available from agricultural supply dealers.) Another good reference source for information about pesticides is the *Farm Chemical Handbook*.

Pesticides can be grouped generally into three toxicity categories: high, moderate and low. Signal words corresponding to the level of toxicity and the caution statement "Keep out of the reach of children" must appear on every label. Three signal words that indicate the level of toxicity of a pesticide include:

Exposure	DANGER High Toxicity	WARNING Moderate Toxicity	CAUTION Low Toxicity	CAUTION Very Low Toxicity
Oral LD_{50}	Less than 50 mg/kg	50–500 mg/kg	500–5000 mg/kg	> 5000 mg/kg
Dermal LD_{50}	Less than 200 mg/kg	200–2000 mg/kg	2000–5000 mg/kg	> 5000 mg/kg
Inhalation LC_{50}	Less than 0.05 mg/l	0.05–0.5 mg/l	0.5 – 2 mg/l	> 2 mg/l
Eye Effects	Corrosive	Irritation persisting for 7 days	Irritation reversible within 7 days	Minimal effects gone in 24 hrs
Skin Effects	Corrosive	Severe Irritation at 72 hours	Moderate irritation at 72 hours	Mild or slight irritation

Signal Word Toxicity Comparison

Pesticides may poison or cause harm to humans by entering the body in one or more of these four ways:

1. Through the eyes
2. Through the skin
3. By inhalation
4. By swallowing

As with most chemicals, exposure to the eyes is the fastest way to become poisoned. The route of entry is the path by which the pesticide enters the body. However, it may not be the point where the injury occurs. For example, a pesticide may enter the body through inhalation, but may have an effect on the central nervous system, and not affect the respiratory system. In this instance, the route of exposure is inhalation through the lungs; the target organ is the central nervous system. Whenever anyone is exposed to a pesticide, it is important to recognize the signs and symptoms of poisoning, so that prompt medical help can be provided. Any unusual appearance or feeling of discomfort of illness can be a sign or symptom of pesticide poisoning. These signs and symptoms may be delayed up to 12 h. When they occur and pesticide contact is suspected, get medical attention immediately.

While emergency responders should never risk life to protect the environment, they certainly shouldn't do anything to damage the environment further than if they had not responded at all. Some pesticide fires may need to be left to burn out on their own to avoid contaminated runoff, which could damage the environment. Many times fire can break down pesticides into less harmful chemicals when they burn. Pesticide labels contain information about what type of environmental damage can occur from the uncontrolled release of the pesticide. Care should be taken to protect the environment in consultation with state and local environmental officials.

Protective Measures

Several protective measures may minimize the effect of toxic materials. Antidotes are available for a small number of toxic materials, but they must be administered immediately after exposure. Your body has the ability to filter out some toxic materials through the normal process of eliminating wastes. Protect yourself from toxic materials by wearing protective clothing and avoiding contact with toxic materials. Practice contamination prevention. Establish zones, deny entry and provide protection to responders and the public (*Firehouse Magazine*).

Synthetic Opioids and the Dangers to Emergency Responders

According to the U.S. Drug Enforcement Administration (DEA), "In the last several years, U.S. Law Enforcement has witnessed a dramatic increase in the availability of dangerous synthetic opioids." A large majority of these synthetic opioids are structural derivatives of the synthetic drug "fentanyl." Fentanyl is a synthetic opioid currently listed as a Schedule II prescription drug by DEA, which mimics the effects of morphine in the human body, but has potency 50–100 times that of morphine. Due to the high potency and availability of fentanyl, both transnational and domestic criminal organizations are increasingly utilizing these dangerous synthetic opioids as an adulterant in heroin and other controlled substances. The presence of these synthetic opioids in the illicit U.S. drug market is extremely concerning as the potency of these drugs has led to a significant increase in overdose incidents and overdose-related deaths throughout the nation.

DEA has issued the following warning to emergency responders including fire, police, EMS and other response personnel (Figure 3.127):

Figure 3.127 DEA has issued a warning to emergency responders including fire, police, EMS and other response personnel who may come into contact with fentanyl and other fentanyl-related substances. (Courtesy DEA)

"There is a significant threat to law enforcement personnel, and other first responders, who may come in contact with fentanyl and other fentanyl-related substances through routine law enforcement, emergency or life-saving activities. Since fentanyl can be ingested orally, inhaled through the nose or mouth, or absorbed through the skin or eyes, any substance suspected to contain fentanyl should be treated with extreme caution as exposure to a small amount can lead to significant health-related complications, respiratory depression, or death."

In the United States during 2016–2017, law enforcement personnel, including SWAT team members, narcotics officers and police dogs, have been exposed to fentanyl and related compounds requiring medical treatment. Fentanyl (also known as Fentanil), $C_{22}H_{28}N_2O$, IUPAC chemical name N-(1-(2-phenylethyl)-4-piperidinyl)-N-phenylpropanamide. Carfentanil (Wildnil®) is an analogue of fentanyl with an analgesic potency 10,000 times that of morphine and is used in veterinary practice to immobilize certain large animals. It has not been approved for use in humans in any form or dose.

Illicit Uses

Fentanyl is abused for its intense euphoric effects. It can serve as a direct substitute for heroin in opioid-dependent individuals. However, fentanyl is a very dangerous substitute for heroin because it is much more potent than heroin and results in frequent overdoses that can lead to respiratory depression and death. Fentanyl patches are abused by removing the gel contents from the patches and then injecting or ingesting these contents. Patches have also been frozen, cut into pieces and placed under the tongue or in the cheek cavity for drug absorption through the oral mucosa. Used patches are attractive to abusers as a large percentage of fentanyl remains in these patches even after a 3-day use. Fentanyl oral transmucosal lozenges and fentanyl injectables are also diverted and abused. Abuse of fentanyl initially appeared in the mid-1970s and has increased in recent years. There have been reports of deaths associated with abuse of fentanyl products.

Clandestine Manufacture

"Drug incidents and overdoses related to fentanyl are occurring at an alarming rate throughout the United States and represent a significant threat to public health and safety," said DEA Administrator Michele M. Leonhart. "Often laced in heroin, Fentanyl and Fentanyl analogues produced in illicit clandestine labs are up to 100 times more powerful than morphine and 30–50 times more powerful than heroin. Fentanyl is extremely dangerous to law enforcement and anyone else who may come into contact with it." In the past 2 years, DEA has seen a significant resurgence in fentanyl-related seizures. According to the National Forensic Laboratory Information System (NFLIS), state and local laboratories reported 3,344 fentanyl submissions in 2014, compared with 942 in 2013. In addition, DEA has identified 15 other fentanyl-related compounds.

Fentanyl is the most potent opioid available for use in medical treatment. It is potentially lethal, even at very low levels. Ingestion of small doses as small as 0.25 mg can be fatal. Its euphoric effects are indistinguishable from morphine or heroin. DEA is concerned about law enforcement coming into contact with fentanyl on the streets during the course of enforcement, such as a buy-walk or buy-bust operation. The current outbreak involves not just Fentanyl but also fentanyl analogues. The current outbreak is wider geographically and involves a wide array of individuals including new and experiences abusers. Carfentanil side effects are similar to those of fentanyl, which include itching, nausea and respiratory depression, which can be life-threatening. Carfentanil is classified as Schedule II Drug by DEA.

CASE STUDY

A 2012 study found evidence supporting the claim that during the 2002 Moscow theater hostage crisis, the Russian military made use of an aerosol form of carfentanil and another similar drug, remifentanil, to subdue Chechen hostage takers. Its short action, easy reversibility, would make it a potential agent for this purpose. Riches et al. found evidence from liquid chromatography–tandem mass spectrometry analysis of extracts of clothing from two British survivors and urine from a third survivor that the aerosol contained a mixture of the two anesthetics, the exact proportions of which the study was unable to determine. Previously, Wax et al. had surmised from the available evidence that the Moscow emergency services had not been informed of the use of the agent but were instructed to bring opioid antagonists. Because of the lack of information provided, the emergency workers did not bring adequate supplies of naloxone or naltrexone (opioid antagonists) to prevent complications in many of the victims, and there were subsequently over 125 confirmed deaths from both respiratory failure and aerosol inhalation during the incident. Assuming that carfentanil and remifentanil were the only active constituents, which have not been verified by the Russian military, the primary acute toxic effect to the theater victims would have been opioid-induced apnea; in this case, mechanical ventilation or treatment with opioid antagonists could have been life-saving for many victims.

Naloxone/Narcan

Naloxone, sold under the brand name Narcan among others, is a medication used to block the effects of opioids, especially in overdose. Naloxone may be combined within the same pill as an opioid to decrease the risk of misuse. When given intravenously, it works within 2 min, and when injected into a muscle, it works within 5 min. The medication may also be used in the nose. The effects of naloxone last about half an hour to an hour. Multiple doses may be required, as the duration of action of most opioids is greater than that of naloxone. Fentanyl has been found in busts involving heroin, morphine, oxycodone and hydrocodone. It has also been found in cocaine and synthetic marijuana, also known as "spice." Law enforcement personnel and drug-sniffing dogs are at risk from exposure to synthetic opioids. Many response organizations that may have to deal with synthetic opioids carry Narcan with them in the event of exposure.

Weapons Grade Narcotic

Carfentanil places law enforcement personnel in even greater risk. A chemical with the size of a grain of salt can kill an officer in less than 3 min. According to DEA, "It is crazy dangerous." Carfentanil is considered a weapons grade chemical. "Terrorists could acquire it commercially as we have seen drug dealers doing," according to Andrew Weber, former U.S. assistant secretary of defense for nuclear, chemical and biological defense systems. In June 2016, 1 kg of carfentanil was seized in Vancouver. That is enough of this potent narcotic for 50 million fatal doses.

Applied Chemistry and Physics 245

Figure 3.128 DEA warns law enforcement officers and other emergency responders to recognize the dangers of synthetic opioids and prepare to prevent contamination. (Courtesy DEA)

Protecting Personnel

DEA warns law enforcement officers and other emergency responders to recognize the dangers of synthetic opioids and prepare to prevent contamination (Figure 3.128). Emergency responders are advised to be vigilant with any suspected fentanyl-related narcotic.

- Exercise extreme caution.
- Don't collect samples without special training and PPE.
- Don't field test or transport samples to the office, and take them directly to testing laboratories equipped to handle entanyls.
- Be aware of symptoms of exposure.
- Seek IMMEDIATE medical attention.
- Administer naloxone as directed.

Personnel suspected of potential exposure based on their duties should carry antidotes with them. These narcotics may look like powered cocaine or heroin, or be cut into them as a filler. Carfentanil may be found in a powder, blotter paper, tablet, patch or spray form.

Fentanyls and analogues are extremely dangerous chemicals but with proper awareness, training, SOP's, personnel protective equipment and antidotes, responders can handle them safely (*Firehouse Magazine*).

Chemistry of Clandestine Methamphetamine Drug Labs

Clandestine drug laboratories continue to present a significant law enforcement and emergency response problem across the United States. These illegal labs contain the chemicals and equipment required to manufacture controlled substances such as methamphetamines (speed, crack, ice, glass and crystal), phenyl-2-propanone (P2P), LSD, PCP (angel dust), MDA/MDPP (Ecstasy), metha-qualude, methcathinone (cat), fentanyl and others. Methamphetamines are by far the most common illegal drugs manufactured in clandestine labs and will be the primary focus of this column. Illegal drug labs have been discovered in homes, apartments, hotel and motel rooms, barns, restaurants, fields, vacant and abandoned buildings, storage facilities and even mobile labs (Figure 3.129).

This is a problem that concerns rural America as well as urban areas. In fact, the more remote the area, drug makers think, the less likely they will be detected by law enforcement. However, the very nature and dangers of clandestine drug labs may cause emergency responders to encounter them by accident or when something goes wrong with the chemicals involved. Other public officials or civilians may also discover clandestine drug labs.

Response personnel should become familiar with the detection clues and hazards of the hazardous materials involved in drug lab operations. The chemicals used are themselves dangerous, and they can produce hazardous by-products and cause fires and explosions. Use of the chemicals can often result in contamination of the area used for the drug lab, which becomes a secondary contamination hazard for response personnel. Illegal drugs can be made with "preferred" or "alternate" chemicals. Some of these chemicals can make the operations more dangerous than others to the operators and the emergency responders.

Figure 3.129 Illegal drug laboratories have been discovered in homes, apartments, hotel and motel rooms, barns, restaurants, fields, vacant and abandoned buildings, storage facilities and even mobile laboratories. (Courtesy DEA).

Applied Chemistry and Physics

Making illegal drugs does not require the sophistication, knowledge or equipment necessary to manufacture chemical and biological terrorist agents. Not all chemicals associated with illegal drug manufacture are regulated, and many are available from local merchants such as pharmacies, hardware stores, supermarkets, discount and convenience stores and agricultural cooperatives. Transactions that occur in these locations involving large purchases of suspected chemicals should be reported to law enforcement. Even better, retailers should be taught to limit the amounts of materials sold, making the drug lab operators work harder to obtain the raw materials needed for drug manufacture. Though not consistent in all locations, some chemicals that may be used in illegal drug manufacture are regulated by government agencies. Manufacturers, distributors and retailers should be aware of precursor chemicals used for illegal drug production.

Anhydrous ammonia is a common hazardous material used to make methamphetamines. It is often stolen from storage tanks on farms and commercial facilities, which may result in leaks and releases that require the response of emergency personnel (Figure 3.130). Frequently, the amounts of ammonia stolen are so small that they are not missed. Thieves then hide stolen ammonia in places where they are not expected to be found, such as the trunks of cars, inside vans and in homes and apartments. Sometimes, thieves use portable propane tanks like those used for barbecue grills, which creates a hazard from ammonia attacking the fittings and valves that in turn may result in a release of the contents. The major problem with the theft of ammonia is that leaks often occur because valves are damaged or left open.

Figure 3.130 Anhydrous ammonia is a common hazardous material used to make methamphetamines. It is often stolen from storage tanks on farms and commercial facilities, which may result in leaks and releases that require the response of emergency personnel.

Detection

Response personnel must know how to recognize the hazards involved in emergencies that may involve clandestine drug labs. When a drug lab is suspected or discovered, it becomes a crime scene, and law enforcement should be contacted immediately. Much like terrorist incident scenes, responders should take care to make sure the scene is safe (free of devices set up to injure response personnel) and take only those actions necessary to save lives and protect exposures.

In addition to all of the other things they should be looking for at an emergency scene, responders should be aware of the potential clues that point to an illegal drug operation. Clues to watch for include blackened out windows, burn pits, stained soil, dead vegetation and multiple over-the-counter-drug containers. Tree kills have been found around locations of methamphetamine labs. Some 150-year-old Ponderosa Pines were killed close to a drug lab in Arizona.

Residents never putting trash out, laboratory glassware being carried into a residence and little or no traffic during the day, but lots of traffic late at night, can also be clues to illegal drug operations. Empty containers from antifreeze, white gas, starting fluids, Freon, lye or drain openers, paint thinner, acetone or alcohol may be clues, especially if they seem out of place for the occupancy. Additional clues are anhydrous ammonia or propane tanks, ceramic or glass containers or other kitchenware with hoses or duct tape, and thermos bottles or other cold storage containers. Also suspect are respiratory masks and filters, dust masks, rubber gloves, funnels, hosing and clamps. Unusual odors may also be present. When residential occupancies contain odors of ammonia, solvents, chemicals, sweet or bitter smells, all should be investigated to determine cause. Manufacturers of illegal drugs often set up booby traps for law enforcement and other responders.

Drug Lab Chemicals

Following is a list of some of the more dangerous and in some cases common hazardous chemicals that may be found in conjunction with clandestine drug lab operations. While the presence of any one of the materials does not automatically indicate a drug operation, the type of location and numbers of chemicals present should be taken into account. The list is not meant to be comprehensive or contain response operational information. It is provided as awareness information for emergency responders. Once chemicals are located at an incident scene, they should be researched using the same methods as other hazardous materials found at a hazmat incident.

- Acetic acid (glacial) is a corrosive organic acid, colorless liquid and a solid below 62°F (glacial means solid at normal temperatures). It has a vinegar-like odor. It is used in the manufacture of P2P,

methamphetamine and amphetamine. Its primary hazards are corrositivity and a strong irritant. In certain concentrations, it can also be flammable.
- Acetic anhydride is a colorless liquid with a strong vinegar-like odor. It can also be used in P2P synthesis. It is corrosive and can cause skin burn. Acetone (dimethyl ketone), a member of the ketone family of hydrocarbon derivatives, is a volatile, highly flammable and colorless liquid solvent with a sweet-type odor. It is a common ingredient in nail polish remover and may be found in beauty salons where nail technicians work. It is not highly toxic, but it is a narcotic. Acetone is used in the manufacture of methamphetamines.
- Anhydrous ammonia is a colorless, lighter-than-air gas with a strong, pungent odor. It is toxic and can be flammable under certain conditions, particularly inside a structure. Ammonia is readily available in rural areas, where it is used as a fertilizer, and has been the target of theft for use in clandestine drug operations. Ammonia in liquid form is reacted with sodium metal that is water reactive.
- Benzene, an aromatic hydrocarbon, is a colorless-to-yellow liquid with an aromatic odor (characteristic of all aromatic hydrocarbons) that is flammable, toxic and known to cause cancer. It is a solvent used in methamphetamine production.
- Ephedrine is composed of odorless white crystals. It can be found in medicines available over the counter. It is one of the primary precursors used in methamphetamine production. Ephedrine is an irritant and mildly toxic.
- Ethanol, an alcohol hydrocarbon derivative, is a clear, colorless liquid solvent used in the production of methamphetamine. It is flammable but only mildly toxic short term (used for drinking alcohol).
- Ethyl ether, an ether hydrocarbon derivative, is colorless with a sweet, pungent odor. Ethyl ether is highly flammable, toxic and used in the manufacture of methamphetamines. While it is the purpose of illegal drug manufacturers to produce drugs quickly and it is unlikely that ethyl ether will be around long enough to become a danger, responders should be aware of the potential of explosive peroxide formation. Where containers of ether are in use for 6 months or longer, these peroxides can form. They are sensitive to heat and shock; any suspected old containers of ether should be treated like potential bombs.
- Formic acid, also an organic acid, is a colorless liquid with a pungent odor. It is used in the process of methamphetamine manufacture. Formic acid is corrosive and toxic. Contact with oxidizing agents may cause explosion.
- Hexane, a colorless liquid with a mild characteristic odor, is a solvent used in the production of methamphetamines. It is a central nervous system toxin and is extremely flammable.

- Hydriodic acid is used by oil refineries to test crude oil for sulfur content. It is the principal chemical in the pseudoephedrine reduction process. It breaks down the pseudoephedrine molecules to create methamphetamines.
- Hydrogen iodide is a colorless gas used as a reagent with red phosphorus in the manufacture of methamphetamines. It is corrosive and an irritant.
- Hydrochloric acid (muriatic acid or pool acid) is a colorless inorganic acid with a pungent odor. It is used in the manufacture of methamphetamines; it is corrosive and has toxic irritating fumes. Gases released during the manufacturing process can be flammable and explosive. It is found in hardware stores.
- Iodine is an element that is solid purple crystals or flakes with a sharp odor. It is used in the synthesis of hydriodic acid. Iodine is toxic by inhalation and ingestion and corrosive. It is sometimes used by ranchers to treat thrush on horse hooves. It is used in the initial stages of the pseudoephedrine/ephedrine cooking process.
- Iodine tincture (solution with alcohol) is a dark red solution with a medicinal odor that is used in the synthesis of hydriodic acid. It is flammable and toxic by inhalation.
- Lead acetate is solid white crystals or brown or gray lumps that are odorless. It is used in P2P synthesis and is hazardous from chronic exposure.
- Lithium aluminum hydride is a solid white-to-gray powder and is odorless. It is corrosive and extremely water reactive. When in contact with water, it will generate explosive hydrogen gas. It is used in the hydrogenation process during methamphetamine production.
- Methyl alcohol (methanol) is an alcohol hydrocarbon derivative and a clear colorless liquid with a characteristic odor. It is a solvent used in the production of amphetamines. Methanol is toxic and flammable and can cause blindness.
- Naphtha is a reddish-brown liquid with an aromatic odor. It is a petroleum distillate solvent used in the manufacture of methamphetamine. The primary hazard is toxicity.
- Phenylacetic acid is an organic acid that is a solid, white, shiny crystal with a floral odor. It is used as a precursor to the synthesis of P2P. It is an irritant and a possible teratogen (causes birth defects).
- Phosgene, a toxic gas, is a by-product of the pseudoephedrine reduction process. When red phosphorus and iodine are heated, the lethal and odorless gas called phosgene is created. This could be dangerous to law enforcement and other responders walking in on an active lab.
- Phosphine is a colorless gas with a fish- or garlic-like odor. It is a product of methamphetamine production. It is highly flammable, reacts explosively with air and is toxic by inhalation.

- Phosphoric acid is an inorganic acid that is hygroscopic (absorbs moisture from the air) colorless crystal. Its primary hazard is as an irritant and it is corrosive. It is used as a precursor in the production of methamphetamine.
- Pseudoephedrine, a white crystalline powder, is a precursor used in the production of methamphetamines. It is an irritant and is toxic by ingestion. Available in over-the-counter decongestants and diet pills, it is used in the pseudoephedrine/ephedrine reduction method. The federal government has a limit of eight packages that can be purchased per person on medications containing pseudoephedrine.
- Red phosphorus is a red-to-violet solid that is odorless. It is used as a catalyst in methamphetamine synthesis. Catalysts are used to control the speed of chemical reactions. Red phosphorus can form phosphine gas during the production process which is toxic by inhalation, is flammable and reacts explosively with air. It can be found at the end of every matchstick in your home and is also used in road flares.
- Ronsonol (lighter fluid) is a reddish-brown liquid with an aromatic odor. It is a petroleum distillate solvent consisting of two solvent naphtha fractions: light aliphatic 95% and medium 5% and Shell Sol RB 100%. Properties are similar to those of naphtha.
- Sodium metal is an element that is a solid, silvery, white metal or crystal and is odorless. It is used in the hydrogenation in methamphetamine synthesis. It is corrosive and extremely water reactive, liberating hydrogen gas.
- Sodium hydroxide (lye) is found as white pellets or flakes that are odorless. It is a reagent used in the manufacture of methamphetamine. It is extremely corrosive. When in contact with metals such as sodium or fire, it can produce explosive hydrogen gas.
- Sulfuric acid (drain cleaner) is a colorless-to-yellow viscous liquid that is generally odorless. It is used in the manufacture of amphetamine and methamphetamines. It is corrosive and may produce corrosive fumes, and is found in battery acid or drain cleaners.
- Toluene is a clear colorless liquid member of the aromatic hydrocarbon family with a benzene-like odor. It is an irritant, highly flammable and a solvent used in manufacture of P2P and methamphetamine. It is found in paint thinners.
- 1,1,2-Trichloro-1,2,2-trifluoroethane (Freon) is a clear colorless liquid with a slight ethereal (ether-like) odor. It is a solvent used to extract D-methamphetamine. It is an irritant and toxic by inhalation.
- White gas is a solvent used to extract methamphetamine. It is a colorless liquid that is flammable. It is used as a fuel for camp stoves and readily available in hardware, discount and home improvement stores.

Homemade Ammonia

Methamphetamine manufacturing is accomplished in several ways, but one that is becoming popular is the use of anhydrous ammonia, sodium or lithium metal and the over-the-counter cold medications pseudoephedrine or ephedrine. Heat may be used but is not required. Other materials required include coffee filters, solvents and other common items easily obtained. (This is also known as the "Nazi" method because the Germans used it during World War II, when German soldiers were given methamphetamines to allow them to keep going on a limited diet.)

Anhydrous ammonia is heavily regulated and difficult to obtain for clandestine use, so drug makers often resort to theft from commercial facilities or farmers. There is, however, another option: liberating anhydrous ammonia from the common garden fertilizers ammonium nitrate and ammonium sulfate. Processing the fertilizer using sodium hydroxide and water liberates ammonia gas. Adding heat to the process speeds up the evolution of ammonia. The liberated ammonia gas is captured and condensed to be used for the methamphetamine manufacture.

Making methamphetamines using this method is not without significant hazard. The uncontrolled release of the ammonia gas, which is flammable, into a closed system is hazardous. The risk of a water reaction with sodium or lithium metal creates an ignition source for the flammable ammonia and hydrogen gas liberated from the water. Ammonium nitrate, an oxidizer, coming into contact with organic fuels can be explosive.

Cleanup Concerns

The cleanup of clandestine drug operations following an investigation by law enforcement personnel is not a function of emergency responders. Cleanup should be coordinated with the state department of the environment using recommended cleanup contractors. Because some aspects of the cleanup may prove dangerous for contractors, emergency personnel may want to keep a crew on-site in case of a fire or release requiring emergency actions.

Personnel entering the contaminated area should only do so using appropriate chemical protective clothing for the hazard present. Contamination may extend beyond the location of the actual drug lab. Ventilation and plumbing systems within hotels, motels and apartment buildings may also be contaminated and require decontamination (Figure 3.131).

When drug labs are discovered, they become miniature hazardous waste cleanup sites. It is estimated that as much as 7 lb of waste chemicals are produced for every pound of methamphetamine processed. Waste from drug labs is also sometimes dumped clandestinely on roadsides, vacant property, fields and wooded areas. Materials dumped

Figure 3.131 Personnel entering the contaminated area should only do so using appropriate chemical protective clothing for the hazard present. Contamination may extend beyond the location of the actual drug laboratory.

to dispose of evidence can include propane containers, empty 2L soda bottles, containers for other materials (such as starter fluid, brake fluid, brake cleaner, lighter fluid, rock salt, acetone or camping stove fuel) or miscellaneous glass and other containers. Investigators at the scene of a suspected clandestine drug-making operation need chemical protective clothing and respiratory protection. Hazmat teams may be called upon to assist because of the dangers of some of the chemicals involved (*Firehouse Magazine*).

Class 7 Radioactive (LF/LR)

A radioactive material is defined as "any material having a specific activity greater than 0.002 microcuries per gram." A specific activity of a radionuclide includes "the activity of the radionuclide per unit mass of that nuclide." Simply stated, a microcurie is a measurement of radioactivity. When a radioactive material emits more than 0.002 µCi/g of material, which is a term of weight, the material is then regulated in transportation by the U.S. Department of Transportation (DOT). Radiation is ionizing energy spontaneously emitted by a material or combination of materials. A radioactive material, then, is a material that spontaneously emits ionizing radiation. There are four types of DOT labels used to mark radioactive packages: radioactive, radioactive I, II and III. These radioactive materials are determined by the radiation level at the package surface. Radioactive III materials are the only radioactive materials that require placards on a transportation vehicle.

The good news about radioactive materials is that they are one of the easiest hazard classes to deal with. Transportation containers for radiation are so well constructed that I am not aware of an accident involving the transportation of radioactive materials where any radioactivity was released. If all hazardous materials containers were as well designed and constructed, our hazmat teams would be out of business. Most incidents we might encounter would be at fixed facilities where radioactive materials are used. If there were a transportation incident where radioactive materials were released, it would be an automobile, a delivery truck or a box truck carrying small packages of radioactive isotopes. These are primarily used for medical purposes and have a very short half-life.

However, don't throw out your radiological meters just yet. We should routinely monitor accident scenes of the above mentioned vehicles to make sure no radiation has been released. It is remotely possible that someone could use radioactive materials in an act of terrorism. It has not happened yet that I know of, so you would be more likely to win the lottery than to encounter clandestine radioactive materials. So don't lose any sleep over it. Just make sure you routinely use the meters and you will be prepared if a career event occurs.

Types of Radiation

There are two types of radiation: ionizing and nonionizing. Ionizing radiation involves particles and "waves of energy" traveling in a wavelike motion. Examples are alpha, beta, gamma, X-ray, neutron and microwave. Nonionizing radiation is also made up of "waves of energy." Examples include ultraviolet, radar, radio, visible light and infrared light. While all radioactive waves travel in a wavelike motion, all radioactivity travels in a straight line.

There are three primary types of radioactive emissions from the nucleus of radioactive atoms. The first is the alpha particle, which looks much like an atom of helium stripped of its electrons with two protons and two neutrons remaining. An alpha particle is a positively charged particle. It is large in size and, therefore, will not penetrate as much or travel as far as beta or gamma radiation. Alpha particles travel 3–4 in and will not penetrate the skin. Complete turnouts, including SCBA, hood and gloves, will protect responders from external exposure. However, if alpha emitters are ingested or enter the body through a broken skin surface, they can cause a great deal of damage to internal organs.

Beta particles are negatively charged and smaller, travel faster and penetrate farther than alpha particles. A beta particle is 1/1,800 the size of a proton or roughly equal to an electron in mass. Beta particles will penetrate the skin and travel from 3 to 100 ft. Full turnouts and SCBAs *will not* provide full protection from beta particles. Particulate radiation results in contamination of personnel and equipment where the particles come to rest. Electromagnetic energy waves, like gamma particles, do not cause contamination.

There is a third type of radioactive particle, but it does not occur naturally. The neutron particle is the result of splitting an atom in a nuclear reactor or accelerator, or it may occur in a thermonuclear explosion. When an atom is split, neutron particles are thrown out. You would have to be inside a nuclear reactor or experience a thermonuclear explosion to be exposed to neutron particles. If you were exposed to neutron particles, you would be quickly vaporized.

Gamma radiation is a naturally occurring, high-energy electromagnetic wave that is emitted from the nucleus of an atom. It is not particulate in nature and has high-penetrating power. Gamma rays have the highest energy level known and are the most dangerous of common forms of radiation. Gamma rays travel at the speed of light or more than 186,000 miles/s and will penetrate the skin, can injure internal organs and pass through the body. No protective clothing can protect against gamma radiation. Shielding from gamma radiation requires several inches of lead, other dense metal or several feet of concrete or earth.

Gamma radiation does not result in contamination because there are no radioactive particles, only energy waves. Examples of other electromagnetic energy waves include ultraviolet, infrared, microwave, visible light, radio and X-ray (Figure 3.132). There is little difference between gamma rays and X-rays; X-rays are produced by a cathode ray tube. To be exposed to X-rays, however, there has to be electrical power to the X-ray machine and the machine has to be turned on. If there is no electrical power, there is no radiation.

Gamma Radiation
Electromagnetic-Energy Waves

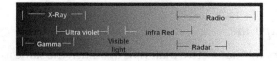

- **Speed = 186,000 miles per second**

- **The energy of gamma is greater than visible light**

Figure 3.132 Examples of other electromagnetic energy waves include ultraviolet, infrared, microwave, visible light, radio and X-ray.

Isotopes

When the nucleus of an element contains more or fewer neutrons than a "normal" atom of that element, it is said to be an isotope of that element. All atoms have 3–25 isotopes; the average is 10 per element. *All isotopes are not radioactive.* Hydrogen has three important isotopes: hydrogen 1, sometimes called protonium, has one proton in the nucleus and no neutrons. Hydrogen 2 has one proton and one neutron in the nucleus and is called deuterium or heavy water. Hydrogen 1 and hydrogen 2 are not radioactive. Hydrogen 3 has one proton and two neutrons in the nucleus and is called tritium. Hydrogen 3 is radioactive. Tritium is used in some "exit" signs, which gives them the glow-in-the-dark ability without batteries or any other electrical source. The signs last approximately 20 years before they have to be replaced. Radiation in the signs being replaced needs to be properly disposed of.

Carbon also has several isotopes. "Normal" carbon is known as carbon 12. Carbon 12 has six protons and six neutrons in the nucleus and is not radioactive. Carbon 13 has six protons and seven neutrons in the nucleus and is not radioactive. Carbon 12 makes up 999 out of 1,000 carbon atoms; the other is carbon 13. Carbon 14 has six protons and eight neutrons and is naturally radioactive. Carbon 14 is a beta emitter that is produced in the atmosphere by the action of cosmic radiation on atmospheric nitrogen. In the process, a proton is forced from the nucleus of nitrogen, which then becomes carbon 14.

The human body is made up of 1/10,000 of 1% of carbon 14. You inhale carbon 14 every time you take a breath. It decays and is replaced in the body, so there is a constant supply. Carbon 14 is sometimes used to determine the age of organic materials, as it takes years to disappear. The amount of carbon 14 can be used to determine how long a person has been dead, or the age of a body or other object. There are three other carbon isotopes that are all man-made: Carbon 11 has six protons and five neutrons, carbon 15 has six protons and nine neutrons and carbon 16 has six protons and ten neutrons; all three are radioactive.

Radiation Exposure

Routes of entry for radioactive materials are much the same as for poisons. However, the radioactive source or material does not have to be directly contacted for radiation exposure to occur. Exposure occurs from the radiation being emitted from the radioactive source. Once a particulate radioactive material enters the body, it is dangerous because the source now becomes an internal source rather than an external one. You cannot protect yourself by time, distance or shielding from a source that is inside your body. Contact with or ingestion of a radioactive material does not make you radioactive. Contamination occurs with radioactive particles, but with proper decontamination, these particles can be successfully removed. After they are removed, they cannot cause any further damage to the body.

Because radiation exposure can be cumulative, there are no truly safe levels of exposure to radioactive materials. Radiation does not cause any specific diseases. Symptoms of radiation exposure may be the same as those of exposure to cancer-causing materials. The tolerable limits for exposure to radiation that have been proposed by some scientists are arbitrary. Scientists concur that some radiation damage can be repaired by the human body. Therefore, tolerable limits are considered acceptable risks when the activity benefits outweigh the potential risks. The maximum annual radiation exposure for an individual person in the United States is 0.1 roentgen equivalent man (REM). Workers in the nuclear industry have a maximum exposure of 5 REM per year. An emergency exposure of 25 REM has been established by the National Institute of Standards and Technology for response personnel. This type of exposure should be attempted under only the most dire circumstances and should occur only once in a lifetime. There are currently efforts to increase this maximum emergency exposure.

Effects of exposure to radiation on the human body depend on the amount of material the body was exposed to, the length of exposure, the type of radiation, the depth of penetration and the frequency of exposure. Cells that are most susceptible are rapidly dividing cells, such as in the bone marrow. Children are more susceptible than adults, and the fetus is the most susceptible. Radiation injuries frequently do not present themselves for quite a long time after exposure. It can be years, or even decades, before symptoms appear. Cancer is one of the main long-term effects of exposure to radiation. Leukemia may take from 5 to 15 years to develop. Lung, skin and breast cancers may take up to 40 years to develop. Varying levels of illness will occur from radiation exposure depending on the dose.

- No detectable symptoms are the result of up to 25 REM.
- Elevated temperature and changes in the blood count occur from 50 REM.
- Nausea and fatigue result from 100 REM.
- Approximately 200–250 REM results in sickness to all exposed and death to some within 30 days.
- Exposure to 500 REM results in the death of half of those exposed in 30 days.
- Exposure to over 600 REM results in the death of all exposed.

Radiation burns are much like thermal burns, although they can be much more severe.

TIME - DISTANCE - SHIELDING

Figure 3.133 Because of the physical characteristics of radioactive materials, protection for emergency responders can be provided by taking a few simple, protective actions.

- First-degree radiation burns result from an exposure of 50–200 radiation absorbed doses (RADs)
- Second-degree burns result from 500 RADs
- Third-degree burns result from 1,000 RADs.

Because of the physical characteristics of radioactive materials, protection for emergency responders can be provided by taking a few simple, protective actions, commonly referred to as time, distance and shielding (Figure 3.133):

Time refers to the length of exposure to a radioactive source and the half-life of a radioactive material. A half-life is the length of time necessary for an unstable element or nuclide to lose one half of its radioactive intensity in the form of alpha, beta and gamma radiation. Half-lives range from fractions of seconds to millions of years. In ten half-lives, almost any radioactive source will no longer put out any more radiation than normal background.

Distance is the second protective measure against radiation. As previously mentioned, radiation travels in a straight line but only for short distances. Therefore, the greater the distance from the radioactive source, the less the intensity of the exposure will be (Figure 3.134). There is a law in dealing with radioactivity known as the "inverse-square law." This means that as the distance from the radioactive source is doubled, the radiation intensity drops off by one-quarter. If the distance is increased ten times, the intensity drops off to 1/100 of the original intensity.

Shielding is the third protective measure against radiation. Shielding simply means placing enough mass between personnel and the radiation, which will provide protection from the radiation (Figure 3.135). In the case of alpha particles, your skin or a sheet of paper will produce enough shielding. Turnouts will provide extra protection. Ingestion is the major hazard of radioactive particles, and wearing SCBA will prevent ingestion. Beta particles require more substantial protection from entering the body. A 1/4 in.-thick piece of aluminum will stop beta radiation. Turnouts will

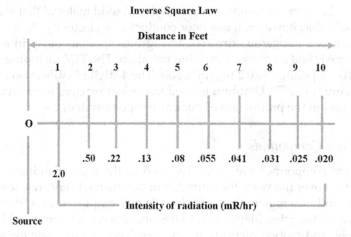

Figure 3.134 Radiation travels in a straight line but only for short distances. Therefore, the greater the distance from the radioactive source, the less the intensity of the exposure will be.

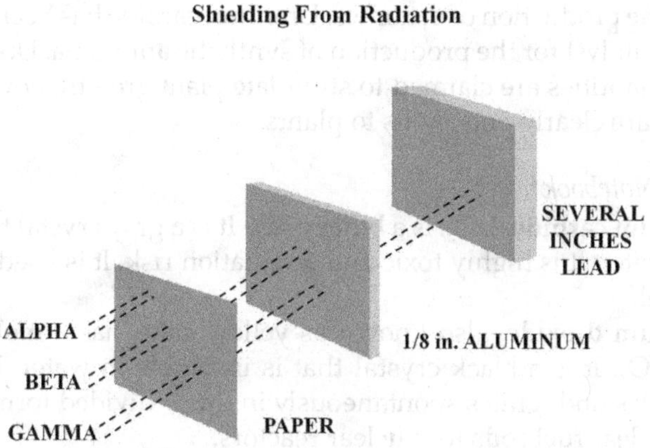

Shielding From Radiation

Figure 3.135 Shielding simply means placing enough mass between personnel and the radiation, which will provide protection from the radiation.

not provide adequate protection. Gamma radiation requires 3–9 in of lead or several feet of concrete or earth.

Chemical Notebook

Uranium, U, is a radioactive metallic element. Uranium has three naturally occurring isotopes: uranium 234 (0.006%), uranium 235 (0.7%) and uranium 238 (99%). Uranium 234 has a half-life of 2.48×10^5 years, uranium 235 has a half-life of 7.13×10^8 years and uranium 238 has a half-life of 4.51×10^9 years. Uranium is a dense, silvery, solid material that is ductile and malleable; however, it is a poor conductor of electricity. As a powder, uranium is a dangerous fire risk and ignites spontaneously in air. It is highly toxic and a source of ionizing radiation. The TLV, including metals and all compounds, is 0.2 mg/m^3 of air. The 4-digit identification number for uranium is 2979. Uranium is used in nuclear reactors to produce electricity and in the production of nuclear weapons systems.

Uranium Compounds

Uranium compounds are primarily used in the nuclear industry. It has been used over the years for a number of commercial ventures, some successful and others not. Uranium dioxide was employed as a filament in series with tungsten filaments for large incandescent lamps used in photography and motion pictures. It has a tendency to eliminate the sudden surge of current through the bulbs when the light is turned on, which extends the life of the bulbs. Some alloys of uranium were used in the production of steel; however, they never proved commercially valuable. Sodium and ammonium diuranates have been used to produce colored glazes in the production of ceramics. Uranium carbide has been suggested as a good catalyst for the production of synthetic ammonia. Uranium salts in small quantities are claimed to stimulate plant growth; however, large quantities are clearly poisonous to plants.

Chemical Notebook

Uranium carbide, UC_2, is a binary salt. It is a gray crystal that decomposes in water. It is highly toxic and a radiation risk. It is used as nuclear reactor fuel.

Uranium dioxide, also known as yellow cake, has a molecular formula of UO_2. It is a black crystal that is insoluble in water. It is a high radiation risk and ignites spontaneously in finely divided form. It is used to pack nuclear fuel rods for nuclear reactors.

Uranium hexafluoride has a molecular formula of UF_6. It is a colorless, volatile crystal that sublimes and reacts vigorously with water. It is highly corrosive and is a radiation risk. The four-digit 4-digit identification number for fissile material containing more than 1% of uranium 235 is 2977; for lower specific activity, the number is 2978. Uranium hexafluoride is used in a gaseous diffusion process for separating isotopes of uranium.

Uranium hydride is a binary salt and has a molecular formula of UH_3 and is a brown–gray-to-black powder that conducts electricity. It is highly toxic and ignites spontaneously in air.

Uranium tetrafluoride, with the molecular formula of UF_4, is a green, nonvolatile, crystalline powder that is insoluble in water. It is highly corrosive and is also a radioactive poison.

Medical uses of radioactive sources include sterilization, implants using radium, scans using iodine and therapy using cobalt. X-rays are used in diagnostic medical procedures. In addition to medical facilities, radioactive materials may be found in research laboratories, educational institutions, industrial applications and hazardous waste sites.

Radium Compounds
Chemical Notebook
Radium, Ra, is a radioactive metallic element. There are 14 radioactive isotopes of radium; however, only radium 226, with a half-life of 1,620 years, is usable. It is a brilliant, white solid that is luminescent and turns black upon exposure to air. Radium is water soluble, and contact with water evolves hydrogen gas. It is in the alkaline-earth metal family and, like calcium, it seeks the bones when it enters the body. It is highly toxic and emits ionizing radiation. Radium is destructive to living tissue.

It is used in the medical treatment of malignant growth and industrial radiography. Compounds formed with radium all have the same hazards as radium itself. Most are used in the treatment of cancer and for radiography in the medical and industrial fields. The compounds are all solids, and the degree of water solubility varies.

Radium bromide is a binary salt and has a molecular formula of $RaBr_2$; it is composed of white crystals that turn yellow to pink. It sublimes at about 1,650°F and is water soluble. The hazards are the same as those of radium. It is used in the medical treatment of cancers.

Radium carbonate, with the molecular formula of $RaCO_3$, is an amorphous, radioactive powder that is white when pure. Because of impurities, radium carbonate is sometimes yellow or pink. It is insoluble in water.

Radium chloride is a binary salt with the molecular formula of $RaCl_2$ and a yellowish-white crystal that becomes yellow or pink upon standing. It is radioactive and soluble in water. It is used in cancer treatment and physical research.

Cobalt, Co, is a metallic element. Cobalt 59 is the only stable isotope. Common isotopes are cobalt 57, cobalt 58 and, the most common, cobalt 60. Cobalt is a steel-gray, shining, hard, ductile and somewhat malleable metal. It has magnetic properties and corrodes readily in air. Cobalt dust is flammable and toxic by inhalation, with a TLV of 0.05 mg/m^3 of air. It is an important trace element in soils and animal nutrition. Cobalt 57 is radioactive. It has a half-life of 267 days.

It is a radioactive poison and used in biological research. Cobalt 58 is also radioactive and has a half-life of 72 days. It is a radioactive poison, and it is used in biological and medical research. Cobalt 60 is one of the most common radioisotopes. It has a half-life of 5.3 years, is available in larger quantities and is cheaper than radium. It is a radioactive poison and used in radiation therapy for cancer and radiographic testing of welds and castings in industry. Compounds of cobalt are not radioactive.

Iodine, I, is a nonmetallic element of the halogen family. There is only one natural stable isotope: iodine 127. There are many artificial radioactive isotopes. Iodine is a heavy, grayish-black solid or granules having a metallic luster and characteristic odor. It is readily sublimed to a violet vapor and insoluble in water. Iodine is toxic by ingestion and inhalation and is a strong irritant to the eyes and skin, with a TLV ceiling of 0.1 ppm in air. Iodine 131 is a radioactive isotope of iodine and has a half-life of 8 days. It is used in the treatment of goiter, hyperthyroidism and other disorders. It is also used as an internal radiation therapy source. Most iodine compounds are not radioactive.

Krypton, Kr, is an elemental, colorless, odorless, inert gas. It is noncombustible, nontoxic and nonreactive; however, it is an asphyxiant gas and displaces oxygen in the air. Krypton 85 is radioactive and has a half-life of 10.3 years. The 4-digit identification number for krypton is 1056 as a compressed gas and 1970 as a cryogenic liquid. These forms of krypton are not radioactive. Radioactive isotopes of krypton are shipped under radioactive labels and placards as required. Its primary uses are in the activation of phosphors for self-luminous markers, detection of leaks and medicine to trace blood flow.

Radon, Rn, is a gaseous radioactive element from the noble gases in family 8 on the periodic table. There are 18 radioactive isotopes of radon, all of which have short half-lives. For example, radon 222 has a half-life of 3.8 days. Radon is a colorless gas that is soluble in water. It can be condensed to a colorless transparent liquid and to an opaque, glowing solid. Radon is the heaviest gas known, with a density of 9.72 g/L at 32°F. It is derived from the radioactive decay of radium. It is highly toxic and emits ionizing radiation. Lead shielding must be used in handling and storage. Radon has appeared naturally in the basements of homes, causing some concern for the residents. The primary uses are in cancer treatment, as a tracer in leak detection, in radiography and in chemical research.

Radioactive materials are often found in transportation. They are heavily regulated, and the containers are well constructed. Most radioactive incidents are not handled by local emergency responders. Agencies other than fire, police, and EMS are responsible for response and handling of radioactive emergencies. Emergency responders must, however, be aware of radioactive materials and know how to protect themselves. Each state has radiological response teams for radioactive emergencies. They may be a part of the emergency management agency, the health department, the department of environment, or some other agency. Federal interests are represented by the U.S. Department of Energy and the Nuclear Regulatory Commission.

Applied Chemistry and Physics 263

Containers for Radioactive Materials

Radioactive materials are shipped and stored in containers which are so well engineered that there has never been a transportation incident where radiation was released (Figure 3.136). In storage areas where radioactive isotopes are stored and used, they are behind lead shields or contained within specially designed lead containers (Figure 3.137) (Hazardous Materials Chemistry for Emergency Responders).

Figure 3.136 Radioactive materials are shipped and stored in containers which are so well engineered that there has never been a transportation incident where radiation was released.

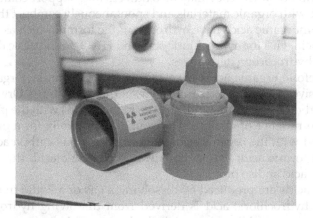

Figure 3.137 In storage areas where radioactive isotopes are stored and used, they are behind lead shields or contained within specially designed lead containers.

Hazard Class 8 Corrosive (HF/LR)

Class 8 materials are liquids and solids that are corrosive. There are no Divisions of corrosives. Next to flammable liquids and gases, corrosives are the next most common hazardous material encountered by emergency responders. There are, however, two types of corrosive materials found in Hazard Class 8: acids and bases. Acids and bases are actually different types of chemicals that are sometimes used to neutralize each other in a spill. They are grouped together in Class 8 because the corrosive effects they have on tissue and metals, if contacted, are much the same. It should be noted that the correct terminology for acids is corrosive and that for bases is caustic. The Department of Transportation (DOT), however, does not differentiate between the two. The DOT definition of corrosive is "a liquid or solid that causes visible destruction or irreversible alterations in human skin tissue at the site of contact, or a liquid that has a severe corrosion rate on steel or aluminum. This corrosive rate on steel and aluminum is 246 inches per year at a test temperature of 131°F."

A chemical definition of an acid from the *Condensed Chemical Dictionary* is "a large class of chemical substances whose water solutions have one or more of the following properties: sour taste, ability to make litmus dye turn red and to cause other indicator dyes to change to characteristic colors, ability to react with and dissolve certain metals to form salts and the ability to react with bases or alkalis to for to form salts."

There are two basic types of acids: organic and inorganic. Inorganic acids are sometimes referred to as mineral acids. Organic acids as a group are generally not as strong as inorganic acids. The main difference between the two is the presence of carbon in the compound. Inorganic acids do not have carbon in the compound. Inorganic acids are corrosive, but they do not burn. They may be oxidizers and support combustion or may react with organic materials and spontaneously combust the organic material. Inorganic acids have hydrogen in the formula such as H_2SO_4 for sulfuric acid, HCl for hydrochloric acid and HNO_3 for nitric acid. Organic acids are hydrocarbon derivatives.

Therefore, organic acids have carbon in the formula. Organic acids are corrosive, may polymerize and some of them may burn. Organic acids have a carbon in the compound, and the name begins with the prefix indicating the number of carbons. For example, the prefix for a one-carbon compound with the organic acids is "form," so a one-carbon acid would be called formic acid; two carbons would be acetic acid; three carbons propionic acid and so on.

Most acids are produced by dissolving a gas or a liquid in water. For example, hydrochloric acid is derived from dissolving hydrogen chloride gas in water. All acids contain hydrogen. This hydrogen is an ion (H^+) and can be measured by using the pH scale (Figure 3.138), which,

pH Scale of Common Materials

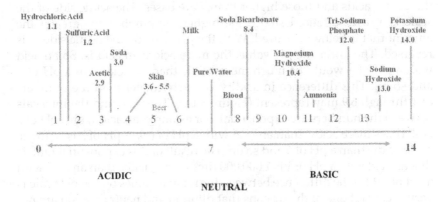

Figure 3.138 All acids contain hydrogen. This hydrogen is an ion (H⁺) and can be measured by using the pH scale.

in simple terms, measures the hydrogen ion concentration of a solution. Acids as a group have high hydrogen ion concentration. Bases have very low hydrogen ion concentrations and high hydroxyl (OH⁻) concentrations. The strength or weakness of an acid or base is the amount of hydrogen ions or hydroxyl ions that are produced as the acid or base is produced. If the hydrogen ion concentration in an acid is high, then the acid is a strong acid. If the hydroxyl concentration is high, then it is a strong base. In both cases, there is almost total ionization of the material dissolved in water to make the strong acid and base. For example, hydrochloric acid is a strong acid with a pH of 1.1; almost all of the hydrogen chloride gas is ionized in the water. If the hydrogen ion concentration is low, then the acid is a weak acid. Acetic acid is a weak acid with a pH of 3; only about 2% ionization has occurred in producing the compound.

Another term associated with corrosives is concentration. Concentration, unlike strength, has to do with the amount of acid that is mixed with water. This is often expressed in terms of percentages. A 98% concentration of sulfuric acid would be 98% sulfuric acid and 2% water. A solution of 50% nitric acid would be 50% nitric acid and 50% water. In the 50% concentration, the solution would have only half of the H⁺ ions than the 100% concentration would have. A 50% concentration of nitric acid would be a solution diluted to 50% of the original acid.

The pH scale measures the acidity or alkalinity of a solution. Acids are acidic and bases are alkaline. Acids have a value on the pH scale of 1–6.9. Materials with a pH value of 7 are considered to be neutral; they are neither acidic nor basic. Bases have values on the pH scale from 7.1 to 14. It is not important that emergency responders know how the pH scale measures corrositivity or the specific values of any given acid or base. It is important, however, for responders to know that numerical values less than 7 are acids and those higher than 7 are bases. The acidic side of the scale is a reverse ratio. Usually, the higher the number, the greater the amount that is being measured. With the pH scale on the acidic side, it is reversed. The lower the pH value, the more acidic an acid is. So an acid with a pH of 1 would be much more acidic than an acid with a pH of 2 and so on. This difference in acidity is much greater than the numbers of 1 through 6.9 may represent (Figure 3.139). The ratio and the intervals between the numbers are exponential. For example, an acid with a pH of 6 is 10 times more acidic than an acid with a pH of 6.9. A pH of 5 is 10 times more acidic than a pH of 6 and so on. The result of this exponential ratio is that an acid with a pH of 1 is 1,000,000 times more acidic than an acid with a pH of 6.9. So the difference between individual values on the pH scale is very great and one of the reasons that dilution and neutralization are not as simple as they might sound.

If the chemical name of a hazardous material is known and it is determined to be a corrosive, looking up the chemical name in reference sources identifies if the material is an acid or a base. It is not necessary for responders to get a pH measurement of the material unless it is to verify the reference information. The use of pH measurements can be useful when a material hasn't been positively identified. The pH measurement can be used to narrow the chemical family possibilities in the identification process. There are a number of ways for emergency responders to measure the pH of a corrosive material. First of all, the proper chemical protective clothing must be worn when working around the corrosive material.

Exponential Logarithm of pH Values

pH Value	H^+ and OH^- Concentration
1	1,000,000
2	100,000
3	10,000
4	1,000
5	100
6	10
7	1
8	10
9	100
10	1,000
11	10,000
12	100,000
13	1,000,000

Figure 3.139 This difference in acidity is much greater than the numbers of 1 through 6.9 may represent.

The simplest and least expensive method of determining pH is the use of pH paper. The pH paper changes color based upon the type and strength of corrosive material that is present. The colored paper is then compared to a chart on the pH paper container. The chart indicates that numerical pH values are much the same as the pH scale. Although not as accurate as a pH meter, the numbers give a "ballpark" measure of the pH of the material. There are also commercially available pH meters from handheld to sophisticated laboratory instruments. This equipment can be expensive and pH paper will be accurate enough for emergency response. To determine whether a material is an acid or a base, litmus paper can be used. Litmus paper turns blue if the corrosive material is a base. If the litmus paper turns red, the corrosive is an acid. The litmus paper does not give actual numerical pH values.

The chemical definition of a base from the *Condensed Chemical Dictionary* is "a large class of compounds with one or more of the following properties: bitter taste, slippery feeling in solution, ability to turn litmus blue and to cause other indicators to take on characteristic colors, and the ability to react with (neutralize) acids to form salts". It is important to note that, while the definition of acids and bases mentions the taste and feeling of the materials, these are dangerous chemicals and can cause damage to tissues upon contact. Therefore, it is **NOT** recommended that responders come into contact with these materials! Ionization occurs with the bases just as with the acids as they are made. Most bases are produced by dissolving a solid, usually a salt, in water.

However, with the bases the ion produced is the hydroxyl ion (OH^-). The base is considered strong or weak depending on the numbers of hydroxyl ions produced as a material is dissolved in water. A large OH^- concentration produces a strong base, and a small OH^- concentration produces a weak base. Sodium and potassium hydroxide are strong bases, and calcium hydroxide (hydrated lime) is a weak base. Bases have a pH from 7.1 to 14 on the scale. The ratio is increasing from 7.1 to 14, with 7.1 being the least basic and 14 the most basic. The amount of basicness between the numerical values on the pH scale is exponential just as with the acids. A base with a pH of 8 is 10 times more basic than that with a pH of 7.1.

Corrosiveness is not the only hazard of Class 8 materials. In addition to being corrosive, they may have other hazards, and they may be poisons, oxidizers or flammables. Many corrosives, especially acids, can be violently water reactive. Contact with water may cause splattering of the acid, produce toxic vapors and produce heat that may ignite nearby combustible materials. Some of the water may be turned to steam. Overpressurization of the corrosives container is possible from contact with water. Many corrosives may also be unstable and reactive. They may explode, polymerize, decompose and produce poisons.

Organic Acid

Organic acids are hydrocarbon derivatives. They can be flammable and toxic, and all are corrosive. Some organic acids burn, whereas others do not. If a flammable placard is found on an acid container or the MC312/DOT412 acid tanker, it is an organic acid. Organic acids are polar; in fact, of all the hydrocarbon derivatives, they are the most polar. They have both the carbonyl **CO** structure like the ketones, aldehydes and esters and the **OH** structure like the alcohol. The formula for a generic organic acid is **COOH**. One radical is attached and the name comes from the radical with an "ic" added and with the ending word "acid." Acetic acid is formed when a methyl radical is attached to the generic COOH of the organic acid. It can be found in varying concentrations from strong, as in glacial acetic acid which is the pure compound (99.8%), to weak solutions known as vinegar. It is a clear, colorless liquid with a pungent odor. It is one of the organic acids that is flammable, and is toxic by inhalation and ingestion in higher concentrations, with a TLV of 10 ppm in air.

Other types of common organic acids include formic acid, acrylic acid, butric acid and propionic acid.

Chemical Notebook

Acetic acid, CH_3COOH, an organic acid, is a clear, colorless corrosive liquid with a pungent odor. It has a pH of 2.9. The glacial form is the pure form without water, which is 99.8% pure. It is flammable with a flash point of 110°F and a flammable range of 4.0–19.9. Acetic acid is water soluble. It is toxic by inhalation and ingestion, with a TLV of 10 ppm in air. It is a food additive at lower concentrations. It is used in the production of plastics, pharmaceuticals, dyes insecticides and photographic chemicals.

Perchloric acid, $HClO_4$, is a colorless, odorless, fuming liquid that is unstable in its concentrated form. It is a strong oxidizing agent and spontaneously ignites when in contact with organic materials. Contact with water produces heat; when shocked or heated, it may detonate. Perchloric acid is toxic by ingestion and inhalation. It is used in the manufacture of explosives and esters, in electropolishing, as a catalyst and in analytical chemistry.

Hydrocyanic acid, HCN, is a very toxic material. It is also a dangerous fire and explosion risk. It has a wide flammable range of 6%–41% in air. It is toxic by inhalation and ingestion and through skin absorption. The TLV of hydrocyanic acid is 10 ppm in air. It is used in the manufacture of acrylonitrile, acrylates, cyanide salts, dyes, rodenticides and other pesticides. It is shipped by a "candy stripe" tank car, which was white with a horizontal stripe around the belly and vertical stripes on each end. Because of the potential for terror attacks on the tank cars, they have been discontinued and the product is shipped as a solid pellet that is reconstituted into HCH when it reaches its destination.

Organic acids may polymerize by exposure to heat or sudden shock. For example, **acrylic acid, C_2H_3COOH**, has a double bond between the carbons that can come apart in a polymerization reaction. Generally, materials that have double bonds are reactive in some manner. If polymerization occurs inside a container, an explosion may occur that can produce heat, light, fragments and a shockwave.

Formic acid, HCOOH, is a colorless, fuming liquid with a penetrating odor. It is soluble in water. As with many of the organic acids, formic acid may burn. It has a flash point of 156°F and a flammable range of 18%–57%. Formic acid has a TLV of 5 ppm in air. It is used in dyeing and finishing of textile, treatment of leather, making of esters, fumigants, insecticides, refrigerants and others.

Propionic acid, C_2H_5COOH, is a colorless, oily liquid with a rancid odor. It is flammable with a flash point of 130°F and a flammable range of 2.9%–12.1%. It is toxic with a TLV of 10 ppm in air. It is used as a mold inhibitor in bread, as a fungicide, as a herbicide, as a preservative for grains, in artificial fruit flavors and in pharmaceuticals. Corrosives may also contact poison materials and produce poison gases as the poisons decompose. The poison vapors could be much more toxic than the corrosive vapors. When acids come in contact with cyanide, hydrogen cyanide gas is produced. Hydrogen cyanide gas is very toxic with a TLV of 10 ppm in air.

When strong corrosives contact flammable liquids, the chemical reaction that occurs may produce heat. The heat produced causes more vapor to be present, and if an ignition source is present, combustion may occur. Corrosives may be strong oxidizers. If they come into contact with particulate combustible solids, spontaneous combustion may occur. Once ignition occurs, the corrosive acts as an oxidizer and accelerates the rate of combustion. Nitric acid in contact with combustible organic materials containing cellulose produces a chemical reaction. This reaction produces nitrocellulose which is a dangerous fire and explosion risk. Toxic vapors may also be produced when the cellulose burns.

Inorganic Acid

Chemical Notebook

Sulfuric acid is an inorganic acid and the most heavily produced and widely used industrial chemical in existence. It is used in batteries for cars and other vehicles. It is a strong corrosive material. It has a pH of 1.2. It is a dense, oily liquid, and is colorless to dark brown depending on purity. The vapors are toxic at 1 ppm in air. It is used in the manufacture of fertilizers, chemicals and dyes; in pigments; as an enchant and a catlyst; in electroplating baths and explosives and many other uses.

Fuming sulfuric acid is called oleum. It is a solution of sulfur trioxide in sulfuric acid. The sulfur trioxide is forced into solution with the sulfuric acid to the point that the solution cannot hold anymore. As soon as the solution is exposed to air, the fuming begins forming dense vapor clouds. It is violently water reactive as are most acids.

Phosphoric acid, H_3PO_4, is an inorganic acid and a white or yellowish, crystalline mass. As a liquid, it is viscous. It is water soluble and absorbs oxygen very readily. Its primary use is in chemical analysis and as a reducing agent.

Nitric acid is an inorganic acid and a colorless, transparent or yellowish, fuming, suffocating, corrosive liquid. It attacks almost all metals. The yellow color results from the exposure of the nitric acid to light. Nitric acid is a strong oxidizer and is soluble in water. It may be found in solutions of 36, 38, 40, 42, degrees B`e and concentrations of 58%–63%, and 95%. Nitric acid is a dangerous fire risk when in contact with organic materials. It is toxic by inhalation and is corrosive to tissue and mucous membranes. The TLV is 2 ppm in air. Nitric acid is used in the manufacture of ammonium nitrate fertilizer and explosives, etching of steel, reprocessing spent nuclear fuel and others. There are two types of fuming nitric acid:

White fuming nitric acid is concentrated with 97.5% nitric acid and less than 2% water. It is a colorless-to-pale yellow liquid which fumes strongly. It is decomposed by heat and exposure to light and becomes red from nitrogen dioxide.

Red fuming nitric acid contains more than 85% nitric acid, 6%–15% nitrogen dioxide and 5% water. Both are toxic by inhalation, strong corrosives and dangerous fire risks, and may explode in contact with reducing agents. They are used in the production of nitro compounds and rocket fuels and as a laboratory reagent.

Hydrochloric acid is an inorganic acid and a colorless or slightly yellow, fuming pungent liquid. It has a pH of 1.1. It is water soluble, a strong corrosive and toxic by ingestion and inhalation. It is used in food processing, in pickling and metal cleaning, as an alcohol denaturant and as a laboratory reagent.

Base

Caustic potash is also known as potassium hydroxide (KOH) and lye. It is a white solid found as pieces, lumps, sticks, pellets or flakes. It is water soluble and may absorb water and carbon dioxide from the air. It is a strong base and toxic by ingestion and inhalation. It is used in soap manufacture, bleaching, fertilizers and herbicides; as an electrolyte in alkaline storage batteries and some fuel cells; and as an absorbent for carbon dioxide and hydrogen sulfide.

Sodium hydroxide, NaOH, is a strong base. It has a pH of 13.0. It is the most important industrial caustic material. It is used in the manufacture of chemicals, as a neutralizer in petroleum refining, in metal etching, in electroplating and as a food additive. Sodium hydroxide is a white,

Applied Chemistry and Physics

deliquescent solid in the form of pellets or beads. It may absorb water and carbon dioxide from the air. It is water soluble and is found in 50% and 73% solutions.

Dilution vs. Neutralization

Often tactics considered when dealing with corrosive materials. Dilution is placing water into the acid to reduce the pH level. The addition of water to a corrosive can create a very dangerous chemical reaction. Acids are very water reactive creating vapors, heat, and splattering. First of all with dilution you must consider the exponential values of the numbers on the pH scale. Just moving the pH from 1 to 2 on the scale will take an enormous amount of water. Dilution may not be a practical approach for large volume spills.

> **Hazmatology Point:** *For example, consider a 2,000 gallon spill of concentrated hydrochloric acid occurs. In order to have enough water to dilute the material to a pH of 6, the following efforts would be required:*
>
> *One 1,000 gpm pumper, pumping 1,000 gpm 24 h/day, 7 days/week, 365 days/year, for 64 years! (The pump operator would be given all of the 29 days of February off.) This would produce 1,440,000 million gallons of water per day! A reservoir (large) would be needed to hold the water. As the process proceeds, it would become necessary to stir the mixture of water and acid to ensure uniformity in the dilution process. So, hundreds of motor boats would need to be brought in to stir the acid/water mix.*
>
> *The idea of dilution may work on small spills, but it will not work well on large spills. Neutralization involves a chemical reaction that works very well under laboratory conditions using small amounts of acids and bases. This was calculated by long time friend and fellow Chemistry Instructor Stan Totten.*

However, when you are in the field facing a large spill of a corrosive material, neutralization may not be feasible. The neutralization reaction, first of all, requires a large amount of neutralizing agent. For the same spill of 2,000 gallons of concentrated hydrochloric acid, it would take 8.7 tons of sodium bicarbonate, 5.5 tons of sodium carbonate or 4.15 tons of sodium hydroxide. The latter would not be recommended because sodium hydroxide is a strong base and would be dangerous to work with by itself without trying to add it to a concentrated acid. There would be a need for a method to apply the neutralizing agent. The reaction that occurs will be a violent one producing heat, vapor and splattering of product. Neutralization may not work well at the scene of an incident with a large spill. The method of choice may turn out to be one of cleaning up the product by a hazardous waste contractor. It may use vacuum trucks, absorbents or jelling materials to accomplish the task (Hazardous Materials Chemistry for Emergency Responders).

CASE STUDY

DENVER, CO, APRIL 4, 1983, RAILCAR LEAK NITRIC ACID

Denver and Rio Grande Western Railroad Company Train Yard Accident Involving Punctured Tank Car, Nitric Acid and Vapor Cloud.

Executive Summary

About 4:00 a.m. mountain standard time on April 3, 1983, a Denver and Rio Grande Western Railroad Company (D&RGW) switch crew was switching 17 cars in the D&RGW's North Yard at Denver, Colorado, when a coupler broke on the 4th car, leading to an undetected separation of 150 feet between the 3rd and 4th cars. The engineer, responding to a hand lamp signal from the foreman, accelerated the locomotive, with a caboose, an empty freight car, and a loaded tank car coupled ahead. The loaded tank car impacted a fourth car at a speed of about 10–12 mph. Upon impact, the end sill of the fourth car (empty boxcar) rode over the coupler of the (loaded tank car) and punctured the tank head. Nitric acid spilled from the car and formed a vapor cloud which dispersed over the area. As a result, 9,000 persons were evacuated from the area; 34 were injured. Damage to railroad property was estimated to be about $341,000.

Probable Cause

The National Transportation Safety Board determines that the probable cause of the accident was the complete failure of a coupler of a box car leading to an undetected separation of cars being switched and the puncturing of a tank car by the end sill of the box car when the coupled cars overtook those which had separated. Contributing to the accident was the lack of a federal regulatory inspection or an industry practice for a periodic inspection to detect defects in hidden car components. Contributing to the severity of the accident was the nature of the released product, insufficient guidelines and absence of preplanning which led to the evacuation of about 9,000 residents and injury to 34 persons.

More than 2,000 people were evacuated from an area near central Denver before dawn this morning after nitric acid spilled from a railroad tank car and sent a plume of deadly gas over the city. Eight people suffered minor injuries after contact with the gas, and three firefighters sustained minor burns from the acid itself, according to the police. It took the Denver Fire Department nearly six hours to contain the 20,000-gallon spill and suppress the deadly clouds of vaporized acid. Until then, the Fire Department advised people to stay out of an area of nearly three square miles, including part of downtown, while firefighters worked to contain the acid spill.

Officials warned that the gas could cause skin irritation and respiratory problems. If inhaled in sufficient quantity, it can be lethal, they said. Still, many people appeared to ignore the hazard. Several churches in central Denver went ahead with Easter morning services as scheduled. The police said 25,000 people lived in the area. Officials had no precise count of the number who left their homes, but it was estimated that at least 2,000 fled or were evacuated. Many waited out the holiday in six Denver schools set aside by the Red Cross as emergency shelters. Stretches of Interstate 25 and Interstate 70, which intersect in the area, were closed until early afternoon. Switchyard Accident Officials said the accident took place about 4 A.M. in the Denver & Rio Grande Western Railroad switching yard when the coupler from one rail car punched a hole of 14 inches by 6 inches in the tank car while switching it.

About 20,000 gallons of nitric acid spilled onto the ground, where it vaporized into thick billowing clouds and was carried over the city by light winds out of the northwest.

The police set off air-raid sirens in the area to warn sleeping residents of the danger and went house to house advising people to leave. "The sirens woke me up and when I looked out the window, I saw this big black cloud," said Bertha Sperry, who lived in a public housing project within a block of the railyard. Mrs Sperry said that she woke her husband and two children, and the family began walking south, away from the smoke. Church opened its breakfast.

Like nearly 200 others, she ended up in a north Denver junior high school, where Red Cross volunteers dispensed juice and coffee. At the Highlands Lutheran Church, across the street, the annual Easter morning breakfast was opened to the evacuees. By midmorning, 300 people had been served. Fire Department workers managed to suppress the acid clouds by blowing soda ash onto the spilled acid to neutralize it. To spread the ash onto the acid, which had spilled over a wide area, firefighters borrowed blowers used to clear snow from the runways at nearby Stapleton International Airport (NTSB Investigation Report).

The main danger of corrosive materials to responders is the contact of these materials to the human body. Corrosive materials destroy living tissue. The destruction begins immediately upon contact with the corrosive material. Many of the strong acids and bases cause severe damage upon contact with the skin. Other materials may not cause any damage for several hours after the exposure. A chemical burn is nine times more damaging than a thermal burn. There are four basic methods of interfering with the chemical action of corrosives on the skin. They are as follows:

- **Physical removal** is a removal of the material which is difficult to accomplish and may leave a residue behind.
- **Neutralization** is a chemical reaction that may be violent and produce heat. This type of reaction on body tissues may cause more damage than they prevent.
- **Dilution** acids are water reactive, diluting suggests being careful not to cause a water reaction that will produce heat and splattering of the acid.
- **Flushing** is the method of choice for corrosive materials. Flushing is flooding, with so much water the water reaction is over come.

Neutralization should not be attempted on personnel wearing chemical suits either, for the same reason as mentioned above. The layer of chemical protection is very thin and the heat from the neutralization may melt the suit or cause burns to the skin below the suit.

Dilution takes a large amount of water to lower the pH to a neutral position. While dilution may be similar to flushing, the intended outcome is different. With dilution the goal is to reduce the pH number as near neutral as possible. With flushing the goal is to remove as much of the material as possible with a large volume of water. Flushing is by far the method of choice and should be started as soon as possible to reduce the amount of chemical damage. Flushing should continue for a minimum of 15 min. Most corrosives are very water soluble in water. This also applies to the eyes. Contact lenses should not be worn at hazmat incident scenes. Contact with acids can "weld" the contact to the eye, which almost always produces blindness. The person being treated may be in a great deal of pain and may have to be restrained during the flushing operation. Treatment after flushing involves the standard first aid for burns.

Corrosive Containers

Corrosives are transported in MC/DOT 312/412 tanker trucks (Figure 3.140). These trucks have a small diameter tank with heavy reinforcing rings around the circumference of the tank from front to rear. The tank diameter is small because most corrosives are very heavy. No other type of hazardous material is carried in this type of tanker. The 312/412 is a corrosive tanker regardless of how it is placarded. The placard may be poison, oxidizer or flammable, but don't forget the

Figure 3.140 If a flammable placard is found on an acid container or the MC312/DOT412 acid tanker, you can be sure that it is an organic acid.

"hidden hazard"; the tank identifies corrosives. Lighter corrosives may also be found in MC/DOT 307 tankers and may be placarded corrosive, flammable, poison and oxidizer. Corrosives may also be found in rail tank cars, intermodal containers and varying sizes of portable containers. Portable containers may range from pint and gallon glass bottles to stainless steel carboys and 55 gallon drums. Some are also shipped in plastic containers.

Emergency responders should have a thorough knowledge of corrosive materials. Next to flammable liquids and gases, corrosives are the next most frequently encountered hazardous materials. Responders should have proper chemical protective equipment and SCBA to deal safely with corrosive materials. Firefighter turnouts will not provide protection from corrosives. The most common places of exposure to responders occur with the hands and feet. The next is inhalation of the vapors. Make sure that the chemical suits chosen to wear are compatible with the corrosive material. No suit will protect you from chemicals forever. They all have breakthrough times. Make sure personnel are rotated to avoid prolonged exposure and they do not contact the material unless absolutely necessary. Safety should be your number 1 concern.

Hydrocarbon and hydrocarbon derivative families all have particular hazards associated with each of them. If responders become familiar with the hazards of the family, they will have an idea of the hazards facing them during an incident, because all compounds in a particular family have similar hazards (Hazardous Materials Chemistry for Emergency Responders).

Hazard Class 9 Miscellaneous Hazardous Materials (LF/LR)

Materials in Class 9 present hazards during transportation but do not meet the definition of any other hazard class. They may be anesthetics, noxious or elevated temperature materials, hazardous substances, hazardous wastes or marine pollutants. These materials may be encountered as solids of varying configurations, gases and liquids. Examples are asbestos, dry ice, molten sulfur and lithium batteries. These materials would carry the Class 9 miscellaneous hazardous materials placard, which is white with seven vertical black strips on the top half. Also included in Class 9 are Other Regulated Materials (ORM-D), Consumer Commodities. They are "materials that present a limited hazard during transportation due to the form, quantity and packaging." Some of these materials, if shipped in tank truck quantities, would fit into another class but because the individual packaging quantities are so small, the hazard is considered limited by the U.S. Department of Transportation (DOT), and they are labeled ORM-D. Generally, ORM-D materials are destined for use in the home, industry and institutions. They are used in small containers, including aerosol cans, in quantities of a gallon or less.

In fires, many small containers can become projectiles as pressure builds up inside from the heat and the containers explode. Aerosol cans can be particularly dangerous because they are already pressurized, and exposure to heat can cause them to explode and rocket from the pressure. Some of them, like shaving cream use propane as a propellant. Those materials found in industry and institutions are usually service products used in cleaning and maintenance rather than in industrial chemical processes. Examples of ORM-D materials include low-concentration acids, charcoal lighters, spray paints, disinfectants and cartridges for small firearms. Even though the containers may be small, the products inside can contaminate responders and kill and injure if not handled properly. No one specific hazard can be attributed to Class 9 materials. The physical and chemical characteristics mentioned in the first eight hazard class articles may be encountered with Class 9 materials. The difference is that the quantities may be smaller or the materials classified as hazardous wastes, which can include almost any of the other hazard classes. With miscellaneous hazardous materials, it is important to obtain more information about the shipment to determine the chemical names and the exact hazards of the materials involved.

Some Class 9 placards on vehicles include 4-digit identification numbers. The corresponding information in the DOT's *Emergency Response Guidebook* (ERG) may not give detailed names of the materials. In such cases, shipping papers or other sources must be consulted to determine the exact hazard of the shipment.

In addition to the Class 9 placard, a second placard may appear next to it with the word HOT. The word may also appear outside of a placard by itself. What it indicates is that the material inside has an elevated temperature that may be a hazard. An elevated temperature material is usually a solid that has been heated until it melts and becomes a molten liquid. The change is in physical state only; the material's chemical characteristics are the same. There may, however, be vapors produced from molten materials that are not present in the solid form. These vapors may be flammable or toxic.

Water in contact with molten materials can cause a violent reaction and instantly turn to steam. If this happens inside a container, the pressure buildup from the steam can cause a boiler-type explosion. The steam, which is a gas, builds up pressure inside a nonpressure container. When the container can no longer withstand the pressure, it fails. The molten material inside may be splattered around by the explosion.

Chemical Notebook

Asbestos—white, gray, green, brown and blue—is an impure magnesium silicate mineral which occurs in fibrous form. It is noncombustible and used extensively as a fire-retardant material until it is found to cause cancer. Asbestos is highly toxic by inhalation of dust particles. The 4-digit identification number for white asbestos is 2590. The primary uses of asbestos are in fireproof fabrics, in brake linings, in gaskets, as a reinforcing agent in rubber and plastics and as a cement reinforcement. Many uses of asbestos are being banned because of the cancer danger of the material.

Ammonium nitrate fertilizers that are not classified as oxidizers are considered miscellaneous hazardous materials. This type of fertilizer contains other materials, including controlled amounts of combustible materials. Mixtures of ammonium nitrate, nitrogen and potash that are not more than 70% ammonium nitrate and don't have more than 0.4% combustible material are included as miscellaneous hazardous materials. Additionally, ammonium nitrate mixtures with nitrogen and potash, with not more than 45% ammonium nitrate, may have combustible material which is unrestricted in quantity. The 4-digit identification number for these mixtures is 2071.

Solid carbon dioxide (dry ice) presents a danger in transportation because of the carbon dioxide gas produced as it warms. This warming is much like water ice melting, although no liquid is formed in the case of dry ice. Dry ice sublimes, going directly from a solid to a vapor or a gas without becoming a liquid. While carbon dioxide gas is nontoxic, it is an asphyxiate and can displace oxygen in the air or in a confined space. Carbon dioxide gas is used as a fire extinguishing agent. The 4-digit identification number for solid carbon dioxide is 1845.

Solutions of formaldehyde, 30%–50% (such as those used in preservatives), are listed as miscellaneous hazardous materials. These solutions are nonflammable, and the toxicity is below the requirements for a poison liquid. However, the material may still be carcinogenic. Formaldehyde solutions usually contain up to 15% methanol to retard polymerization. The 4-digit identification number for nonflammable solutions is 2209.

Polychlorinated biphenyls (PCBs) are composed of two attached benzene rings, with at least two chlorine atoms in the compound. PCBs has been widely used in industry since 1930 because of their stability; however, it was this same stability that led to their downfall. They are highly toxic, colorless liquids with a specific gravity of 1.4–1.5, which is heavier than water. They are not biodegraded and remain as an ecological hazard through water pollution. The manufacture of PCBs was discontinued in the United States in 1976. The material that remains is considered hazardous waste and is shipped as a miscellaneous hazardous material. According to Environmental Protection Agency (EPA) statistics, from 1988 to 1992, PCBs accounted for 3,586 accidental releases, resulting in 34 deaths. This was the number 1 chemical involved in accidental releases during that period.

Batteries

Batteries containing lithium are listed as miscellaneous hazardous materials. The storage batteries are composed of lithium, sulfur, selenium, tellurium and chlorine. The 4-digit identification numbers assigned to them depend on the use and composition of the batteries. Lithium batteries contained in equipment has the number 3091. Batteries with liquid or solid cathodes, not in any kind of equipment, are given the 4-digit identification number 3090.

Solid Materials

Also listed in CPR 49 hazmat tables are solid materials, fish meal or fish scrap that has been stabilized. Fish meal is subject to spontaneous heating. These materials are given the 4-digit identification number 2216. Castor beans, castor meal or flakes may also undergo spontaneous heating. The 4-digit identification number is 2969. Other materials include cotton (wet, 1365), polystyrene beads (2211), self-inflating lifesaving appliances (2990, not self-inflating (3072), environmentally hazardous liquids or substances (3077), hazardous waste/liquid (3082) and hazardous waste/solid (3077). Hazards also are presented by self-propelled vehicles with internal combustion engines or electric storage batteries, and electric wheelchairs with spillable or nonspillable batteries.

Chemical Notebook

Titanium dioxide, TiO_2, is a white powder and has the greatest hiding power of all white pigments. It is noncombustible; however, it is a powder and, when suspended in air, may cause a dust explosion, if an ignition source is present. It is not listed in the DOT Hazardous Materials Table and is not considered hazardous in transportation by the DOT. The primary uses are as white pigment in paints, paper, rubber and plastics, cosmetics, welding rods and radioactive decontamination of the skin.

Sodium silicate, Na_2OSiO_2, also known as water glass, is the simplest form of glass. It is found as lumps of greenish glass soluble in steam under pressure, white powders of varying degrees of solubility or cloudy or clear liquids. It is noncombustible; however, when the powdered form is suspended in air, it could cause a dust explosion if an ignition source is present. The glass form could also create a hazard to responders in an accident. It is not listed as a hazardous material in the DOT Hazardous Materials Table. The primary uses are as catalysts, soaps, adhesives and flame retardants and for water treatment, bleaching, waterproofing.

Bisphenol A, $(CH_3)_2C(C_6H_4OH)_2$, is made up of white flakes that have a mild phenolic odor. It is insoluble in water. Bisphenol A is combustible with a flash point of 175°F. It is not listed in the DOT Hazardous Materials Table. It is used in the manufacture of epoxy, polycarbonate, polysulfone and polyester resins, and as a flame retardant and a fungicide.

Urea (carbamide), $CO(NH_2)_2$, is composed of white crystals or powder and is almost odorless, with a saline taste. It is soluble in water and decomposes before reaching its boiling point. Urea is noncombustible. The primary uses of urea are in fertilizers, animal feed, plastics, pharmaceuticals and cosmetics, and as a stabilizer in explosives and flame-proofing agents. Urea appears to be both a ketone and an amine by structure and molecular formula; however, it is neither, nor does it have any of the characteristics of either family.

Hot materials such as asphalt can cause serious thermal burns if contacted with parts of the body.

> *Hazmatology Point:* The TV program "Rescue 911" highlighted a rescue operation that involved hot asphalt. In that incident, a dump truck being used to haul solid hot asphalt to patch roadway holes collided with a car. As a result of the collision, the load of hot asphalt was dumped into the car, covering the driver to the point that he could not escape. Before rescuers could remove the driver, he suffered severe second- and third-degree thermal burns from the hot asphalt. This same hazard exists with all elevated temperature materials. Responders should work carefully around transportation vehicles that have the HOT placard or the word HOT on the container.

Molten Materials

Two molten materials are specifically listed in the hazardous materials tables in 49 CPR: molten aluminum and molten sulfur. Molten aluminum has a 4-digit identification number of 9260. When referenced in the *ERG*, it refers to Guide 77 for hazards of the material. (Guide 77 was an addition to the 1993 version of the *ERG*. Molten aluminum is the only material that refers to this guide.) The guide indicates that the material is above 1,300°F and will react violently with water, which may cause an explosion, and release a flammable gas.

The molten material in contact with combustible materials may cause ignition if the molten material is above the ignition temperature of the combustible.

For example, gasoline has an average ignition temperature of around 800°F. Diesel fuel has an average ignition temperature of around 400°F. In an accident, gasoline or diesel fuel could be spilled. The molten material could be an ignition source for the gasoline or diesel fuel. When contacting with concrete on a roadway or at a fixed facility, molten materials could cause spalling and small pops. This could cause pieces of concrete to become projectiles. Contact with the skin would cause severe thermal burns. No personnel protective clothing will adequately protect responders from contacting with molten materials.

There are other molten materials that are not as hot as molten aluminum. Molten sulfur refers you to Guide 32 in the *ERG*. Molten sulfur, carrying the 4-digit number 2448, may ignite combustible materials with which it comes into contact. It has a melting point of about 245°F. Molten sulfur in transportation would be above that temperature, but not as hot as molten aluminum. Contact would still cause severe thermal burns, and the vapor is toxic.

Hot asphalt in liquid form can also cause combustion of combustible materials and severe thermal burns. Asphalt that has been heated to become a hot liquid refers you to Guide 27 in the *ERG* for hazard information. When referencing Guide 27, the information will indicate that the materials may be HOT. Asphalt has a boiling point of >700°F and a flash point of >400°F. The ignition temperature is 905°F.

Fires involving asphalt should be fought with care. Water may cause frothing, as it does with all combustible liquids with flash points above 212°F. This does not mean that water should not be used; just be aware that the frothing may be violent and the water contacting the molten material may cause a reaction. Asphalt has a 4-digit identification number of 1999 for all forms.

Emergency responders should have a thorough understanding of the physical and chemical characteristics of hazardous materials, including parameters of combustion, water and air reactivity, incompatibilities with other materials and the effects of temperature and pressure.

Applied Chemistry and Physics 281

Responders should have the same level of understanding of hazardous materials as they do of firefighting, EMS protocols and law enforcement procedures. All emergency response incidents have the potential to involve hazardous materials (Hazardous Materials Chemistry for Emergency Responders).

Applied Physics (Physical Characteristics)

Physical changes involve alterations of the physical state of the chemical but do not produce a new substance. They do not alter the chemical makeup of a material, for example, the physical transformation from a liquid to a gas or a liquid to a solid. Physical properties or applied physics includes specific gravity, vapor pressure, boiling point, vapor density, melting point, solubility, flash point, fire point, autoignition temperature, flammable range, heat content and pH (Hazardous Materials Chemistry for Emergency Responders).

Combustion Analysis

Up to this point, most of the discussions of this volume have centered on the chemical characteristics of hazardous materials. This section begins looking at the physical characteristics of some of these materials and is called "applied physics." Flammable liquids and gases make up the largest number of hazardous materials incidents that responders will encounter. Just look at the fuels used to power internal combustion engines and heat our homes and businesses, cook our food, run our trains and fly our planes. Flammable liquids and gases are everywhere in our society. Responders should have at least a basic understanding of the physical characteristics of these materials. Things such as flash point, boiling point, ignition temperature, flammable range, vapor pressure, vapor density, volatility, polarity, miscibility and the effects of temperature on flammable liquids and gases, to name a few. Flammable liquids can be divided into some basic families based upon use and physical characteristics. For our purposes, in "applied chemistry," we call them the fuel family, hydrocarbon derivatives and the animal/vegetable oils, which undergo slow oxidation or spontaneous combustion (FEMA/NFA).

Hydrocarbons

Hydrocarbons have flammability as a primary hazard. Different materials have different flammable ranges. The only way to determine if a material is present in a mixture with air that is in its flammable range is to use a meter to monitor the lower explosive limit of the material. Monitoring instruments check for a percentage of the lower explosive limit. The rule of thumb according to the EPA is that when you reach 25% of the lower explosive limit, it becomes too dangerous for personnel to proceed any further. There is currently talk within OSHA that wants to raise the threshold above 25%.

Hydrocarbon derivatives that have flammability as one of their primary hazards include the ether, amine, alcohol and aldehydes. Alkane, alkene and alkyne hydrocarbons, and the aromatic hydrocarbons, also known as the BTX fraction, make up the fuel families. Many of the compounds in the fuel family are mixtures such as gasoline, diesel fuel, fuel oil and aviation fuels. These materials are mixtures of two or more of the hydrocarbons with other chemicals as additives. There are also pure compounds among the fuel family members. For example, propane, butane, pentane and octane are pure compounds. The rule of thumb for determining a mixture or pure compound involves the ability to draw a structural formula for the compound. A structure can be drawn for pure compounds; however, there are no structures for mixtures. Animal/vegetable oils include such compounds as linseed oil, cottonseed oil, corn oil, lard and other cooking oils.

Emergency responders should become familiar with the physical characteristics of flammable and combustible liquids and know where to get the necessary information from reference books and other sources.

Boiling Point

One of the very basic characteristics of a flammable liquid or liquefied gas is its boiling point. Boiling point of a liquid is the temperature at which the vapor pressure of the liquid overcomes the atmospheric pressure. As the atmospheric pressure is overcome, the vapors from a liquid start to move farther away from a spill, or if inside a container, the pressure in the container starts to increase. The boiling point decreases with altitude. The sea-level boiling point of water is 212°F; however, in Denver, the boiling point decreases to about 202°F (Figure 3.141). That decrease is because the atmospheric pressure is lower at higher altitudes; therefore, it is easier for a vapor coming from a liquid to overcome the atmospheric pressure. It also takes less energy.

Boiling Point of Water and Atmospheric Pressure (psi) Vs. Altitude

Altitude	Atmospheric Pressure	Boiling Point
Sea Level	14.7 psi	212° F
3300 Feet	13.03 psi	203° F
5280 (1 mile)	12.26 psi	201° F
10,000 Feet	10.17 psi	194° F
13,000 Feet	9.00 psi	188° F

Figure 3.141 The sea-level boiling point of water is 212°F; however, in Denver, the boiling point decreases to about 202°F.

Physical characteristics establish whether a flammable liquid or family of liquids have high or low boiling points. First of all, liquids with a high carbon and hydrogen content tend to have higher boiling points. They are heavier, which is referred to as the "size" or molecular weight of a compound. For example, propane is a three-carbon alkane hydrocarbon, with eight hydrogen atoms. Propane has a boiling point of around –40°F and is a gas at normal temperatures and pressures. Pentane, on the other hand, has five carbon atoms and 12 hydrogen atoms. Propane has a boiling point of about 94°F, and it is a liquid at normal temperatures and pressures. You can easily see the effect that size has on the boiling point.

Vapor Pressure

Boiling point is related to vapor pressure and vapor content, although the relationship is opposite in nature. Materials with low boiling points and flash points will have high vapor pressure and high vapor content. The vapor pressure of a liquid is defined in the *Condensed Chemical Dictionary* as "the characteristic at any given temperature of a vapor in equilibrium with its liquid or solid form. This pressure is often expressed in millimeters of mercury, mm Hg."

In simple terms, vapor pressure is the pressure exerted by the liquid against the atmospheric pressure. When the pressure of the vapor is greater than the atmospheric pressure, vapor will spread beyond an open container or an open spill. If liquid is in a container, vapor pressure is the pressure exerted by the liquid vapors on the sides of the container. For example, when a gas is liquefied, the only thing keeping it a liquid is the pressure in the tank; the liquid is already above its boiling point. The pressure inside the container is the atmospheric pressure in that container. It can be much higher than the atmospheric pressure outside the container. For example, if the pressure in the tank is 50 psi, the atmospheric pressure in that tank remains 50 psi regardless of conditions outside the tank.

Vapor Content

Vapor content is the amount of vapor that is present in a spill or open container. The lower the boiling point and the flash point of a liquid, the more vapor there will be. The parallel relationship of boiling point and flash point is comparable to the opposite relationship of vapor pressure and vapor content shown in the diagram, using a seesaw to illustrate the up-and-down and opposing relationship.

As shown in Figure 3.145 when the boiling point and flash point are low, vapor content and vapor pressure are high. When the boiling point and flash point are high, vapor pressure and vapor content are low. The lower the boiling point or flash point of a liquid, the higher the vapor content at a spill and the higher the vapor pressure inside a container. If a liquid is above its boiling point temperature, there is likely to be more vapor moving farther away from a spill. If the liquid is below its boiling point temperature, there will be some vapor above the surface of the liquid, but it will not travel far. If the flammable liquid in a container is below its boiling point, the vapor content and vapor pressure in the container will be low. If the flammable liquid in a container is above its boiling point, the vapor content and vapor pressure in the container will be high.

Vapor Density

Vapor density is a physical characteristic that affects the travel of vapor; it is the weight of a vapor compared to the weight of air. Vapor density is usually determined in the reference books by dividing the molecular weight of a compound by 29, which is the assumed molecular weight of air. Air is given a weight value of 1, which is used to compare the vapor density of a material. If the vapor of a material has a density greater than 1, it is considered heavier than air. Heavier-than-air vapor will lie low to the ground and collect in confined spaces and basements. This can cause problems because many ignition sources are in basements, such as hot water heaters and furnace pilot lights. If the vapor density is less than 1, the vapor is considered to be lighter than air, so it will move up and travel farther from the spill.

Another term associated with vapor is volatility. It is the tendency of a solid or a liquid to pass into the vapor state easily. This usually occurs with liquids that have low boiling points. A volatile liquid or solid will produce significant amounts of vapor at normal temperatures, creating an additional flammability hazard. The vapor produced by a volatile liquid is affected by wind, vapor pressure, temperature and surface area. Temperature always causes an increase in vapor pressure and vapor content in an incident. The more vapor pressure in a container, the greater the chance of container failure. The more vapor content, the farther the vapor may travel away from a spill.

Polarity

Another characteristic that affects the boiling point is polarity. Polarity causes the boiling points of polar liquids to have higher boiling points than nonpolar liquids of the same or similar molecular weight. Polarity is a kind of "glue" that causes molecules within a compound to stick together. In order for a liquid to boil, not only does it have to overcome atmospheric pressure, but you have to overcome polarity that causes the molecules to want to stick together. In order to accomplish this, more energy in the form of heat has to be applied to the liquid to get it to boil, which results in the compound having a higher boiling point.

Applied Chemistry and Physics

The flammable liquid families can further be divided into those that are polar and those that are nonpolar. Polarity not only has an effect on the physical characteristics of combustion but determines the type of fire extinguishing agent you will need to use to fight fires. Water is made up of two atoms of hydrogen and one atom of oxygen. As a compound, it has an atomic weight of 18 atomic mass units, 1 from each of the hydrogen atoms and 16 from the oxygen atom. The average molecule of air weighs about 29 atomic mass units. So in effect, water is lighter than air! How can this be? Water is a liquid at normal temperatures and pressures. It should be a gas judging by its weight. Well, the answer is polarity. Water is a polar compound with a boiling point of 212°F at sea level. Polarity causes water, which is smaller in weight than air, to have a high boiling point, and thus, water exists as a liquid rather than a gas. It's a good thing that water is a liquid and not a gas or life as we know that it might not be possible!

Of the families we have discussed, alcohol, ketone, ester, aldehyde and organic acid are polar compounds. Hydrocarbons, aromatics, alkyl halide, nitro, peroxide and ether are nonpolar. Now to set the record straight, all materials may have some degree of polarity, but for our purposes, here we can make general statements about groups of hazardous materials, and it will work for "applied physics."

Molecular Weight

In order to compare the weights of compounds, it is necessary to determine the molecular weight of the elements in the compound. Information is found on the periodic table of elements for determining the molecular weight of an element in a compound. You locate the atomic weight of the element on the table and round it off to the nearest whole number. For example, carbon weighs 12.011 atomic mass units, and hydrogen weighs 1.0079 atomic mass units. Carbon would be rounded to 12 and hydrogen to 1. Oxygen weighs 15.9994 atomic mass units and would be rounded to 16.

If you compare ethyl alcohol with an atomic weight of 46 and ethyl chloride (an alkyl halide) with an atomic weight of 64, you might expect the ethyl chloride to have a higher boiling point because it weighs more. However, the boiling point of ethyl chloride is about 54°F, and the boiling point of ethyl alcohol is much higher at 173°F. That is the effect of polarity. Alcohol is polar and the alkyl halide is nonpolar; polarity causes the boiling point of the alcohol to be higher.

Branching

A third factor that affects the boiling point is isomerism or branching. The definition of an isomer is a compound that has the same formula but a different structure. For example, butane, sometimes referred to as straight chained or normal butane, has a molecular formula of C_4H_{10}. Isobutane also has a molecular formula of C_4H_{10}; however, isobutane is branched and therefore has a different structure than normal butane. The isomer of butane is indicated in the formula by placing a small "i" in front of the formula (i-C_4H_{10}).

Isomers of a compound have the effect of lowering the boiling point of the compound. For example, butane is used as a home heating fuel and has a boiling point of around 31°F. Because of this relatively high boiling point, butane cannot be used in some colder parts of the country, as a fuel, where the temperature goes below 32°F in the winter time. However, if butane is branched, the boiling point is lowered to approximately 12°F, which increases the areas where it can be used as a fuel.

Sublimation

Certain solid materials may undergo sublimation, a process in which they can go directly from a solid to a vapor or a gas without ever becoming a liquid. These types of materials can have flash points and other characteristics of combustion and sometimes called flash point solids. Moth balls and camphors are examples of solids that go directly from a solid to a gas without becoming a liquid.

Pyrophoric Materials

Flammable liquids can also be pyrophoric, which means that they will spontaneously combust when exposed to air. Alkyl aluminum and boranes are pyrophoric liquids. Phosphorus is a pyrophoric solid. When in contact with air, it spontaneously combusts. Several major incidents have occurred over the years when phosphorus was exposed to air in two train derailments and one box truck incident (Hazardous Materials Chemistry for Emergency Responders).

CASE STUDIES

GETTYSBURG, PA, MARCH 22, 1979, PHOSPHOROUS TRUCK FIRE AND EXPLOSION

During the night of March 22, 1979, a truck transporting phosphorus though the town of Gettysburg, PA, stopped on Buford Avenue just west of the downtown business district when the driver noticed smoke coming from the cargo space of the truck. The truck was hauling white phosphorus in 55 gallon drums (the same white phosphorus that was in the tank car in the previous incident). Phosphorus being air reactive is shipped under water to keep the phosphorus from igniting. One of the barrels developed a leak, and water covering the phosphorus dropped below the level of the material in the drum. At that point, the phosphorus ignited and burned, impinging on other drums on board. This repeated like a chain reaction until the entire truck was ablaze on the streets of Gettysburg. Gettysburg Volunteer Fire Department responded and the firefighters were told by dispatchers not to put water on the fire. So, exposures were protected and the fire allowed to burn itself out. A couple of buildings received surface damage from the radiant heat.

By the next morning, the fire had gone out and cleanup crews were on-site to clean up the remaining phosphorus. As the phosphorus would be uncovered, it would again ignite and sand had to be brought in to prevent the phosphorus from reaching the air. Several persons are injured when 55 gallon drums of white phosphorus exploded during the cleanup from a fire that destroyed the truck that carried them. Parts of the drums were rocketed several hundred feet.

It is likely the water heated up in the drums from the heat of the fire and turned the water to steam causing the drums to open up explosively from the steam pressure built up inside. Phosphorus is shipped in ordinary 55 gallon drums meant for liquids. The drums have no ability to withstand pressure increases inside the drum and will fail rather easily, which is likely what happened that day. Several area fire departments provided ambulances and other apparatuses through mutual aid (*Firehouse Magazine*).

> **Hazmatology Point:** *Remember that 55 gallon drums are used for many chemicals including acids, poisons, flammable liquids and solids. They do not carry gases because they are liquid containers. Anytime 55 gallon drums are exposed to fire, expect pressure buildup in the containers, and they will rocket when the pressure causes the container to fail.*

BROWNSON, NE, APRIL 2, 1978, TRAIN DERAILMENT PHOSPHORUS FIRE AND EXPLOSION

Brownson is located is an unincorporated village about 5 miles west of Sidney, NE, on US Highway 30. Union Pacific 142 was eastbound with 101 cars when the derailment occurred. Thirty-one cars left the tracks including a liquid tank car carrying white (yellow) phosphorus. Phosphorus is an air-reactive material that ignites spontaneously in air at 86°F. The phosphorus car turned over and approximately half of the water covering the phosphorus leaked from the tank. Immediately the phosphorus ignited and started to burn vigorously. No water supply was available so the decision was made to allow the fire to burn. In consultation with NFPA action guides and CHEMTREC, they were advised that there was no hazard of the tank car exploding. White phosphorus is very toxic and should not be allowed to come into contact with the skin. Dense white clouds of toxic fumes evolved from the burning tank which can attack the lungs.

Firefighters were given three options by CHEMREC to deal with the burning tank car. They could deluge the tank with large volumes of water. This was not possible because of the remote location of the derailment and no water source readily available. They could bury the car with sand and keep the sand wet, which would block the phosphorus from getting to the air. Lastly, the option chosen was to let the tank burn. Wind speed and direction changed around 11:00 a.m. Winds were gusting to 30 miles/h, blowing the toxic cloud toward Sidney. Evacuation of the north side of Sidney was undertaken as a precaution. The fire continued to burn until 11:38 a.m.

Without warning, the tank car exploded and parts of the tank car were propelled 2,400 ft away. Even though firefighters were in close proximity, no one was killed in the blast. Six railroad employees were injured (*Firehouse Magazine*).

Hazmatology Point: *Everyone was correct in terms of the phosphorus not causing an explosion in the tank car. Then what caused the explosion? Water the very thing they didn't have enough of to fight the fire ended up causing the explosion. Phosphorus being air reactive was shipped under water in the tank, thus keeping the phosphorus from spontaneously combusting. When the water leaked off, the phosphorus ignited and continued to burn all the while heating the remaining water in the tank.*

When water is heated to the point of boiling, it becomes a gas; steam. Since this was a liquid tank, it was not designed to hold pressure. When the steam pressure built up to the point the tank could no longer hold it, the tank came apart explosively. No, this was not a BLEVE. This was a simple boiler explosion. When dealing with hazardous materials incidents, be sure to consider all of the actors. In this case, in addition to the phosphorus, the tank and the water were also actors, which precipitated the explosion.

Volatility

Some liquids are volatile, which means that they readily produce vapor, and they can become a fire hazard if all of the conditions for combustion are just right. For example, gasoline is volatile and produces vapor at almost any ambient temperature, because its boiling point is –40° below zero. Ethyl alcohol (ethanol), on the other hand, is not volatile and has a boiling point of 172°F.

Flash Point

One of the most important physical characteristics of a flammable liquid is its flash point. Flash point is the minimum temperature to which a liquid must be heated to produce enough vapor to allow a vapor flash to occur (if an ignition source is present). Remember that it is the vapor that burns, not the liquid. Flash point is a measurement of the liquid temperature. Therefore, even if the ambient temperature is not at the flash point temperature of the liquid, the liquid may have been heated to its flash point by some external heat source. For example, the radiant heat from the sun, heat from a fire or heat from a chemical process may heat the liquid to its flash point. If an ignition source is present, fire can occur and probably will.

If a flammable liquid is not at its flash point temperature, combustion cannot occur. So, when researching chemicals in reference sources, the flash point temperature is the first physical characteristic responders should look for. Reference books used to research chemical characteristics in hazardous materials emergencies may show different flash point values. There are different tests used to determine the flash point of a liquid. Tests use open- and closed-cup testing apparatus, which often produce somewhat different temperature values. Open-cup flash point tests try to simulate conditions of a flammable liquid in the open, such as a spill from a container onto the ground. The open-cup test usually results in a higher flash point temperature for the same flammable liquid than the closed-cup method. Responders should use the lowest flash point value given.

Flash Point Solids

The flash point solids are a small group of materials. Examples include paraformaldehyde, naphthalene (moth balls) and camphor. Flash point solids sublime (i.e., change directly into a vapor without passing through the liquid state). As a result, these materials have flash points and ignite in a manner similar to combustible liquids. Once ignited, the materials will melt and flow, like a flammable liquid.

Fire Point

Fire point is the temperature to which the liquid is heated that produces enough vapor for ignition to occur after the vapor flash occurs and continues. The fire point temperature is 1°–3° higher than the flash point temperature. The fire point is so close to the flash point temperature that it really isn't much of a concern to emergency responders. If a liquid is at its flash point in a spill or leak, then it is very likely to also be at its fire point and prepare for a fire!

I have only heard of two occasions when there was a vapor flash from vapors being ignited and no fire occurred after the vapors burned off (Figure 3.142). A gasoline tanker was filling the underground storage tanks of a service station early one morning. The vent pipes for the underground tanks were not up to code and were just 3 ft above the ground. The code requires that the vents be 3 ft above the nearest roofline. As a result of the improper vents, the vapors, being heavier than air, were collected on the ground. When a soda machine's compressor came on, it provided the ignition source and a vapor flash occurred.

In another instance, vapors from a gas tank on a snowmobile ignited and engulfed the operator, but there was only vapor present, and once the vapor burned off, the fire went out. That was a vapor flash. The operator was not injured and his clothes were not burned (Figure 3.143). So don't be too concerned about the fire point, find out what the flash point of the liquid is and determine what the temperature of the liquid is. If the flash point is close to the ambient temperature, or there is some source of radiant heat present that may have heated the liquid, take proper precautions because the temperature of the liquid may be at its flash point.

Figure 3.142 The author has only heard of two occasions when there was a vapor flash from vapors being ignited and no fire occurred after the vapors burned off.

Applied Chemistry and Physics   291

Figure 3.143 That was a vapor flash. The operator was not injured and his clothes were not burned.

Ignition Temperature

Another term associated with combustion, which is sometimes misunderstood, is ignition temperature, also known as autoignition temperature. The ignition temperature is the minimum temperature to which a material must be heated to cause autoignition without the need for an ignition source to be present. In other words, a flammable liquid is heated from an outside heat source and autoignites when its ignition temperature is reached (Figure 3.144). In order for an ignition source to ignite a material at its flash point temperature, the ignition source must be at the ignition temperature of the liquid.

For example, if you placed gasoline in a container and used a lit cigarette to ignite the gasoline, you would be wasting your time. The temperature of a lit cigarette is around 400°F, and the average ignition temperature of gasoline is around 800°F; therefore, the cigarette is not hot enough to ignite the gasoline. On the other hand, if you placed diesel fuel

IGNITION TEMPERATURES OF COMMON COMBUSTIBLES

Wood	392° F
#1 Fuel Oil	444° F
Paper	446° F
60 Octane Gas	536° F
Acetylene Gas	571° F
Wheat Flour	748° F
Corn	752° F
Propane Gas	871° F

Figure 3.144 When a flammable material is heated from an outside heat source, it autoignites when its ignition temperature is reached.

in a container and used a lit cigarette to ignite the diesel fuel, you might have a fire on your hands! The ignition temperature of diesel fuel is about the same as the temperature of the lit cigarette. When a pan of animal/vegetable oil is placed on a stove to cook food, often the burner is turned up too high and the oil ignites. This is because the ignition temperature of most animal/vegetable oils is in the range of 400°F–500°F. The stove burner is able to reach the autoignition temperature of the oil without much trouble and ignite the oil if the temperature is turned up too high.

Relationship of Physical Characteristics

Boiling point and flash point have a parallel relationship. When the boiling point of a compound is low, the flash point is also low. Compounds that have high boiling points also have high flash points. If a flammable liquid that has a low boiling point and a correspondingly low flash point, the quantity of vapor produced by the compound will be high. If the liquid is spilled on the ground, the vapor will travel farther away from the spill.

This is referred to as the vapor content of the spill. Inside a closed container, the vapor will increase the pressure inside the container. This is the vapor pressure within the container of liquid. If the boiling point and flash point are high, then the vapor content and vapor pressure will be low. Another term associated with combustion is the heat output when the vapor from a liquid burns. Large compounds have high heat output and small compounds have low heat output.

There is a relationship between boiling point, flash point, ignition temperature, vapor content, vapor pressure and heat output. Compounds that have high boiling points tend to have high flash points. Those that have low boiling points tend to have low flash points. The numeric values are different, but the ratio holds true. For example, butane has a boiling point of around 31°F and a flash point of –76°F. Butyl alcohol has a relatively high boiling point of about 273°F and a flash point of 95°F, which is also high compared to butane. Some additional terms that are associated with the boiling point and flash point include vapor content, vapor pressure heat output and ignition temperature. Compounds that have high boiling points and high flash points tend to have high heat output, low vapor pressure, low vapor content and low ignition temperatures. This concept can be illustrated by using a "seesaw" or "teeter totter" (depending on which part of the country you are from). Figure 3.145 is an illustration showing the relationship of combustion characteristics.

If a flammable liquid is at its flash point temperature, and the ignition source is at or above the ignition temperature of the liquid, there is still one more physical characteristic that must be correct for combustion to occur. This characteristic is referred to as the flammable range or explosive limits.

Physical Relationships

BP = Boiling Point   VP = Vapor Pressure
FP = Flash Point   VC = Vapor Content
IT = Ignition Temp   HO = Heat Output

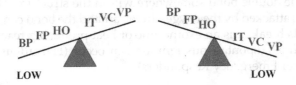

Figure 3.145 An illustration showing the relationship of combustion characteristics.

Flammable Range or Explosive Limits

The flammable range is expressed on a scale from 0% to 100%. Between 0% and 100%, a flammable liquid vapor will encounter the proper mixture of air and fuel to complete the combustion process.

Each vapor has an upper (UEL) and a lower (LEL) explosive limit (Figure 3.146). Above the UEL, the mixture is too rich to burn, when there is enough fuel for combustion to occur, but not enough oxygen. Below the lower explosive limit, the mixture is too lean to burn, when there is enough oxygen, but not enough fuel. Somewhere between the LEL and the UEL, there is a proper mixture of fuel and oxygen for combustion to occur.

Most hydrocarbon fuel family compounds have a flammable range of 1%–12%. Alcohol, ether and aldehyde compounds generally have wide flammable ranges. Alcohol ranges from 1% to 36%, ether from 2% to 48%

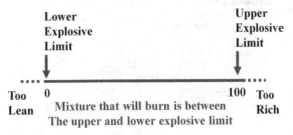

Figure 3.146 Each vapor has a UEL and a LEL.

and aldehyde from 3% to 55% (Figure 3.147). Acetylene has an extremely wide range from 2% at the lower end to between 80% and 100% at the upper end (Figure 3.148). Materials with wide flammable ranges are dangerous because they will burn leaner and combustion may occur inside the container.

Large hydrocarbon compounds such as animal/vegetable oils have a characteristic double bond somewhere within the structure. This double bond can be attacked by the oxygen in the air and the bond can be broken. When bonds break, it is an exothermic or heat-producing reaction. If the heat is confined, spontaneous ignition can occur (Hazardous Materials Chemistry for Emergency Responders).

**TYPICAL FLAMMABLE RANGES
OF FLAMMABLE LIQUID FAMILIES**

Fuel Family	1 to 8%
Aromatic hydrocarbons	1 to 7%
Ketones	2 to 12%
Esters	1 to 9%
Amines	2 to 14%
Alcohols	1 to 36%
Ethers	2 to 48%
Aldehydes	3 to 55%
Acetylene	2 to 85%

Figure 3.147 Alcohol ranges from 1% to 36%, ether from 2% to 48% and aldehyde from 3% to 55%.

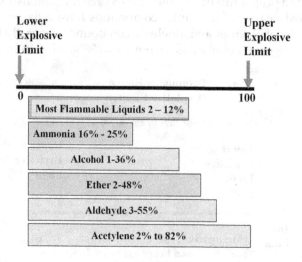

Figure 3.148 Acetylene has an extremely wide range from 2% at the lower end to between 80% and 100% at the upper end.

CASE STUDIES

PHILADELPHIA, PA

The One Meridian Plaza Building fire in Philadelphia several years ago, which resulted in the deaths of three firefighters, was started by linseed oil-soaked rags that were improperly stored, and they spontaneously combusted starting the fire (*Firehouse Magazine*).

VERDIGRIS, OK

I personally investigated a fire in an aircraft hanger that also resulted from improperly stored rags soaked with linseed oil. You could see the characteristic "V" pattern on the wall pointing right back to the box of rags.

The process of spontaneous combustion with animal/vegetable oils is very slow, which occurs over period of hours. Fires have also occurred in commercial laundries and restaurants while cleaning rags that were used for cleaning up animal/vegetable oils spontaneously combusting. Rags that are laundered and then dried in a dryer, and stored in laundry carts or on shelves, allow the heat to be confined within the rags. This laundry process does not remove all of the animal/vegetable oils and spontaneous combustion can occur.

Fuel family mixtures, such as gasoline, diesel fuel, fuel oil, aviation fuels, motor oils and grease, do not have any double bonds in the compounds.

These types of compounds cannot undergo spontaneous combustion and start fires because there is no double bond to break to produce heat required for spontaneous combustion to occur. These materials on rags have been blamed for causing fires in the past when it was not possible for them to be responsible.

Comparison of physical characteristics of gasoline, diesel fuel and ethanol							
Product	Boiling point (°F)	Flash point (°F)	Ignition temperature (°F)	Flammable range	Miscibility water	Specific gravity	Vapor density
Gasoline	85–437	–45	>530	1.4–7.6	No	0.70–0.78	3–4
Ethanol	122–176	61.8	685	3.3–19.0	Yes	0.79	1.6
Diesel fuel	320–700	125	500	0.3–10.0	No	0.81–0.88	>3

Specific Gravity

Specific gravity is to water as vapor density is to air. Specific gravity is a relationship of the weight of a material compared to water or another liquid (Figure 3.149). Water is given a weight value of 1.0. If a flammable liquid has a specific gravity greater than 1.0, it is heavier than water, and the flammable liquid will sink to the bottom in a water spill. If the flammable liquid has a specific gravity less than 1.0, it will float on top of the water.

The specific gravity of a hazardous material is very important in a water spill because it will determine what kinds of tactics are necessary to contain the spill. Another term associated with liquids is miscibility. If a liquid is miscible in water, it will mix with the water making clean-up very difficult. If the liquid is not miscible with water, then it will form a separate layer. The layer will form on top of the water or at the bottom of the water depending on the specific gravity of the liquid.

All of the conditions must be just right for combustion to occur. The liquid must be at its flash point, the air/vapor mixture must be at the appropriate flammable range and the ignition source must be above the ignition temperature of the liquid. There are also materials that have very wide flammable ranges. This fact makes them very dangerous to emergency responders. Some wide flammable range materials can burn inside the container, because they burn very rich.

Critical Temperature and Pressure

There are three terms that are important in understanding the liquefaction of gases by pressure. The first is critical point. This has to do with whether a gas will exist as a gas or as a liquid. Two factors come into

Specific Gravity

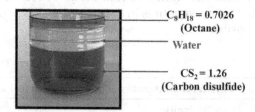

C_8H_{18} = 0.7026 (Octane)

Water

CS_2 = 1.26 (Carbon disulfide)

The relative density of a solid or liquid compared to water

Figure 3.149 Specific gravity is to water as vapor density is to air. Specific gravity is a relationship of the weight of the flammable liquid compared to water or another liquid.

play at the critical point to make this determination. Gases that are liquefied, such as propane, butane, and liquefied petroleum gases (LPGs), are pressurized within a container at their critical temperature and pressure. LPGs are mixtures of isobutane, propane, propylene and butylene. Critical temperature is the maximum temperature that a liquid (in this case a liquefied gas) can be heated and still remain a liquid.

For example, butane has a critical temperature of 305°F. As more heat is added, more of the liquid vaporizes. When a liquid is at its critical temperature, no amount of pressure can keep the liquid from becoming a gas. The critical pressure is the maximum pressure required to liquefy a gas that has been cooled to a temperature below its critical temperature. Butane has a critical pressure of 525 psi. In order to liquefy butane, it would have to be cooled below 305°F and pressurized to 525 psi to become a liquefied compressed gas.

Once created, the liquefied compressed gases remain under pressure to keep them in the liquid state, and they are shipped and stored at ambient temperatures. Because liquefied gases have large liquid-to-gas expansion ratios, they may present special hazards to emergency responders. For example, a very small amount of a liquid leaking from a container can form a very large gas cloud because of the large expansion. Larger leaks will produce large vapor clouds. This increases the danger of flammability if an ignition source is present and of asphyxiation or toxicity when vapor clouds form (Hazardous Materials Chemistry for Emergency Responders).

Hazards of the "Invisible Force"

Powerful magnets are in use at research universities, medical centers, imaging centers and industry. These magnets produce an "invisible force" that cannot be detected by the human senses but can be hazardous to firefighters and other emergency personnel if they do not understand the dangers that are present. In terms of the "invisible force" or magnetic field itself, there are no ill health effects from short-term exposure. However, this magnetic field produced by medical magnetic resonance imaging (MRI) and research nuclear magnetic resonance (NMR) devices can create an unexpected hazard to response personnel. Magnets attract metal objects. Because of the power of large magnets, metal objects brought into an MRI or NMR room can be "ripped" from the hands or body of the person(s) with the metal objects and drawn into the magnet at 45 miles/h (Figure 3.150).

A small child receiving an MRI at a medical facility was killed when the magnet pulled an oxygen tank into the patient area of the unit hitting the child in the head. Fire extinguishers with metal tanks, firefighter tools, oxygen tanks, SCBA with metal tanks or metal parts of the regulator or harness, belt buckles, steel toed shoes or boots and anything else metal can be dangerous if taken into the magnet rooms. The NMR magnets are more powerful than the electromagnets used to move old cars at the junkyard. Magnets can also damage certain types of credit cards, ID cards, watches and calculators if taken into the magnet room. They may also affect pacemakers and other metal implants in the human body.

Pay attention to warning signs placed on doors and in the area of powerful magnets. Read the directions on these signs and take them seriously. These magnets are always turned on and cannot easily or quickly be shut down during an emergency situation. Be careful when entering magnet rooms as there may be many obstacles such as tanks, piping, wires and other apparatuses associated with the magnet that can create trip and fall hazards as well as overhead dangers. When the magnet room is constructed, great care is taken to make that sure none of the construction materials contain ferrous (iron) metals that would be impacted by the magnet force. Sprinkler piping has to be copper or plastic. Fire extinguishers located in or near these magnet rooms need to have containers that are not ferrous metals. Specially designed extinguishers are commercially available.

Operation of one of these high-powered magnets requires the use of cryogenic liquid helium and liquid nitrogen. Both of these liquids are very cold, and when in contact with the liquid, can result in frostbite or solidification of body parts. There is no protective equipment that can be worn to protect the skin from the effects of the cold liquids. The gases produced as the liquids warm can cause asphyxiation to those present by displacing the oxygen in the area in the event of a release. Wearing SCBA will protect the respiratory system and prevent asphyxiation. Magnet rooms are

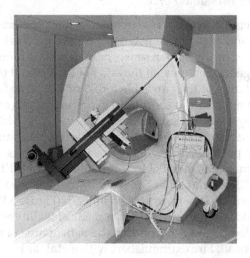

Figure 3.150 Because of the power of large magnets metal objects brought into an MRI or NMR room can be "ripped" from the hands or body of the person(s) with the metal objects and drawn into the magnet at 45 miles/h.

equipped with low-oxygen warning monitors to warn of an oxygen deficiency. The only way to "shut down" the magnet is to activate the Quench button located in the magnet control room. This evacuates the nitrogen and helium from the magnet and in effect shuts it down. Quenching should only be done under the advice and consent of the magnet operator as it is an expensive operation to refill the magnet. Magnets provide an important function in medical treatment and research. Response personnel should be aware of the potential hazards of these powerful magnets and learn how to work safely around them during an emergency to avoid serious injury or death (*Firehouse Magazine*).

Glossary

Absorption A route of exposure. It occurs when a toxic material contacts the skin and enters the bloodstream by passing through the skin.

Accidental explosion An unplanned or premature detonation/ignition of explosive/incendiary material or material possessing explosive properties. The activity leading to the detonation/ignition has no criminal intent and is primarily associated with legal, industrial or commercial activities.

Acid Any of a class of chemical compounds with aqueous solutions that turn litmus paper red (has a pH less than 7) or react with and dissolve certain metals or react with bases to form salts. (2) A compound capable of transferring a hydrogen ion in solution. (3) A molecule or ion that combines with another molecule or ion by forming a covalent bond with two electrons from other species.

Acid, corrosive A material that usually contains an H^+ ion and is capable of dehydrating other materials.

Acute exposure The adverse effects resulting from a single dose or exposure to a material. Ordinarily used to denote the effects observed in experimental animals.

Acute toxicity Any harmful effect produced by a single short-term exposure that may result in severe biological harm or death.

Aerosol The dispersion of very fine particles of a solid or liquid in a gas, fog, foam or mist.

Agent dosage The concentration of a toxic vapor in the air multiplied by the time that the concentration is present or the time that an individual is exposed ($mg\text{-}min/m^3$).

Alcohol foam A type of foam developed to suppress ignitable vapors on polar solvents (those miscible in water). Examples of polar flammable liquids are alcohols and ketones.

Alkaline Any compound having the qualities of a base. Simplified, a substance that readily ionizes in aqueous solution to yield hydroxyl (OH⁻) anions. Alkalis have a pH >7 and turn litmus paper blue.

Alpha particle A form of ionizing radiation that consists of two protons and two neutrons.

Ambient temperature The normal temperature of the environment.

ANFO An ammonium nitrate and fuel oil mixture, commonly used as a blasting agent. The proportions are determined by the manufacturer or user. It is commonly mixed with the addition of an "enhancer," such as magnesium or aluminum, to increase the rate of burn.

Anhydrous Describing a material that contains no water (water-free).

Antidote A material administered to an individual who has been exposed to a poison in order to counteract its toxic effects.

Arsenical Pertaining to or containing arsenic; a reference to the vesicant lewisite.

Asphyxia Lack of oxygen and interference with oxygenation of the blood. Can lead to unconsciousness.

Asphyxiant A vapor or gas that can cause unconsciousness or death by suffocation (lack of oxygen). Most simple asphyxiants are harmful to the body when they become so concentrated that they reduce (displace) the available oxygen in air (normally about 21%) to dangerous levels (18% or lower). Chemical asphyxiants, like carbon monoxide (CO), reduce the blood's ability to carry oxygen or, like cyanide, interfere with the body's utilization of oxygen.

Asphyxiation Asphyxia or suffocation. Asphyxiation is one of the principal potential hazards of working in confined spaces.

Atmospheric container A type of container that holds products at atmospheric pressure (760 mm).

Atom The smallest unit into which a material may be broken by chemical means. In order to be broken into any smaller units, a material must be subjected to a nuclear reaction.

Atomic weight (at. wt.) The relative mass of an atom. Basically, it equals the number of protons plus the number of neutrons.

Autoignition A process in which a material ignites without any apparent outside ignition source. In the process, the temperature of the material is raised to its ignition temperature by heat transferred by radiation, convection, combustion or some combination of all three.

Atropine An anticholinergic used as an antidote for nerve agents to counteract excessive amounts of acetylcholine. It also has other medical uses.

Bacteria Single-celled organisms that multiply by cell division and can cause disease in humans, plants or animals. Examples include anthrax, cholera, plague, tularemia and Q fever.

Base A chemical compound that reacts with an acid to form a salt. The term is applied to the hydroxides of the metals, certain metallic oxides and groups of atoms containing one or more hydroxyl groups (OH⁻) in which hydrogen is replaceable by an acid radical. *See* Alkaline.

Beta particle A form of ionizing radiation that consists of either electrons or positrons.

Biohazard The organisms that have a pathogenic effect on life and the environment, and can exist in normal ambient environments. These hazards can represent themselves as disease germs and viruses.

Biological agent A living organism, or the materials derived from it, that causes disease in or harms humans, animals or plants, or causes deterioration of material. Biological agents may be found as liquid droplets, aerosols or dry powders. A biological agent can be adapted and used as a terrorist weapon, such as anthrax, tularemia, cholera, encephalitis, plague and botulism. There are three different types of biological agents: bacteria, viruses and toxins.

Blasting agent A material designed for blasting that has been tested in accordance with Section 173.114a (49 CFR). It must be so insensitive that there is little probability of accidental explosion or going from burning to detonation.

BLEVE *See* Boiling liquid, expanding vapor, explosion.

Blister agent A chemical agent, also called a vesicant, that causes severe blistering and burns to eyes, skin, and tissues of the respiratory tract. Exposure is through liquid or vapor contact. Also referred to as mustard agent. Examples include mustard and lewisite.

Blood agent A chemical agent that interferes with the ability of blood to transport oxygen and causes asphyxiation. These substances injure a person by interfering with cell respiration (the exchange of oxygen and carbon dioxide between blood and tissues). Common examples are hydrogen cyanide and cyanogen chloride.

Blood asphyxiant A chemical that is absorbed by the blood and changes or prevents the blood from flowing or carrying oxygen to cells. An example is carbon monoxide poisoning.

Boiling liquid, expanding vapor, explosion (BLEVE) The explosion and rupture of a container caused by the expanding vapor pressure as liquids in the container become overheated.

Boiling point At this temperature, the vapor pressure of a liquid now equals the surrounding atmospheric pressure (14.7 psi at sea level).

BTU British Thermal Unit. The amount of heat required to raise 1 lb of H_2O—1°F at sea level.

Carcinogen A material that causes cancer in either humans or animals; because it causes cancer in animals, it is considered capable of causing cancer in humans.

Caustic (1) Burning or corrosive. (2) A hydroxide of a light metal. Broadly, any compound having highly basic properties. A compound that readily ionizes in aqueous solution to yield OH^- anions, with a pH >7, and turns litmus paper blue. *See* Alkaline; Base.

Central nervous system (CNS) In humans, the brain and spinal cord, as opposed to the peripheral nerves found in the fingers, etc.

Chemical agent There are five classes of chemical agents, all of which produce incapacitation or cause serious injury or death: nerve agents, blister agents, blood agents, choking agents and irritating agents. A chemical substance used in military operations intended to kill, seriously injure, or incapacitate people through its physiological effects.

Chemical burn A burn that occurs when the skin comes into contact with strong acids, strong alkalis or other corrosive materials. These agents literally eat through the skin and, in many cases, continue to do damage as long as they remain in contact with the skin.

Chemical properties A property of matter that describes how it reacts with other substances.

Chemical reaction A process that involves the bonding, unbonding or rebonding of atoms. A chemical change takes place that actually changes substances into other substances.

Chemical reactivity A process whereby substances are changed into other substances by the rearrangement, or recombination, of atoms.

CHEMTREC Chemical Transportation Emergency Center operated by the Chemical Manufacturers Association. Provides information or assistance to emergency responders. Contacts the shipper or producer of the material for more detailed information, including on-scene assistance when feasible. Can be reached 24 h a day by calling 1-800-424-9300.

CHLOREP Chlorine Emergency Plan operated by the Chlorine Institute. A 24-h mutual-aid program. Response is activated by a CHEMTREC call to the designated CHLOREP's geographical-sector assignments for teams.

Choking agents These agents exert their effects solely on the lungs and result in the irritation of the alveoli of the lungs. Agents cause the alveoli to constantly secrete watery fluid into the air sacs; this condition is called pulmonary edema. When a lethal amount of a choking agent is received, the air sacs become so flooded that the air cannot enter and the victim dies of anoxia (oxygen deficiency); also known as dry drowning.

Chronic Applies to long periods of action, such as weeks, months or years.

Chronic effect An adverse health effect on a human or animal body with symptoms that develop slowly or recur frequently due to the exposure of hazardous chemicals.

Chronic exposure Repeated doses or exposure to a material over a relatively prolonged period of time.

Closed-cup tester A device for determining flash points of flammable and combustible liquids, utilizing an enclosed cup or container for the liquid. Recognized types are the Tagliabue (Tag) Closed Tester, the Pensky–Martens Closed Tester and the Setaflash Closed-Cup Tester.

CNS *See* Central nervous system.

Combustibility The ability of a substance to undergo a rapid chemical combination with oxygen, with the evolution of heat.

Combustible dust A particulate material that, when mixed in air, burns or explodes.

Combustible liquid A term commonly used for liquids that emit burnable vapors or mists. Technically, a liquid whose vapors will ignite at a temperature of 100°F or above but below 93.3°C (200°F).

Compound A substance composed of two or more elements that have chemically reacted. The compound that results from the chemical reaction is unique in its chemical and physical properties.

Compressed gas Any material or mixture having in the container an absolute pressure exceeding 40 psi at 70°F or, regardless of the pressure at 70°F, having an absolute pressure exceeding 104 psi at 130°F; any flammable material having a vapor pressure exceeding 40 psi absolute at 100°F as determined by testing. Also includes cryogenic or "refrigerated liquids" (DOT) with boiling points lower than −130°F at 1 atmosphere.

Concentration The amount of a material that is mixed with another material.

Concentration (corrosives) In corrosives, the amount of acid or base compared to the amount of water present. Corrosives have "strength" and "concentration." *See* Strength.

Contaminant (1) A toxic substance that is potentially harmful to people, animals and the environment. (2) A substance not in pure form.

Corrosive A chemical that causes visible destruction of or irreversible alterations in living tissue by chemical action at the site of contact; a liquid that causes a severe corrosion rate in steel. A corrosive is either an acid or a caustic (a material that reads at either end of the pH scale).

Corrosive material (DOT) A material that causes the destruction of living tissue and metals.

Covalent bond A chemical bond in which atoms share electrons to form a molecule.

Critical pressure The pressure required to liquefy a gas at its critical temperature.

Critical temperature The temperature above which a gas cannot be liquefied by pressure.

Cryogenic burn Frostbite; damage to tissues as the result of exposure to low temperatures. It may involve only the skin, extend to the tissue immediately beneath it or lead to gangrene and loss of affected parts.

Cryogenic cylinder An insulated metal cylinder contained within an outer protective metal jacket. The area between the cylinder and the jacket is normally under vacuum. The cylinders range in size from a Dewier (similar to a small thermos) up to 24 inches diameter and 5 ft in length. Examples of materials found in these types of cylinders are argon, helium, nitrogen and oxygen.

Cryogenic liquid A liquid with a boiling point below –130°F.

Cylinder A container for liquids, gases or solids under pressure. Ranges in size from aerosol containers found at home, such as spray deodorant, to the cryogenic (insulated) cylinders for nitrogen that can be approximately 24 inches diameter and 5 ft in length. Pressure ranges from a few pounds to 6,000 pounds per square inch.

Dangerous When Wet Materials that when exposed to water allow a chemical reaction to take place and often produce flammable or poisonous gases, heat and a caustic solution. An example is sodium.

Decomposition Separation of larger molecules into separate constituent and smaller parts.

Decomposition (chemical) A reaction in which the molecules of a chemical break down to its basic elements, such as carbon, hydrogen or nitrogen, or to more simple compounds. This often occurs spontaneously, liberating considerable heat and often large volumes of gas.

Decontamination The physical or chemical process of reducing and preventing the spread of contamination from persons and equipment used at a hazardous materials incident.

Deflagration Explosion, with rapid combustion, up to 1,250 ft/s.

Detonating cord A flexible cord containing a center cord of high explosives used to detonate other explosives with which it comes into contact.

Detonation An explosion at speeds above 1,250 ft/s and many times over 3,300 ft/s.
Detonator Any device containing a detonating charge that is used for initiating detonation in an explosive. This term includes, but is not limited to, electric and nonelectric detonators (either instantaneous or delayed) and detonating connectors.
Dewier container A small (less than 25 gallons) container used for temporary storage or handling of cryogenic liquids.
Dilution The application of water to water-miscible hazardous materials. The goal is to reduce the hazard of a material to safe levels by reducing its concentration.
Dose The accumulated amount of a chemical to which a person is exposed.
DOT U.S. Department of Transportation. Regulates transportation of materials to protect the public as well as fire, law and other emergency response personnel.
Dry bulk A type of container used to carry large amounts of solid materials (more than 882 lb or 400 kg). It can be placed either on or in a transport vehicle or vessel constructed as an integral part of the transport vehicle.
Dyspnea Shortness of breath, a subjective difficulty or distress in breathing, usually associated with disease of the heart or lungs; occurs normally during intense physical exertion or at high altitudes.
Element A substance that cannot be broken down into any other substance by chemical means.
Empirical formula Describes the *ratio* of the number of each element in the molecule, but not the exact number of atoms in the molecule.
Emulsification The process of dispersing one liquid in a second immiscible liquid. The largest group of emulsifying agents includes soaps, detergents and other compounds whose basic structure is a paraffin chain terminating in a polar group.
Endothermic A process or chemical reaction that is accompanied by absorption of heat.
Enterotoxin A cytotoxin specific for the cells of the intestinal mucosa.
Erythema Red area of the skin caused by heat or cold injury, trauma or inflammation.
Etiologic agent The living organisms or their toxins that contribute to the cause of infection, disease or other abnormal condition.
Evaporation A process in which liquid becomes vapor as more molecules leave the vapor than return.
Exothermic reaction A chemical reaction that liberates heat during the reaction.
Expansion ratio The amount of gas produced from a given volume of liquid escaping from a container at a given temperature.

Explosion The sudden and rapid production of gas, heat, noise and many times a shock wave, within a confined space.

Explosive (DOT) Any chemical compound or mixture whose primary function is to produce an explosion.

Explosives, high Explosive materials that can be used to detonate by means of a detonator when unconfined (e.g., dynamite).

Explosives, low Explosive materials that deflagrate rather than detonate (e.g., black powder, safety fuses and "special fireworks" as defined by Class 1.3 Explosives).

Explosive limits See Flammable limits.

Febrile Denoting or relating to fever.

Fire point The lowest temperature at which the vapor above the liquid ignites and continues to burn, usually a few degrees above the flash point.

Flammable gas A gas that at ambient temperature and pressure forms a flammable mixture with air at a concentration of 13% by volume or less; a gas that at ambient temperature and pressure forms a range of flammable mixtures with air greater than 12 percentage points by volume, regardless of the lower explosive limit.

Flammable limits The range of the percentages of vapor mixed with air that are capable of ignition, as opposed to those mixtures that have too much or too little vapor to be ignited. Also called explosive limits.

Flammable liquid A liquid that gives off readily ignitable vapors. Defined by the NFPA and DOT as a liquid with a flash point below 100°F (38°C).

Flammable range The percentage of fuel vapors in air where ignition can occur. It has an upper and a lower limit.

Flammable solid A solid (other than an explosive) that ignites readily and continues to burn. It is liable to cause fires under ordinary conditions or during transportation through friction or retains heat from manufacturing or processing. It burns so vigorously and persistently as to create a serious transportation hazard. Included in this class are spontaneously combustible and water-reactive materials. An example is white phosphorus.

Flash back The ignition of vapors and the travel of the flame from the ignition source back to the liquid/vapor release source.

Flash point The minimum temperature at which a liquid gives off vapor within a test vessel in sufficient concentration to form an ignitable mixture with air near the surface of the liquid.

Foam A firefighting material consisting of small bubbles of air, water and concentrating agents. Chemically, the air in the bubbles is suspended in the fluid. The foam clings to vertical and horizontal surfaces and flows freely over burning or vaporizing materials.

Foam puts out a fire by blanketing it, excluding air and blocking the escape of volatile vapor. Its flowing properties resist mechanical interruption and reseal the burning material.

Formula A combination of the symbols for atoms or ions that are held together chemically.

Freezing point The temperature at which a material changes its physical state from a liquid to a solid.

Frothing A foaming action caused when water, turning to steam in contact with a liquid at a temperature higher than the boiling point (212°F), picks up a part of a viscous liquid.

Gas A formless fluid that occupies the space of its enclosure. It can settle to the bottom or top of an enclosure when mixed with other chemicals. It can be changed to its liquid or solid state only by increased pressure and decreased temperature.

Gastrointestinal tract (GI tract) The entire digestive canal from the mouth to the anus.

Halogens A chemical family that includes fluorine, chlorine, bromine and iodine.

Halons Halogenated hydrocarbons (containing the elements such as F, Cl, Br or I) used to suppress or prevent combustion.

Hazard class One of nine classes of hazardous materials as categorized and defined by the DOT in 49 CFR.

Hazmat foam A special vapor-suppressing mix that can be applied to liquids or solids to prevent off-gassing.

High-expansion foam A detergent-based foam (low water content) that expands at a ratio of 1,000 to 1.

High-order explosion Materials that require moderate heat and reducing agents to initiate combustion.

Hypergolic materials Materials that ignite upon contact with one another.

Hypergolic reaction The immediate spontaneous ignition when two or more materials are mixed.

Hypotension Subnormal arterial blood pressure.

Hypoxemia Subnormal oxygenation of arterial blood; short of anoxia.

IDLH Immediately dangerous to life and health. The maximum levels to which a healthy worker can be exposed for 30 min to a chemical and escape without suffering irreversible health effects or escape impairing symptoms.

Ignition temperature The minimum temperature at which a material will ignite without a spark or flame present. This is also the temperature the ignition source must be.

Immiscible A matter that cannot be mixed. For example, water and gasoline are immiscible.

Incompatibility The inability to function or exist in the presence of something else, such as when a chemical will destroy the container.

Inert A material that under normal temperatures and pressures does not react with other materials.

Inhibited A substance that has had another substance added to prevent or deter its reaction with either other materials or itself (polymerization). Usually used to deter polymerization.

Inhibitor A substance that is capable of stopping or retarding a chemical reaction. To be technically useful, it must be effective in low concentration (i.e., to stop polymerization).

Initiator The substance or molecule (other than reactant) that initiates a chain reaction, as in polymerization.

Inorganic Pertaining to or composed of chemical compounds that do not contain carbon as the principal element (except carbonates, cyanides and cyanates). A matter other than plant or animal.

Inorganic peroxides Inorganic compounds containing an element at its highest state of oxidation (such as sodium peroxide) or having the peroxy group –O–O– (such as perchloric acid).

Ionic bond A chemical bond in which atoms of different elements transfer (exchange) electrons. As the electrons are exchanged, charged particles known as ions are formed.

Ionizing radiation High-energy radiation, such as an X-ray, that causes the formation of ions in substances through which it passes (gamma rays). Excessive amounts of ionizing radiation will cause permanent genetic or bodily damage.

Irritant A noncorrosive material that causes a reversible inflammatory effect on living tissue by chemical action at the site of contact.

Latent period Specifically in the case of mustard agent, the period between the exposure and the onset of signs and symptoms; otherwise, an incubation period.

Lethal chemical agent An agent that may be used effectively in a field concentration to produce death.

Lacrimation Secretion and discharge of tears.

LC_{50} Lethal concentration 50 or median lethal concentration. The concentration of a material in air that, based on laboratory tests (respiratory route), is expected to kill 50% of a group of test animals when administered as a single exposure in a specific time period.

LD_{50} Lethal dose 50. The single dose of a substance that causes death of 50% of an animal population from exposure to the substance by any route other than inhalation.

Liquid A substance that is neither a solid nor a gas; a substance that flows freely like water.

Low-order explosion Materials that require excessive heat and reducing agents to initiate combustion.

Low-pressure container A container designed to withstand pressures from 5 to 100 psi.

Lower explosive limit The lowest concentration of gas or vapor (% by volume in air) that burns or explodes if an ignition source is present at ambient temperatures.

Mechanical foam A substance introduced into the water line by various means at a 6% concentration. Air is then introduced to yield foam consisting generally of 90 volumes of air, 9.4 volumes of water and 0.6 volumes of foam liquid. It uses hydrolyzed soybean, fish scales, hoof and horn meal, and peanut or corn protein as a base.

Median incapacitating dosage (ICT_{50}) The volume of a chemical agent vapor or aerosol inhaled that is sufficient to disable 50% of exposed, unprotected people (expressed as mg-min/m^3).

Median lethal dosage (LCT_{50}) The dosage of a chemical agent vapor or aerosol inhaled that is lethal to 50% of exposed, unprotected people (expressed as mg-min/m^3).

Median lethal dosage (LD_{50}) The amount of liquid chemical agent expected to kill 50% of a group of exposed, unprotected individuals.

Melting point The degree of temperature at which a solid substance becomes a liquid, especially under a pressure of 1 atmosphere.

Miosis A condition where the pupil of the eye becomes contracted (pinpointed), impairing night vision.

Miscible Mixable in any and all proportions to form a uniform mixture. Water and alcohol are miscible; water and oil are not.

Molecular formula Shows the *exact number* of each atom in the molecule.

Molecular weight The sum of the atomic weights of all the atoms in a molecule.

Molecule The smallest possible particle of a chemical compound that can exist in the free state and still retain the characteristics of the substance. Molecules are made up of atoms of various elements that form the compound.

Monomer A simple molecule capable of combining with a number of like or unlike molecules to form a polymer. It is a repeating structure unit within a polymer.

Mutagen A material that induces genetic changes (mutations) in the DNA of chromosomes. Chromosomes are the "blueprints" of life within individual cells.

Myalgia Muscle pain.

Mydriasis Dilation of the pupil.

Narcosis General and nonspecific reversible depression of neuronal excitability, produced by a number of physical and chemical agents, usually resulting in stupor rather than in anesthesia.

Necrosis Cell or tissue death due to disease or injury.

Nerve agent A substance that interferes with the central nervous system. Exposure is primarily through contact with the liquid (skin and eyes) and secondarily through inhalation of the vapor.

Neutralization A chemical reaction used to remove H^+ ions from acidic solutions and OH^- ions from basic solutions. The reaction can be violent and usually produces water, a salt, heat and, many times, a gas.

Nonpersistent agent An agent that remains in the target areas for a relatively short period of time. The hazard, predominantly vapor, exists for minutes or, in exceptional cases, hours after dissemination of the agent. As a general rule, the duration is less than 12 h.

Normal A solution that contains one equivalent of solute per liter of solution.

NRC National Response Center, a communications center for activities related to response actions, located at Coast Guard headquarters in Washington, DC. The toll-free number, (800) 424-8802, can be reached 24 h a day for reporting actual or potential pollution incidents.

NRT National Response Team, consisting of representatives of 14 government agencies (DOD, DOI, DOT/RSPA, DHS, USCG, EPA, DOC, FEMA, DOS, USDA, DOJ, HHS, DOL, NRC and DOE), is the principal organization for implementing the National Contingency Plan. When the NRT is not activated for a response action, it serves as a standing committee to develop and maintain preparedness, to evaluate methods of responding to discharges or releases, to recommend needed changes in the response organization and to recommend revisions to the NCP.

Odor A quality of something that affects the sense of smell; fragrance.

Odor threshold The greatest dilution of a sample with odor-free water to yield the least definitely perceptible odor.

Open-cup tester A device for determining flash points of flammable and combustible liquids, utilizing an open cup or a container for the liquid. Recognized types are the Tagliabue (Tag) Open-Cup Apparatus and the Cleveland Open-Cup Apparatus.

Organic A material that comes from living plants or animals, such as waste or decay products. Distinguished from mineral matter. Organic chemistry deals with materials that contain the element carbon (C).

Organic peroxides Any organic compound containing oxygen (O) in the bivalent –O–O– structure and that may be considered a derivative of hydrogen peroxide, where one or more of the hydrogen atoms have been replaced by organic radicals.

Organophosphate A compound with a specific phosphate group that inhibits acetylcholinesterase. Used in chemical warfare and as an insecticide.

Oxidizer A material that gives up oxygen easily, removes hydrogen from another compound or attracts negative electrons (such as chlorine or fluorine), thus enhancing the combustion of other materials.

Oxidizing agent A material that gains electrons from the fuel during combustion.

Oxygen deficient Defined by OSHA as ambient air containing less than 19.5% oxygen concentration.

Oxygen enriched Defined by OSHA as ambient air containing above 24% oxygen concentration.

2-PAM CL Pralidoxime chloride (Protopam®). An antidote to organophosphate poisoning which might result from exposure to nerve agents or some insecticides. The drug, which helps restore an enzyme called acetylcholinesterase, must be used in conjunction with atropine to be effective. Restores normal control of skeletal muscle contraction (relieves twitching and paralysis).

Papule A small, circumscribed, solid elevation on the skin.

PEL Permissible exposure limit. A term used by OSHA for its health standards covering exposures to hazardous chemicals. PEL generally relates to legally enforceable TLV limits.

Persistency An expression of the duration of effectiveness of a chemical agent, dependent on physical and chemical properties of the agent, weather, method of dissemination and terrain conditions.

Persistent agent An agent that remains in the target area for longer periods of time. Hazards from both vapor and liquids may exist for hours, days or, in exceptional cases, weeks or months after dissemination of the agent. As a general rule, the duration is greater than 12 h.

Photophobia Morbid dread and avoidance of light. Photosensitivity, or pain in the eyes with exposure to light, can be a cause.

pH The "power of hydrogen." A measure of the acidity or basicity of a solution, that is, of the concentration of H^+ or OH^- ions in solution. The scale ranges from 0 to 14, with 7 being neutral.

Physical properties A property of matter that describes only its condition, not the way it reacts with other substances. Examples are size, density, color and electrical conductivity.

Poison Any substance (solid, liquid, or gas) that, by reason of an inherent deleterious property, tends to destroy life or impair health.
Polar-solvent liquids The liquids that mix (are miscible with water).
Polymer A long chain of molecules having extremely high molecular weights made up of many repeating smaller units called monomers or comonomers.
Polymerization A chemical reaction in which small molecules combine to form larger molecules. A hazardous polymerization is a reaction that takes place at a rate that releases large amounts of energy which can cause fires or explosions or burst containers. Materials that can polymerize usually contain inhibitors that can delay the reaction.
Powder A solid reduced to dust by pounding, crushing or grinding.
ppm (parts per million) Parts of vapor or gas per million parts of contaminated air by volume at 25°C and 1 torr pressure.
Pressure vessel A tank or a container constructed so as to withstand the interior pressure greater than that of the atmosphere.
psi Pounds per square inch.
Presynaptic Pertaining to the area on the proximal side of a synaptic cleft.
Pruritus Synonym: itching.
Ptosis (pl. Ptoses) In reference to the eyes, drooping of the eyelids.
Pustule A small, circumscribed elevation of the skin containing pus and having an inflamed base.
Pyrogenic Causing fever.
Pyrolysis A chemical decomposition or breaking apart of molecules produced by heating in the absence of air.
Pyrophoric A material that ignites spontaneously in air below 130°F (54°C). Occasionally caused by friction.
Pyrophoric gas A gaseous material that spontaneously ignites when exposed to air under ambient conditions. An example is trimethyl aluminum.
Pyrophoric liquid A liquid material that spontaneously ignites when exposed to air under ambient conditions.
Pyrophoric solid A solid material that spontaneously ignites when exposed to air under ambient conditions. An example is phosphorus.
RAD Radiation-absorbed dose.
Radiation Ionizing energy, either particulate or wave, that is spontaneously emitted by a material or combination of materials.
Radioactive material (DOT) A material that emits ionizing radiation.
Radioactivity Any process by which unstable nuclei increase their stability by emitting particles (alpha or beta) or gamma rays.

Glossary

Rate of explosion Rate of decomposition measured in feet per second in relation to the speed of sound. If subsonic, the rate is described as a deflagration. If supersonic, the rate of decomposition is defined as a detonation.

Reducing agent A substance that gives electrons to (and thereby reduces) another substance.

Respiratory asphyxiant A material that prevents or reduces the available oxygen necessary for normal breathing. Divided into simple and chemical asphyxiants.

Respiratory dosage This is equal to the time in minutes an individual is unmasked in an agent cloud multiplied by the concentration of the cloud.

Retinitis Inflammation of the retina.

Rhinorrhea A runny nose.

Roentgen The amount of ionization that occurs per cubic centimeter of air.

Routes of exposure Ways in which chemicals get in contact with or enter the body. These are inhalation, absorption, ingestion or injection.

SADT *See* Self-accelerating decomposition temperature.

Safety fuse A flexible cord containing an internal burning medium by which fire or flame is conveyed at a uniform rate from the point of ignition to the point of use, usually a detonator.

Safety-relief valve A safety-relief device containing an operating part that is held normally in a position closing a relief channel by spring force and is intended to open and close at a predetermined pressure.

Self-accelerating decomposition temperature (SADT) Organic peroxides or other synthetic chemicals that decompose at ambient temperature, or react to light or heat, resulting in a chemical breakdown. This releases oxygen, energy and fuel in the form of rapid fire or explosion. To ensure stabilization, these materials must be kept in a dark or refrigerated environment.

Sensitizer A substance that on the first exposure causes little or no reaction in humans or test animals, but that on the repeated exposure may cause a marked response not necessarily limited to the contact site.

Septic shock Shock associated with septicemia caused by Gram-negative bacteria.

Shigellosis Bacillary dysentery caused by bacteria of the genus *Shigella*, often occurring in epidemic patterns.

Simple asphyxiant A material that replaces the amount of oxygen admitted into the body without further damage to tissue or poisoning. Examples are nitrogen and carbon dioxide.

Slurry A pourable mixture of solid and liquid.

Solid The state of matter having definite volume and rigid shapes. Its atoms or molecules are restricted to vibration only.

Solubility The ability of a substance to form a solution with another substance.

Solution The even dispersion (mixing) of molecules of two or more substances. The most commonly encountered solutions involve mixing of liquids and liquids or solids and liquids.

Solvent A substance, usually a liquid, capable of absorbing another liquid, gas or solid to form a homogeneous mixture.

Specific gravity The weight of a solid or liquid substance compared to the weight of an equal volume of water; the specific gravity of water equals 1.

Spontaneous combustion A process by which heat is generated within the material by either a slow oxidation reaction or microorganisms.

Spontaneous ignition Ignition that can occur when certain materials, such as tung oil, are stored in bulk, resulting from the generation of heat, which cannot be readily dissipated; often heat is generated by microbial action.

Spontaneous ignition temperature *See* Ignition temperature.

States of matter Any of three physical forms of matter: solid, liquid or gas.

Strength (acid/base) The amount of ionization that occurs when an acid or a base is dissolved in a liquid.

Stridor A high-pitched, noisy respiration, like the blowing of the wind; a sign of respiratory obstruction, especially in the trachea or larynx.

Subacute exposure (1) Less than acute. (2) Of or pertaining to a disease or other abnormal condition present in a person who appears to be clinically well. The condition may be identified or discovered by means of a laboratory test or by radiologic examination.

Sublimation The direct change of state from solid to vapor.

Subscripts Identify the number of atomic weights of the element present in the molecule.

Symbol Letters used to identify each element. The symbol for an element represents a definite weight (1 atomic weight) of that element.

Systemic toxicity Poisoning of the whole system or organism, rather than poisoning that affects, for example, a single organ.

Target organ The primary organ to which specific chemicals cause harm. Examples are the lungs, liver or kidneys.

Temperature Measure of the vibratory rate of a molecule.

Teratogen A material that affects the offspring when a developing embryo or fetus is exposed to that material.

Thermal burn Pertaining to or characterized by heat.
TLV Threshold limit value, *estimated* exposure value, below which no ill health effects *should* occur to the individual.
Toxemia A condition in which toxins produced by cells at a local source of infection or derived from the growth of microorganisms are contained in the blood.
Toxic Harmful, poisonous.
Toxicity The ability of a substance to cause damage to living tissue, impairment of the central nervous system, severe illness or death when ingested, inhaled or absorbed by the skin.
Toxins Toxic substances of natural origin produced by an animal, plant or microbe. They differ from chemical substances in that they are not man-made. Toxins may include botulism, ricin and mycotoxins.
TTL Threshold toxic limit, *estimated* exposure value, below which no ill health effects *should* occur to the individual.
Upper explosive limit The maximum fuel-to-air mixture in which combustion can occur.
Uricant A chemical agent that produces irritation at the point of contact, resembling a stinging sensation, such as a bee sting. For example, the initial physiological effects of phosgene oxime (CX) upon contact with a person's skin.
Vapor density The weight of a vapor or gas compared to the weight of an equal volume of air; an expression of the density of the vapor or gas calculated as the ratio of the molecule weight of the gas to the average molecule weight of air, which is 29. Materials lighter than air have vapor densities less than 1.
Vapor pressure The pressure exerted by a saturated vapor above its own liquid in a closed container. Vapor pressure reported on Material Safety Data Sheet (MSDS) is in millimeters of mercury at 68°F (20°C), unless stated otherwise.
Vapors Molecules of liquid in air; moisture, such as steam, fog and mist, often forming a cloud suspended or floating in the air, usually due to the effect of heat upon a liquid.
Variola Syn: smallpox.
Vesicants Chemical agents, also called blister agents, that cause severe burns to eyes, skin and tissues of the respiratory tract. Also referred to as mustard agents. Examples include mustard and lewisite.
Vesicles Blisters on the skin.
Virus The simplest type of microorganism, lacking a system for its own metabolism. It depends on living cells to multiply and cannot live long outside of a host. Types of viruses are smallpox, Ebola, Marburg and Lassa fever.

Violent reaction The action by which a chemical changes its composition near or exceeds the speed of sound, often releasing heat and gases.
Viscosity The measurement of the flow properties of a material expressed as its resistance to flow. Unit of measurement and temperature are included.
Vomiting agent A compounds that causes irritation of the upper respiratory tract and involuntary vomiting.
Water-reactive material A material that decomposes or reacts when exposed to moisture or water.
Water solubility The ability of a substance to mix with water.

Bibliography

Burke, Robert, Anthrax Scares: The Bomb Threat of the 90s, *Firehouse Magazine*, July 1, 1999.
Burke, Robert, Chemical Notebook: Phosphorus, *Firehouse Magazine*, November 1, 2019.
Burke, Robert, Chemistry of Clandestine Methamphetamine Drug Labs, *Firehouse Magazine*, December 31, 2003.
Burke, Robert, *Counter Terrorism for Emergency Responders*, Third Edition, CRC Press, Taylor & Francis Group, Boca Raton, FL, 2017.
Burke, Robert, Counter Terrorism for Emergency Responders, Third Edition, CRC Press, Taylor & Francis Group, Boca Raton, FL, 2018.
Burke, Robert, Cryogenic Liquids, *Firehouse Magazine*, November 30, 1996.
Burke, Robert, DOT Releases 2016 ERG, *Firehouse Magazine*, September 1, 2016.
Burke, Robert, Emergency Response to Agricultural Ammonia Releases, *Firehouse Magazine*, July 1, 2014.
Burke, Robert, Five-Step Field-Test Process for Ruling Out Anthrax Spores, *Firehouse Magazine*, September 1, 2004.
Burke, Robert, Handling Anhydrous Ammonia Emergencies, *Firehouse Magazine*, February 28, 2002.
Burke, Robert, *Hazardous Materials Chemistry for Emergency Responders*, Third Edition, CRC Press, Taylor & Francis Group, Boca Raton, FL, 2017.
Burke, Robert, Hazmat Studies: Safe Response to Aerial Crop-Spraying Accidents, *Firehouse Magazine*, February 1, 2012.
Burke, Robert, *Hazmatology: The Science of Hazardous Materials*, Volume One, CRC Press, Taylor Francis Group, Boca Raton, FL, 2020.
Burke, Robert, *Hazmatology: The Science of Hazardous Materials*, Volume Two, CRC Press, Taylor Francis Group, Boca Raton, FL, 2020.
Burke, Robert, Inside the Houston Hazmat Team, *Firehouse Magazine*, February 1, 2019.
Burke, Robert, Lessons Learned from Anhydrous Ammonia Incident, *Firehouse Magazine*, April 1, 2017.
Burke, Robert, NMR and MRI Medical Scanners: Surviving the "Invisible Force", *Firehouse Magazine*, May 1, 2012.
Burke, Robert, Plastics & Polymerization: What Firefighters Need To Know, *Firehouse Magazine*, February 28, 1999.
Burke, Robert, Safe Response to Aerial Crop-Spraying Accidents, *Firehouse Magazine*, February 1, 2012.

Burke, Robert, Tactical Response to Incidents Involving Crude Oil, *Firehouse Magazine*, September 1, 2015.
Burke, Robert, The Dangers of Synthetic Opioids, *Firehouse Magazine*, September 1, 2017.
Burke, Robert, The Facts About Anthrax, *Firehouse Magazine*, December 31, 2001.
Burke, Robert, The Phenomenon of Spontaneous Combustion, *Firehouse Magazine*, October 31, 2003.
Burke, Robert, The Phenomenon of Spontaneous Combustion, *Firehouse Magazine*, October 31, 2003.
Burke, Robert, The Waverly Propane Explosion 25th Anniversary: What Has Changed? *Firehouse Magazine*, January 31, 2003.
Burke, Robert, Understanding Chlorine, *Firehouse Magazine*, February 29, 2004.
Chemical Safety Board, https://www.csb.gov/.
Code of Federal Regulations (CFR), Title 49 section 172.101, https://www.law.cornell.edu/cfr/text/49/part-172.
DEA, Drug Labs in the United States, https://www.dea.gov/clan-lab.
DOT Reader, http://dot111.info/category/disasters/aliceville-al/
DOT, *Department of Transportation Emergency Response Guide Book*, 2016.
EPA, A-Z Index for Pesticide Topics, https://www.epa.gov/pesticides/z-index-pesticide-topics.
FBI, WMD Unit, https://www.fbi.gov/investigate/wmd.
FEMA, National Fire Academy, Chemistry for Emergency Response Instructor Guide, 2017.
FEMA, National Fire Academy, Hazardous Materials Incident Management, Instructor Guide, 2014.
Inside Climate News, https://insideclimatenews.org/.
International Association of Firefighters (IAFF), Train Derailment in Miamisburg, OH (Lessons Learned), https://www.iafc.org/topics-and-tools/resources/resource/miamisburg-ohio-train-derailment.
Mercapman Site, Gas Odorization History: From Water to Natural Gas, April 15, 2016, https://naturalgasodorization.com/gas-odorization-history/.
NASA, NASA Safety Center, System Failure Case Study, From Rockets to Ruins, the PEPCON Ammonium Perchlorate Plant Explosion, November 2012, Volume 6, Issue 9, Henderson NV May 4, 1988 PEPCON Explosions, www.NASA.gov.
Nebraska State Fire Marshal, Investigation Report, Midwest Farmers COOP, Tecumseh, NE, Anhydrous Ammonia Release, March 20, 2014.
NFPA, Flammable and Combustible Liquids Code, NFPA 30, https://www.nfpa.org/codes-and-standards/all-codes-and-standards/list-of-codes-and-standards/detail?code=30.
NTSB Investigation Report, Derailment of CN Freight Train U70691-18 With Subsequent Hazardous Materials Release and Fire, Cherry Hill IL, June 19, 2009, https://www.ntsb.gov/investigations/AccidentReports/Pages/RAR1201.aspx.
NTSB Investigation Report, https://www.ntsb.gov/Investigations/AccidentReports/Pages/RAR8510.aspx
NTSB, NTSB Safety, DOT-111 and CPC-1232, Tank Cars, https://www.bing.com/search?q=NTSB+Safety%2C+DOT-111+and+CPC-1232%2C+Tank+Cars&cvid=402cd00b77924633986a1915172cd226&PC=U531
Reed, Brian, Nitroglycerin Explosion of 1927, *Butler County Historical*, accessed July 17, 2020, https://butlerhistorical.org/items/show/41.
Wikipedia.org, https://en.wikipedia.org/wiki/Lac-M%C3%A9gantic_rail_disaster.

Index

acetic acid 96, 170, 221, 248, 264, 265, 269
acetone peroxide 95
acetonitrile 218
acrylonitrile 84, 165, 218, 268
alcohol 82, 163, 180,
aldehyde 74, 78, 83, 282
alkali metals 31, 38, 205
alkaline earth metal 31, 205
alkyl halide 74, 159, 218, 285
aluminum alkyls 195, 203
aluminum nitrate 210
aluminum phosphide 195
amine 74, 160, 282
ammonium chlorate 210
ammonium nitrate 3, 40, 53, 86, 98
    ammonium nitrate incidents 1947-2013
        102, 104, 106, 107, 108, 109, 209,
        277, 300
ammonium perchlorate 53, 204
ammonium picrate 95, 193
anhydrous ammonia 3, 40, 120, 137, 150, 247, 249, 252
    historic incidents 15, 142
animal vegetable oils 196, 201, 202, 281, 292, 294, 295
anthrax 227
    case study 232
    cutaneous 228
    diagnosis 230
    FBI guidelines 233
    gastrointestinal 230
    inhalation 229
    oropharyngeal 230
    treatment 230
    vaccine 230
anthrax scares 231
    "The Bomb Threat of the 21st Century" 231
applied chemistry 1

applied physics 281
argon 33, 132, 152, 304
atomic number 26, 27, 30
atomic weight 26, 27, 30, 285, 300

barium azide 193
benzoyl peroxide 75, 82, 213
black powder 85, 94, 98, 104, 112, 213, 306
boiling point 25, 131, 281, 282, 284, 301
branching hydrocarbon compounds 66, 285
bromine 21, 27, 32, 159, 207, 210, 219
butadiene 61, 69, 94, 167
butane 65, 120, 124, 125, 147, 186, 282, 285
butylene 124, 125, 197

calcium carbide 19, 39, 48, 70, 193, 206
calcium chlorite 210
camphor 22, 193, 286, 289
carbon 21, 26, 35, 48, 57, 59, 65, 256
carbon dioxide 56, 136, 277
carfentanil 242, 243, 244, 245
case studies
    Aliceville, AL, November 8, 2013 Crude Oil 187
    Brownson, NE, April 2, 1978, phosphorus 10, 199, 287
    Campbell County, TN, December 14, 1895, dynamite 101
    Cherry Valley, IL, June 17, 2009, ethanol 182
    Crosby, TX, August 29, 2017, organic peroxide 212
    Gettysburg PA, March 23, 1979, phosphorus 11, 198, 286
    Henderson, NV, May 4, 1988, ammonium perchlorate 209
    Indiana, Kentucky, and Tennessee, October 30, 1998, anthrax 232

319

case studies (*cont.*)
    Kansas City, MO, November 29, 1988, ammonium nitrate (ANFO) 209
    Lac-Megantic, Quebec, Canada July 6, 2013 crude oil 190
    Marshall's Creek, PA, June 26, 1964, dynamite, nitro carbo nitrate blasting agent 102
    Moscow, Russia, October 23, 2002, carfentanil and another similar drug, remifentanil 243
    New York City NY, February 26, 1993, ammonium nitrate 105
    Oklahoma City, OK, April 19,1995, ammonium nitrate 106
    Philadelphia, PA, February 23, 1991, Linseed Oil Soaked Rags 197, 295
    Richmond, IN, April 6, 1968, black powder 103
    San Francisco, CA, May 22, 1895, nitroglycerine 99
    Shamokin, PA, November 5, 1888, nitroglycerine 100
    Spencer, NC, October 2, 1908, powder magazine 102
    Syracuse, NY, August 30, 1841, dynamite 98
    Texas City, TX, April 16, 1947, ammonium nitrate 102
    Van Nuys, CA, December 17, 1998, anthrax 232
    Verdigris, OK, 1982, linseed oil soaked rags 295
    Vestal, NY, June 10, 1901, dynamite 101
    Waco, GA, June 1971, dynamite 104
    West Texas, TX, April 17, 2013, ammonium nitrate 107
    Woodland Hills, CA, December 14, 1998, anthrax 232
carbon monoxide 26, 121, 145
chemical activator 207
chemical explosions 109, 110
    chemical 109
    physical 109
chemical explosion types 111
    detonation 111
    deflagration 111
Chemistry of fire 158
chlorine 126, 159, 193, 207, 210, 219, 222, 238
    historic incidents 129, 130, 131, 232
chromic acid 63, 210
clandestine drug labs 245
    anhydrous ammonia, home made 252

    chemicals 248
    clean up 252
    detection 248
    methamphetamine 252
cobalt naphthenates 194
combustion analysis 281
combustible dust 40, 114, 116, 196, 303
common hazardous materials xiii, 3, 5, 40, 46
compounds 21, 26, 27, 28, 34, 26, 55, 76, 80, 92, 218, 260, 261, 282, 285, 292
cone roof tank 174
confined space 135, 284
    gases 144, 277
containers 8, 146, 152, 156, 171, 213, 263, 274
    bulk 116, 151, 156
    compressed gas 120
    container damage 11
    cryogenic 131, 213
    fixed facility 157, 174
    highway 213
    flammable liquid 157
    railroad 213
    tube bank 154
covalent bond 34, 36, 37, 55, 304
critical pressure 125, 297, 304
critical temperature 125, 297
crude oil 66, 182, 185, 187
    response challenges 182
cryogenic liquid 20, 33, 120, 304
cyanide 51, 218

diazoninitrophenol (DDNP) 96
diethyl zinc 194
diphenyl trisulfice 219

ebola 224
elements 21, 25, 26
ester 38, 74, 76, 84
ethanol 163, 171, 178, 295
    emergency response to spills & leaks 178
ether 74, 94, 160
ether peroxide 213
ethylene 69, 166
ethyl ether 94, 160, 249
expansion ratio 134, 297, 305
explosion 84, 111
    dust 114
    nuclear 118
    overpressure effects 114
    yield vs order 112
explosion effects, primary 112
    blast pressure 112

Index 321

fragmentation 113
thermal wave 113
explosive effects, secondary 112
explosion phases 112
   negative 113
   positive 113
explosives, forbidden 86
explosive families 86
   azide 88
   Fulminate 86, 87
   inorganic 86
   nitro 88
explosive incidents, selected 1841-1900's 98, 99, 100, 101, 102, 103, 105, 106, 107
explosive limits 293
explosives tetrahedron 110

families of compounds 46
families of elements 31
fentanyl 241
   clandestine manufacture 243
fire point 281, 290
fixed facility 153, 157, 174
   bulk 174
flammable liquid 157
flammable range 293
flashpoint Flash point solids 289
fluorine 21, 28, 32, 35, 259, 207, 210
formic acid 221, 269
fusee 194

halogen 219, 159, 307
hazard classes 84
   Hazard Class 1 Explosives 84
   Hazard Class 2 Compressed Gases 119
   Hazard Class 3 Flammable Liquids 157
   Hazard Class 4 Flammable Solids 193
   Hazard Class 5 Oxidizers 206
   Hazard Class 6 Poisons and Infectious Substances 214
   Hazard Class 7 Radioactive 253
   Hazard Class 8 Corrosive 264
   Hazard Class 9 Miscellaneous Hazardous Materials 276
hazmat elements 28
hazardous materials incident actors 8
   B.L.E.V.E. 16
   chemicals 8
   container damage 11
   geography 19
   incompatible chemicals 8
   incident caused container damage 11
   inert materials 9

   post incident container damage 15
   weather 18
hazardous materials incident Xiii, 4, 6
helium 33, 131, 132
horizontal tanks 177
hydrocarbon derivative compounds, complex 80
hydrocarbon derivative families 72, 76
   alcohol 82, 163
   aldehyde 83, 162
   alkyl halide 159, 219
   amine 160
   cyanide 218
   ester 164
   ether 160
   isocyanate 218
   ketone 81, 161
   nitro 88
   organic acid 170, 221, 268
   organic peroxide 206, 211, 212
   sulfur compounds 218
hydrocarbon families 63, 57
   alkane 59, 63
   alkene 63, 68, 75
   alkyne 63, 69
   aromatic 70
hydrocarbon naming 58, 59
   IUPAC 59
   trivial 58
hydrogen 28, 31, 32, 35, 48, 57, 105, 120, 127, 136, 256, 264
hydrogen cyanide 51, 218
hydrogen peroxide 95, 206, 213
hydrogen sulfide 74, 121
hypergolic combustion 207, 307
hypergolic propellant 207

ignition temperature 281, 291, 292, 307
immiscible 39, 307
Incident Frequency/Risk Model 42
inert gases 31, 152
internal floating roof tank 176
Invisible Force 297
ionic bond 34, 308
inorganic acid 39, 264, 269
into the atom 29
introduction 2
iso butane 66, 125, 285
iso cyanate 218
isopropyl ether 94

ketone 81, 161
krypton 133, 162

lead styphnate 96
liquefied compressed gases 124, 197
liquefied petroleum gas 12, 40, 57, 124, 297

magnesium 32, 25, 194, 205
matches 193, 194, 211
MC/DOT-306/406 171, 172
MC/DOT-307-407 172, 173
MC/DOT-331 13, 126, 147, 148
MC/DOT-338 135, 136, 152, 153
MC/DOT-412 170, 172, 173, 213, 214, 268, 274, 275
mercury fulminate 85, 96, 97
mercury II fulminate 86
metal 10, 21, 31, 32, 34, 35, 38, 46
metal nitrate 210
methane 64, 146
    hazards 146
methyl acrylate 221
methyl cyanide *see* acetonitrile
methyl tert-butyl ether (MTBE) 94
methyl mercaptan 219
military chemical agents 221
    blister 301, 302, 315
    chocking 302
    nerve 221, 302, 310
mixtures 34, 37, 295, 306
miscibility 39, 40, 281, 295
molecular weight 283, 284, 285

naloxone 244
narcan 244
neon 33, 132, 133
nitric acid 210, 270, 272
nitrogen 8, 21, 28, 35, 47, 53, 109, 127, 132, 135, 137, 161
nitromethane 89
nitroglycerine 89, 91, 100
nitromannite 96
noble gases 30, 33, 35, 55
nonmetals 31, 207
nonsalt compounds 46

open floating roof tank 175
organic acid 21, 76, 78, 79, 170, 221, 268
organic peroxide 206, 212, 311
    types 212
opioids, synthetic 241, 245
oxygen 21, 28, 29, 127, 135, 146, 311
oxygen deficiency 136

perchloric acid 213
periodic table of elements 21, 25, 26, 27, 28, 29, 31, 33, 34, 38, 39, 46, 55, 57, 126, 132
pentaborane 195
pentaerythritol tetranitrate (PETN) 95
petroleum 12, 20, 40, 57, 67, 124, 137, 174, 183, 186, 187, 189, 197, 201, 250, 297
pesticide 216, 219, 222, 235, 136, 237, 238, 239, 240, 241
    chlorophenol 238
    organochlorine 238
    organophosphates 238
    protective measures 241
    signal word 240
phosphorus 10, 44, 193, 196, 198, 199, 204, 221, 288
phosphorous pentasulfide 206
physical characteristics relationships 292
picric acid 85, 93, 95, 193, 194
plastics 164
    combustion products 168
    hazards to responders 169
polar compounds 38, 20, 76, 78, 162, 285
polarity 76, 77, 78, 79, 162, 163, 220, 281, 284
polymerization 9, 68, 76, 84, 164, 165, 167, 195, 211, 212
polymer 9, 10, 69, 89, 164, 165, 167, 195, 211, 212
potassium metal (K) 51, 94, 95
potassium nitrate 211
potassium sulfide 105, 204
propane 3, 12, 65, 66, 67, 120, 124, 125, 137, 147, 148, 149, 151, 181, 183, 185, 188, 247
propionic acid 171, 221, 164, 168, 169
pyrophoric materials 193, 196, 286, 312
propylene 69, 94, 124, 125, 297
prussic acid *see* hydrogen cyanide

railcars 153, 154, 181, 188, 213
    non-pressure 181, 188, 213
    pressure 153, 154
RDX, Cyclonite 95, 97
recognition primed decision model 41
remifentanil 243, 244
routes of exposure 84, 122, 123, 216, 313
    absorption 123, 217
    inhalation 123, 216
    ingestion 123, 217
    injection 123, 217

Index

salt 34, 35, 38, 39, 46
salt families 46
   binary 47
   binary oxide 49
   peroxide 49
   hydroxide 50
   cyanide 51
   oxysalts 51
   ammonium salt 53
self-accelerating decomposition temperature (SADT) 211, 212, 313
silver acetylide 87
site specific hazardous materials 40
sodium hydride 196, 205
sodium nitrate 94, 210
sodium peroxide 63, 193, 211, 308
solubility 34, 38, 39, 281, 314
specific gravity 281, 295, 296, 314
spontaneous combustion 196, 197, 199, 202, 314
states of matter 21, 22, 314
styrene 72, 73, 167, 168, 213
sublimation 22, 136, 194, 286, 314
sulfur compounds 74, 218, 219, 221

target organs 216, 122, 297, 314
titanium 194
toxic affects 122, 216
   acute 122
   chronic 122
   sub-acute 122
toxicity 74, 76, 83, 121, 123, 127, 159, 163, 214, 217, 222, 239, 240
   dermal 214, 240
   inhalation 214
   oral 214

toxicological terms 121
   asphyxiant 64, 313
   asphyxiate 121
   carcinogen 121, 215, 302
   concentration 265, 303
   corrosive 121, 264, 303, 304
   IDLH 129, 217, 307
   irritant 122, 215, 308
   mutagen 215, 309
   PEL 123, 145, 311
   sensitizer 215, 313
   TLV 123, 315
   TLV-ceiling 123
   TLV-STEL 123
   routes of exposure 123, 216, 313
   target organ 122, 216, 314
   teratogen 122, 216, 314
transitional metal 31
triacetone triperoxide (TATP) 95
trinitrobenzene (TNB) 92
trinitrophenol (picric acid) 85, 93, 193, 194
trinitrotoluene (TNT) 85, 93, 193, 194

urea nitrate 194

vinyl acetate 164, 221
vinyl chloride 213
vinyl cyanide *see* acrylonitrile
vinyl sulfide 219
volatility 284, 289
VX 222

xenon 132, 133
   xenon compounds 134

zirconium 194